Issue

Throu... rent,
and fu... ing
lands...
overv...
ecolo...
contri... s
practi... uatic
as we...
enter...
conse...

JOHN...
The a... s
emph...
envir... ssor
Wien... years
of cla... al
worlc

MICH... ces at
the U... in
land s... te to
impr... rce
planr... e of
incor

Cambridge Studies in Landscape Ecology

Series editors

Professor John Wiens *The Nature Conservancy*
Dr. Peter Dennis *Macaulay Land Use Research Institute*
Dr. Lenore Fahrig *Carleton University*
Dr. Marie-Josée Fortin *University of Toronto*
Dr. Richard Hobbs *Murdoch University, Western Australia*
Dr. Bruce Milne *University of New Mexico*
Dr. Joan Nassauer *University of Michigan*
Professor Paul Opdam *Alterra Wageningen*

Cambridge Studies in Landscape Ecology presents synthetic and comprehensive examinations of topics that reflect the breadth of the discipline of landscape ecology. Landscape ecology deals with the development and changes in the spatial structure of landscapes and their ecological consequences. Because humans are so tightly tied to landscapes, the science explicitly includes human actions as both causes and consequences of landscape patterns. The focus is on spatial relationships at a variety of scales, in both natural and highly modified landscapes, on the factors that create landscape patterns, and on the influences of landscape structure on the functioning of ecological systems and their management. Some books in the series develop theoretical or methodological approaches to studying landscapes, while others deal more directly with the effects of landscape spatial patterns on population dynamics, community structure, or ecosystem processes. Still others examine the interplay between landscapes and human societies and cultures.

The series is aimed at advanced undergraduates, graduate students, researchers and teachers, resource and land-use managers, and practitioners in other sciences that deal with landscapes.

The series is published in collaboration with the International Association for Landscape Ecology (IALE), which has Chapters in over 50 countries. IALE aims to develop landscape ecology as the scientific basis for the analysis, planning, and management of landscapes throughout the world.The organization advances international cooperation and interdisciplinary synthesis through scientific, scholary, educational and communication activities.

Also in the series:

J. Liu and W. W. Taylor (eds.) *Integrating Landscape Ecology into Natural Resource Management*

R. Jongman and G. Pungetti (eds.) *Ecological Networks and Greenways*

W. A. Reiners and K. L. Driese *Transport Processes in Nature*

EDITED BY

JOHN A. WIENS
THE NATURE CONSERVANCY

MICHAEL R. MOSS
THE UNIVERSITY OF GUELPH

Issues and Perspectives in Landscape Ecology

PUBLISHED BY THE PRESS SYNDICATE OF THE UNIVERSITY OF CAMBRIDGE
Cambridge, New York, Melbourne, Madrid, Cape Town, Singapore, São Paulo

CAMBRIDGE UNIVERSITY PRESS
The Edinburgh Building, Cambridge CB2 2RU, UK

http://www.cambridge.org
Information on this title: http://www.cambridge.org/9780521830532

First published 2005

Printed in the United Kingdom at the University Press, Cambridge

Typeface 10/14pt. Lexicon   *System* Advent 3b2 8.07f [PND]

A catalog record for this book is available from the British Library

Library of Congress Cataloging in Publication data

ISBN-13 978-0-521-83053-2 hardback
ISBN-10 0-521-83053-2 hardback

ISBN-13 978-0-521-53754-1 paperback
ISBN-10 0-521-53754-1 paperback

Contents

Contributors

JACK AHERN
Department of Landscape Architecture and Regional Planning, University of Massachusetts, Amherst, MA 01003, USA

THOMAS R. CROW
USDA Forest Service, North Central Research Station, Grand Rapids, MN 55744, USA

SAMUEL A. CUSHMAN
Department of Natural Resources Conservation, University of Massachusetts, Amherst, MA 01003, USA (present address: US Forest Service, RMRS, PO Box 8089, Missoula, MT 59807, USA)

DONALD A. DAVIDSON
School of Biological and Environmental Sciences, University of Stirling, Stirling FK9 4LA, UK

HENRI DÉCAMPS
Centre National de la Recherche Scientifique, 29 rue Jeanne Marvig, 31055 Toulouse, France

HAZEL R. DELCOURT
Department of Ecology and Evolutionary Biology, University of Tennessee, Knoxville, TN 37996, USA

PAUL A. DELCOURT
Department of Ecology and Evolutionary Biology, University of Tennessee, Knoxville, TN 37996, USA

LENORE FAHRIG
Ottawa–Carleton Institute of Biology, Carleton University, 1125 Colonel By Drive, Ottawa, Ontario K1S 5B6, Canada

KATHRYN FREEMARK
National Wildlife Research Centre, Canadian Wildlife Service, Environment Canada, Ottawa, Ontario K1A 0H3, Canada

ALISA L. GALLANT
Raytheon ITSS, Inc., EROS Data Center, Sioux Falls, SD 57198, USA

ROY HAINES-YOUNG
Centre for Environmental Management, School of Geography, University of Nottingham, Nottingham NG7 2RD, UK

RICHARD J. HOBBS
School of Environmental Science, Murdoch University, Murdoch, WA 6150, Australia

ROLF A. IMS
Institute of Biology, University of Tromsø, N-9037 Tromsø, Norway

ROB H. G. JONGMAN
Alterra Green World Research, Wageningen University, PO Box 47, NL-6700 AA Wageningen, The Netherlands

ANTHONY W. KING
Environmental Sciences Division, Oak Ridge National Laboratory, Oak Ridge, TN 37831, USA

FRANS KLIJN
WL/Delft Hydraulics, PO Box 177, NL-2600 MH Delft, the Netherlands

THOMAS R. LOVELAND
US Geological Survey, EROS Data Center, Sioux Falls, SD 57198, USA

JOHN A. LUDWIG
Savannas Cooperative Research Centre and CSIRO Sustainable Ecosystems, PO Box 780, Atherton, QLD 4883, Australia

RALPH MAC NALLY
Australian Centre for Biodiversity: Analysis, Policy and Management, School of Biological Sciences, PO Box 18, Monash University, VIC 3800, Australia

CHRIS MARGULES
Rainforest Cooperative Research Centre and CSIRO Sustainable Ecosystems, PO Box 780, Atherton, QLD 4883, Australia

KEVIN MCGARIGAL
Department of Natural Resources Conservation, University of Massachusetts, Amherst, MA 01003, USA

DAVID J. MLADENOFF
Department of Forest Ecology and Management, University of Wisconsin–Madison, Madison, WI 53706, USA

MICHAEL R. MOSS
Faculty of Environmental Sciences, University of Guelph, Guelph, Ontario N1G 2W1, Canada

JOAN IVERSON NASSAUER
School of Natural Resources and Environment, University of Michigan, Ann Arbor, MI 48103, USA

ZEV NAVEH
Faculty of Civil and Environmental Engineering, Lowdermilk Division of Agricultural Engineering, Technion Institute of Technology, Haifa 3200, Israel

RONALD P. NEILSON
USDA Forest Service, Pacific Northwest Research Station, Corvallis, OR 97331, USA

R. V. O'NEILL
Environmental Sciences Division, Oak Ridge National Laboratory, Oak Ridge, TN 37831, USA

JÁN OT'AHEL'
Institute of Geography, Slovak Academy of Sciences, Štefánikova 49, 814 73 Bratislava, Slovak Republic

BAS PEDROLI
Alterra Green World Research, Wageningen University, PO Box 47, NL-6700 AA Wageningen, the Netherlands

NANCY POLLOCK-ELLWAND
Faculty of Environmental Design and Rural Development, University of Guelph, Guelph, Ontario N1G 2W1, Canada

JØRUND ROLSTAD
Norwegian Forest Research Institute, Høgskoleveien 12, N-1430 Ås, Norway

H. H. SHUGART
Department of Environmental Sciences, University of Virginia, Charlottesville, VA 22901, USA

IAN A. SIMPSON
School of Biological and Environmental Sciences, University of Stirling, Stirling FK9 4LA, UK

JERZY SOLON
Institute of Geography and Spatial Organization, Polish Academy of Sciences, 00–818 Warsaw, Twarda 51/55, Poland

MICHAEL F. THOMAS
School of Biological and Environmental Sciences, University of Stirling, Stirling FK9 4LA, UK

JANA VERBOOM
Department of Landscape Ecology, Alterra Green World Research, Wageningen University, PO Box 47, NL-6700 AA Wageningen, the Netherlands

JAMES E. VOGELMANN
Raytheon ITSS, Inc., EROS Data Center, Sioux Falls, SD 57198, USA

WIEGER WAMELINK
Department of Landscape Ecology, Alterra Green World Research, Wageningen University, PO Box 47, NL-6700 AA Wageningen, the Netherlands

JOHN A. WIENS
The Nature Conservancy, 4245 North Fairfax Drive, Suite 100, Arlington, VA 22203, USA

KIMBERLY A. WITH
Division of Biology, Kansas State University, Manhattan, KS 66506, USA

I. S. ZONNEVELD
Enschede, the Netherlands

Preface

In a broad sense, landscape ecology is the study of environmental relationships in and of landscapes. But what are "landscapes"? Are they heterogeneous mosaics of interacting ecosystems? Particular configurations of topography, vegetation, land use, and human settlement patterns? A level of organization that encompasses populations, communities, and ecosystems? Holistic systems that integrate human activities with land areas? Sceneries that have aesthetic values determined by culture? Arrays of pixels in a satellite image? Depending on one's perspective, landscapes are any or all of these, and more. Landscape ecology is therefore a diverse and multifaceted discipline, one which is at the same time integrative and splintered.

The promise of landscape ecology lies in its integrative powers. There are few disciplines that cast such a broad net, that welcome – indeed, demand – insights from perspectives as varied as theoretical ecology, human geography, land-use planning, animal behavior, sociology, resource management, photogrammetry and remote sensing, agricultural policy, restoration ecology, or environmental ethics. Yet this diversity carries with it traditional ways of doing things and different perceptions of the linkages between humans and nature, and these act to impede the cohesion that is necessary to give landscape ecology conceptual and philosophical unity.

The contributions we have collected here do not produce that cohesion, but they demonstrate with remarkable clarity the elements from which we must forge this unification. Individually and collectively, they provide glimpses into the varied ways that landscape ecologists think about landscapes and about what landscape ecology is (or isn't). The contributions are *essays,* rather than traditional book chapters or reviews. We solicited essays from individuals in many countries and with many backgrounds, and the essays therefore express a diversity of perspectives, approaches, and styles, often in highly individualistic ways. We have edited the contributions sparingly, believing that it is in the spirit of essays to be somewhat idiosyncratic. Although we have grouped essays together in broad thematic areas, they are independent of one another and can (or perhaps should) be read in any order. Readers looking for stylistic consistency or an overarching central theme to this collection will be disappointed, but those who wish to sample the varied flavors of landscape ecology and obtain a glimpse of the future of the discipline will, we hope, be rewarded.

This collection grew out of an earlier set of essays that were invited as part of the Fifth World Congress of the International Association for Landscape Ecology (IALE), held in Snowmass, Colorado in 1999. That collection was distributed to registrants at the Congress and had limited distribution. With the encouragement of Alan Crowden of Cambridge University Press, we asked the contributors to that original collection to revise and update their essays, and we added several contributions in areas that were under-represented in the original collection. The essays presented here are therefore considerably more than a repackaging of old essays in new binding.

Production of this collection was aided by the United States Geological Survey, the University of Massachusetts, Colorado State University, IALE, and The Nature Conservancy. Cynthia Botteron and Vicki Fogel Mykles were instrumental in bringing a vision into a finished product for the Snowmass Congress. The assistance of Robert J. Milne of Wilfrid Laurier University, Ontario, was critical in bringing parts of this volume to fruition. But most of all, we thank the essayists, who came back to revise their contributions after several years or who produced new essays in the spirit of essays rather than research papers. Enjoy their thinking and perspectives!

PART I

Introductory perspectives

1

When is a landscape perspective important?

What is landscape ecology?

Although the definition of landscape ecology has been dealt with extensively (some would say ad nauseam) in the landscape ecological literature, there remains confusion among other ecologists as to exactly what landscape ecology is and, particularly, what its unique contribution is to ecology as a whole.

Ecology is the study of the interrelationships between organisms and their environment (Ricklefs, 1979). The goal of ecological research is to understand how the environment, including biotic and abiotic patterns and processes, affects the abundance and distribution of organisms (Fig. 1.1). This includes indirect effects such as the effect of an abiotic process (e.g., fire) on a biotic process (e.g., germination), which in turn affects the abundance and/or distribution of an organism. Processes considered are typically at a "local" scale, that is, at the same scale or smaller than the scale of the abundance/distribution pattern of interest.

Landscape ecology, a subdiscipline of ecology, is the study of how landscape structure affects the abundance and distribution of organisms (Fig. 1.2). Landscape ecology has also been defined as the study of the effect of pattern on process (Turner, 1989), where "pattern" refers specifically to landscape structure. The full definition of landscape ecology is, then, the study of how landscape structure affects (the processes that determine) the abundance and distribution of organisms. In statistical parlance, the "response" variables in landscape ecology are abundance/distribution/process variables, and the "predictors" are variables that describe landscape structure. Again, this includes indirect effects such as the effect of a biotic process (e.g., herbivory) on landscape structure, which in turn affects the abundance and/or distribution of the organisms of interest.

Issues and Perspectives in Landscape Ecology, ed. John A. Wiens and Michael R. Moss. Published by Cambridge University Press.
© Cambridge University Press 2005.

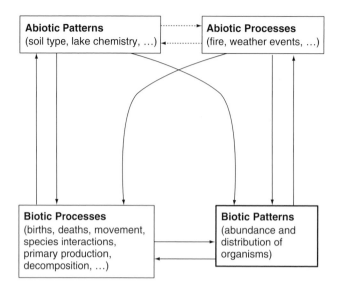

FIGURE 1.1
The study of ecology. Solid lines represent ecological interactions. The goal of ecological research is to understand how abiotic and biotic patterns and processes affect the abundance and distribution of organisms.

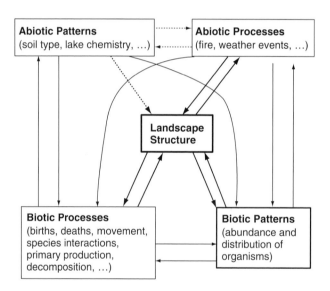

FIGURE 1.2
The study of landscape ecology. Dark solid lines represent landscape ecological interactions. The goal of landscape ecological research is to understand how landscape structure affects the abundance and distribution of organisms.

What is landscape structure?

The above definition raises the question, "What is landscape structure or pattern?" "Structure" and "pattern" imply spatial heterogeneity. Spatial heterogeneity has two components: the amounts of different possible entities (e.g., different habitat types) and their spatial arrangements. In landscape ecology these have been labeled landscape "composition" and "configuration," respectively. The amount of forest or wetland, the length of forest edge, or the density of roads are aspects of landscape composition. The juxtaposition of different landscape elements and measures of habitat fragmentation per se (independent of habitat amount) are aspects of landscape configuration (McGarigal and McComb, 1995).

What is a landscape-scale study?

A landscape ecological study asks how landscape structure affects (the processes that determine) the abundance and/or distribution of organisms. To answer this, the response variable (process/abundance/distribution) must be compared across different landscapes having different structures (Brennan *et al.*, 2002). This imposes a fundamentally different design on a landscape-scale study than on a traditional ecological study. Each data point in a landscape-scale study is a single landscape. The entire study is comprised of several non-overlapping landscapes having different structures (Fig. 1.3).

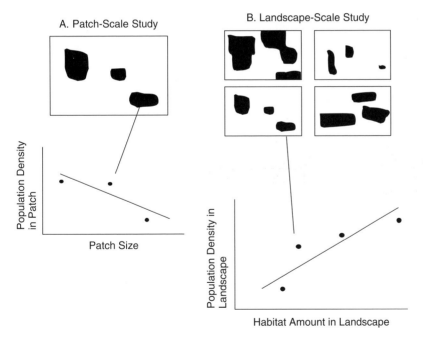

FIGURE 1.3

(A) Patch-scale study: each observation represents the information from a single patch (black areas). Only one landscape is studied, so sample size for landscape-scale inferences is one. (B) Landscape-scale study: each observation represents the information from a single landscape. Multiple landscapes, with different structures, are studied. Here, sample size for landscape-scale inferences is four.

A landscape-scale study therefore has the following attributes: (1) individual data points in the study represent individual landscapes, i.e., the landscape is the observational unit; and (2) the size of a landscape depends on the scale at which the response variable responds to landscape structure. This typically depends on the scale at which the organism(s) in question move about on the landscape, or the typical scale of the process of interest. Note that the landscape is not a level of biological organization (King, this volume, Chapter 4). In fact, a landscape-scale study can be conducted at the individual, population, community, or ecosystem level of biological organization. In the following I provide two hypothetical examples of landscape-scale studies: the first is at the individual level and the second is at the population level.

Example 1. Individual-level study

Consider a researcher who is interested in identifying the factors that determine the fledging success rate of a particular bird species. The usual approach to this would be to locate a number of nests and their associated territories. For each nest, response variables measured might be the number of young fledged or proportion of eggs taken by predators, and the predictor variables might be availability of food in the territory or density of predators in the territory.

To include a landscape perspective in this study, the researcher would determine whether the landscape context of a territory (i.e., the landscape structure of the region surrounding each territory) affects the number of young fledged or the proportion of eggs taken by predators in that territory. This will require a completely different study design.

First, the researcher must determine a reasonable maximum size for individual landscapes. This is done by asking at what scale (s)he expects no effect of landscape structure on the response variables. This will generally depend on movement scales of the organisms in the study. For example, if the predator has a daily movement range of 3 km, then each landscape should be at least 3 km in radius. The researcher must then locate individual territories that are spaced far enough apart such that non-overlapping landscapes of this size can be delineated around them.

Predictor variables in the study will then include both the original predictor variables (local availability of food, local density of predators) and new predictor variables that describe the structure of the landscape surrounding each territory. These variables might include compositional variables (e.g., amount of wetland, amount of forest) and configurational variables (e.g., fragmentation and juxtaposition of habitat types). Optimally, the landscape structural variables should be measured at several scales to determine the size of landscape unit that has the greatest effect on the response variables.

Example 2. Population-level study

In the above example the researcher is interested in the factors that determine a process (fledging success) which has an assumed effect on bird abundance/distribution. An ecologist may also examine directly the factors determining abundance/distribution at a population level. For example, one might ask, "What factors determine presence/absence of this frog species in different ponds?" Variables such as pond size or presence/absence of fish in the ponds might be considered.

The fact that multiple ponds are studied does not render this a landscape-scale study (Fig. 1.3A). In a landscape-scale study, the landscape context of each pond would need to be determined. A new set of ponds would be identified for the landscape-scale study. These ponds would need to be spaced far enough apart that non-overlapping landscapes could be delineated around them. As above, a reasonable maximum landscape size would need to be determined. This might be based on the maximum between-population dispersal distances of the frog species in question.

Predictor variables in the study again include both the original predictor variables (pond size, presence/absence of fish) and new predictor variables that describe the structure of the landscape surrounding each pond. These variables might include compositional variables (e.g., amount of forest, amount of road surface) and configurational variables (e.g., fragmentation, juxtaposition of various landscape elements). Again, the landscape structural variables should be measured for several different landscape sizes, to determine the size of landscape unit that has the greatest effect on the response variables (e.g., Findlay and Houlahan, 1997; Pope *et al.*, 2000).

When is a landscape perspective necessary?

It should be clear from the preceding that a landscape perspective is necessary whenever landscape structure can be expected to have a significant effect on the response variable (abundance/distribution/process) of interest. This leads to the somewhat frustrating catch-22 that one must conduct a landscape-scale study in order to determine whether a landscape perspective is necessary. Practically speaking, this implies that a landscape perspective is always necessary. However, we expect that there must be some, if not many, situations in which landscape structure does not have a large effect on the response variable of interest. In retrospect, this tells us that a landscape perspective was not necessary for that problem. Avoiding a landscape-scale study when one is not necessary will be time- and money-saving. Can we delineate some circumstances in which a landscape perspective is not necessary?

When is a landscape perspective not necessary?

Probably the most straightforward situation in which a landscape perspective is not necessary is when a sufficient proportion of variation in the response variable can be explained with local variables only. The definition of "sufficient" will, of course, depend on the purpose of the study. One might argue that the rarity of landscape-scale studies (as defined above) in the ecological literature suggests that the proportion of variation explained by local variables is high in most cases. However, we know this is not the case. Reasons for the lack of landscape-scale studies are discussed in the following section.

It may also be possible to identify circumstances in which at least certain components of a landscape perspective can be ignored. For example, most studies that have examined the effects of landscape structure on ecological responses have found large effects of landscape composition (reviewed in Fahrig, 2003). In contrast, modeling studies suggest that there are many situations in which landscape configuration has little or no effect on abundance and/or distribution of organisms, such as when the landscape structure itself is highly dynamic or when the amount of habitat on the landscape is above a certain level (Fahrig, 1992, 1998; Flather and Bevers, 2002).

Impediments to landscape-scale studies

The impact of landscape structure has been largely ignored in ecology, mainly because of the perceived difficulty of conducting broad-scale studies. This constraint is disappearing with the increasing availability of remotely sensed data, allowing much easier measurement of landscape structural variables.

The main constraints that must now be overcome are cultural constraints within the discipline of ecology. For example, many ecologists view a "landscape-scale" study as simply a study that covers a large area. If a study including several patches of forest is "large" to that researcher, (s)he may call it a landscape-scale study; however, it is more correctly termed a "patch-scale" study (Fig. 1.3A). As I argue above, a landscape-scale study is one that examines the effect of landscape context on a response variable. It answers the question, "Does the structure of the landscape in which this observation is imbedded affect its value?" This can only be answered by comparing the response variable across several landscapes with different structures (Fig. 1.3B).

Probably a greater hindrance to true landscape-scale studies is the current emphasis in ecology on experimental studies. By definition, landscape ecological studies look at the effect of a pattern (landscape structure) on a response. Judicious choice of landscapes with contrasting structures can result in a pseudo-experimental design, termed a "mensurative experiment" (McGarigal and Cushman, 2002; e.g., Trzcinski *et al.*, 1999). In contrast, manipulative experimentation at a landscape scale (i.e., multiple experimental landscapes) is generally not possible. Where landscape-scale studies have been conducted, large effects of landscape structure (especially landscape composition) have been found. Inability to apply "in vogue" experimental methods to landscape ecological studies is no reason to ignore these effects or to avoid the landscape perspective.

Acknowledgments

I thank the Landscape Ecology Laboratory at Carleton for helpful discussions and comments, particularly Dan Bert, Julie Bouchard, Julie Brennan, Neil Charbonneau, Tom Contreras, Stéphanie Duguay, Jeff Holland, Jochen Jaeger, Maxim Larivée, Michelle Lee, Rachelle McGregor, Shealagh Pope, Lutz Tischendorf, and Rebecca Tittler.

References

Brennan, J. M., Bender, D. J., Contreras, T. A., and Fahrig, L. (2002). Focal patch landscape studies for wildlife management: optimizing sampling effort across scales. In *Integrating Landscape Ecology into Natural Resource Management*, ed. J. Liu and W. W. Taylor. Cambridge: Cambridge University Press, pp. 68–91.

Fahrig, L. (1992). Relative importance of spatial and temporal scales in a patchy environment. *Theoretical Population Biology*, 41, 300–314.

Fahrig, L. (1998). When does fragmentation of breeding habitat affect population survival? *Ecological Modelling*, 105, 273–292.

Fahrig, L. (2003). Effects of habitat fragementation on biodiversity. *Annual Review of Ecology and Sysrematics*, 34, 487–515.

Findlay, C. S. and Houlahan, J. (1997). Anthropogenic correlates of species richness in southeastern Ontario wetlands. *Conservation Biology*, 11, 1000–1009.

Flather, C. H. and Bevers, M. (2002). Patchy reaction-diffusion and population abundance: The relative importance of habitat amount and arrangement *American Naturalist*, 159, 40–56.

McGarigal, K. and Cushman, S. A. (2002). Comparative evaluation of experimental approaches to the study of habitat fragmentation effects. *Ecological Applications*, 12, 335–345.

McGarigal, K. and McComb, W. C. (1995). Relationships between landscape structure and breeding birds in the Oregon coast range. *Ecological Monographs*, 65, 235–260.

Pope, S. E., Fahrig, L., and Merriam, H. G. (2000). Landscape complementation and metapopulation effects on leopard frog populations. *Ecology*, 81, 2498–2508.

Ricklefs, R. E. (1979.) *Ecology*. New York, NY: Chiron Press.

Trzcinski, M. K., Fahrig, L., and Merriam, G. (1999). Independent effects of forest cover and fragmentation on the distribution of forest breeding birds. *Ecological Applications*, 9, 586–593.

Turner, M. G. (1989). Landscape ecology: the effect of pattern on process. *Annual Review of Ecology and Systematics*, 20, 171–197.

2

Incorporating geographical (biophysical) principles in studies of landscape systems

The geographical and biological roots of landscape ecology are in Central and Eastern Europe. Here landscape has always been treated in a holistic manner, starting from von Humboldt (1769–1859), who defined landscape as a holistic characterization of a region of the earth. In 1850 Rosenkranz defined landscapes as hierarchically organized local systems of all the kingdoms of nature. The term "landscape ecology" was introduced by Troll in the late 1930s. He proposed that the fundamental task of this discipline be the functional analysis of landscape content as well as the explanation of its multiple and varying interrelations. Later he modified the definition by referring to Tansley's concept of the ecosystem. In this approach, landscape ecology is the science dealing with the system of interconnections between biocenoses and their environmental conditions in definite segments of space (Richling and Solon, 1996).

A further impulse to the development of landscape ecology was provided by the concepts drawn up in the 1950s within vegetation science. Particularly worthy of emphasis here is the work of Tüxen (1956), which introduced the concept of potential natural vegetation, as well giving rise to that of dynamic circles of plant communities; of Dansereau (1951), who was the first to apply the landscape concept in biogeography; and of Whittaker (1956), whose gradient analysis approach remains as important as ever.

It was only later that a landscape-based conceptualization was brought into animal ecology, although as early as the 1930s Soviet ecologists were emphasizing the influence of the combination of patch types on rodent control. But the real beginning of a landscape approach to the study of animal population dynamics was made in the 1970s, in the wake of Hansson's (1979) work on the importance of landscape heterogeneity for the ecology of small mammals.

Notwithstanding the widespread claims regarding the integrated nature of landscape ecology, historical reasons ensure that there remain differences in the

Issues and Perspectives in Landscape Ecology, ed. John A. Wiens and Michael R. Moss. Published by Cambridge University Press.

attitudes taken by researchers and in the concepts they apply. These differences are so far-reaching that some workers speak straightforwardly of bioecology and geoecology as separate branches of landscape ecology (Leser and Rodd, 1991).

The present disparities in research approaches, and the lack of cohesion between the many concepts applied, point to the need for a new theoretical synthesis within the framework of landscape ecology. As a contribution to this goal, I aim here to recall certain geographical regularities and principles which are now often forgotten in the course of detailed analyses, but which may provide a good basis for wider generalization of both a methodological and theoretical nature.

Space as the main subject of landscape ecology analysis

Irrespective of the precise aim of a study, which is formulated according to need, the subject of analysis each time is geographical space. Space may be understood in two ways: (1) in its entirety, together with its attributes, features, and dynamics; and (2) as an arena characterized solely by geometrical features, upon which abiotic and biotic processes (including the life histories of organisms) are played out.

Space, understood in a holistic manner, may be analyzed in various ways. Two classic approaches are most often distinguished – the structural and the functional. The structural approach deals with spatial scope, including (1) the topic approach, which concentrates on vertical structure and the links between components, and (2) the choric approach, wherein the subjects are territorial landscape structures or geocomplexes. The functional approach can be divided into (1) a process-related approach that analyzes the factors governing the behaviour of geocomplexes, and (2) a dynamic approach that studies the dynamics and evolution of geocomplexes (Richling and Solon, 1996).

The following remarks relate first and foremost to the topic and choric approaches, which should, it would seem, be treated as basic and preliminary to the geographical and ecological functional analysis of the landscape.

The principle of the hierarchical ordering of geocomponents

The simplest breakdown of the natural environment is defined by the geospheres (i.e., lithosphere, hydrosphere, atmosphere, and biosphere). In detailed studies, especially those related to a definite location or a small surface treated as a homogeneous area, a classification into geocomponents can be applied, with distinctions drawn between rocks, air, water, soil, vegetation, and animals.

Geocomponents exist in a mutual interrelationship and interact with each other in a hierarchically ordered way. It is commonly stated that the leading role is played by the bedrock, the most conservative of all the geocomponents and the one least susceptible to change. Hydroclimatic components occupy a subordinate position in this hierarchy and they, in turn, determine the edaphic and biotic components (soils, vegetation, and the animal world).

The place of climate in this perspective depends upon the scale of the approach. For the natural environment as a whole, climate is the superior component. In detailed studies, though, local climate or local modifications of macroclimate are functions of the character of rocks and surface relief, of the abundance and character of surface waters, and the depth of groundwater, as well as of kinds of soils and vegetation.

The non-nested hierarchical ordering of geocomponents (Allen and Starr, 1982) implies that superior components set constraints on the feasible states of subordinated components. A similar idea has also been formulated in the field of ecology, known as Shelford's general law of tolerance (see, for example, Odum, 1971). According to this principle, each geocomponent of a given place is limited by (among other things) two groups of environmental conditions. The first group includes those factors that cannot be influenced by a given geocomponent. The second group includes local environmental conditions that can be modified over timescales similar to those in which the geocomponent changes. When considering vegetation as the geocomponent in question, the first group encompasses macroclimate, parent rock, and topography. Light accessibility, soil humidity, and the organic matter content of soil belong to the second group, along with available surface area.

The distinction between hierarchically ordered independent versus labile environmental factors is relative, and depends upon the temporal and spatial scales of analysis. For instance, when we consider the plant cover of the earth through geological time, the chemical composition of the atmosphere is a labile factor, modified by living organisms. On the other hand, at the level of an individual in a population of short-lived annuals, almost all of the characteristics of the environment remain beyond control.

The principle of the relative discontinuity of the natural environment

A long-lasting conflict among geographers and ecologists concerns the continuity or non-continuity of the natural environment. Proponents of the concept of continuity (including Gleason, Ramiensky, and Whittaker among the plant ecologists, along with many climatologists and hydrologists) ascribe a major role in the shaping of the natural environment to gradient-related and independent changes in different abiotic geocomponents, and in the individualistic responses of different species. Those favoring the concept of non-continuity (including Clements and Braun-Blanquet among the plant ecologists, and most physical geographers in Europe) stress the existence of clear causal linkages between abiotic geocomponents, biocoenotic interdependences between organisms, and the role of plant communities in creating and buffering the environment.

From today's perspective, however, this dispute would seem to be a groundless one, as it takes no account of the influence of at least two factors: (1) the spatial extent and resolution of a study; and (2) the precision of measurements made and the number of analyzed features of the geocomponent. In reality, the boundaries of a geocomplex (patch) are only of significance in relation to a given scale of study. Even a relatively discrete patch boundary between two areas becomes more and more like a continuous gradient as one progresses to a finer and finer resolution.

There are several consequences of this general principle of relative discontinuity. First, ecotones and ecoclines represent a widespread phenomenon, rather than something exceptional, as was once believed. Second, it is not possible to speak of an ecotone in isolation, as the concept only makes sense when related to a defined feature or a group of features. Third, the greater and more diversified the anthropogenic impact in the landscape, the stronger the manifestation of a patch mosaic and the less visible the gradient-related differentiation. And finally, the definitions and criteria used to distinguish a class of spatial unit (a geocomplex) determine the spatial dimension in which the identification of the unit makes sense. In analyses that include both larger and much smaller areas, there is a blurring of the characteristics of geocomplexes, with the larger areas mainly including units of an intermediate nature, while the small areas are gradient-related transitional zones between neighboring geocomplexes.

Adoption of the principle of relative discontinuity of the natural environment allows theoretical models of the landscape to be treated as a series of progressive simplifications of reality. In such a conceptualization, the island–ocean model of MacArthur and Wilson (1967) is simplest in character. Here there are only two categories of object: ocean (with the value of 0) and island (with the value of 1). The patch–corridor model of Forman and Godron (1986) is characterized by the occurrence of three categories of object with values 0, p ($1 > p > 0$), and 1. The spatial-mosaic model has a large, though finite, number of objects belonging to a variable (but also finite) number of value classes. Finally, the gradient models (including the diffusional and gravitational variants often applied in geographical studies) are characterized by an infinite number of analyzed objects (points), with the indicator capable of taking on an infinite number of values in the interval between 0 and 1.

Each of these theoretical models requires its own methods of data collection and analysis. However, there is now a possibility (although not a very widely used one) for a single procedure common to all the models to be applied, with no a-priori assumptions being made with regard to any of them. Such independence is ensured by grid models or cellular-automata models (Wolfram, 1984). This approach is also compatible with both pixel-based remote-sensed imagery and with quadrat-based field observations.

The principle of the delimitation of partial geocomplexes

In accordance with the principle of the relative discontinuity of the natural environment, it is accepted that geocomponents can form natural spatial units – geocomplexes. According to a popular definition, a geocomplex is a relatively closed segment of nature constituting a whole on account of the processes taking part within it and the interrelationships among its components. One should note, however, that in the delimitation of comprehensively understood natural spatial units, it is not possible to account for all components and the interactions between them. None of the systems for the delimitation and classification of geocomplexes is entirely holistic.

Mutual relations of various systems of units can be determined solely on the basis of the theory of partial geocomplexes. Partial geocomplexes (Haase, 1964) reflect the variability of individual geocomponents with respect to the differentiation of the natural environment as a whole. Hence, a basis for their delimitation is provided by studies referring to a given geocomponent, albeit with due consideration given to relations between this component and the remaining geocomponents. The smallest partial units are called morphotopes, climatopes, hydrotopes, biotopes, and pedotopes. Each of these terms designates an area which is homogeneous from a given point of view.

It should be emphasized clearly that, in the early days, both the concept of partial geocomplexes and the closely related concept of the geosystem (Sochava, 1978) assumed an objectivity and a reality to the existence of geocomplexes. In the light of the principle of the relative discontinuity of the natural environment, this view gave rise to much unnecessary polemic. Today, basic spatial units are more likely to be identified on the basis of an objective function. In other words, instead of "discovering" objectively existing geosystems, spatial units are "constructed" according to need. Such an approach, which is entirely in accord with the concept of the partial geocomplex, may also justify a systemic conceptualization under which reality is the so-called "systemic material," while the creation of systems (e.g., geocomplexes) depends on the integrating function adopted (Richling and Solon, 1996). If the life requirements of a given species are accepted as an integrating function, then habitat patches should be defined relative to an organism's perception of the environment. In this case, landscape (heterogenous geocomplex) size would differ among organisms because each organism defines a mosaic of habitat or resource patches differently and on different scales.

The principle of partial geocomplexes gives rise to two additional points. First, from the formal point of view, all criteria distinguishing partial geocomplexes (landscapes and elements thereof) are of equal value – there are no better or worse ones, only ones that are more or less suitable from the point of view of a stated goal. Second, in analyzing landscape structure on the basis of the geocomplexes identified according to different criteria, different answers to the same questions are obtained. This is particularly true of assessments of the diversity and stability of the landscape (Solon, 2000), as well as of the linkage between its biotic and abiotic components.

Finally, the principle of partial geocomplexes is in agreement with the idea that landscape structure can be understood as a superimposition of three partly independent spatial hierarchies: abiotic, biotic, and anthropogenic (e.g., Cousins, 1993; Perez-Trejo, 1993; Barthlott *et al.*, 1996, 1999; Farina, 2000). According to this idea, it is possible to distinguish at least three perspectives in landscape ecology: (1) the human, when landscape elements are distinguished, grouped, and analyzed as meaningful entities for human life; (2) the geographic, focused on spatial and functional relationships between landscape elements and components, distinguished according to their abiotic character; and (3) the biological (both geobotanical and animal approaches), when space is analyzed at an object-specific scale (for example, species-specific) and major account is taken of object sensitivity and requirements. One of the main tasks of landscape ecology is to integrate the above perspectives into one theoretical system.

The principle of equivalence of the bottom-up and top-down approaches to spatial division

In physical geography, there has long been a prevailing view that spatial division on the basis of these two methods is equally proper and equivalent. It is purely by convention that the top-down approach tends to be applied more often for the division of large areas, and the bottom-up approach where detailed analysis of small areas is required.

Recently, however, concerns have been expressed that, in the case of self-organizing spatial systems, the bottom-up approach is the only proper one. In this case, the top-down approach violates two basic features of biological phenomena: individuality and locality. Ignoring locality obscures the factors that might contribute to spatial and temporal dynamics. According to this view, to say that a system is self-organized means that it is not governed by top-down rules, although there might be global constraints on each individual geocomponent (Perry, 1995).

The principle of the compound and temporally variable potential of a geocomplex

In accordance with the classic anthropocentric definition, the potential of a geocomplex is given by all of the resources whose exploitation is of interest to humankind (Neef, 1984). This definition may easily be generalized for any selected group of organisms using different resources and attributes of the environment. From the point of view of such a selected group of organisms, it is possible to speak generally of several partial potentials. First, one may consider the self-regulating and resistance potential and the capacity to counteract changes in the structure and nature of functioning of the geocomplex (landscape or elements thereof) that are induced by natural stimuli (particularly exploitation by the given group of organisms) or those of anthropogenic origin. Second, there is the resource-utilitarian potential, manifested in the ability of the landscape to meet the energy and material needs of the defined group of organisms. This may be considered in relation to the following sub-potentials:

- the food-related; i.e., the ability to produce organic matter of appropriate quality and quantity
- the concealment-related; i.e., the ability to supply the appropriate number of shelters or places in which shelters may be constructed
- the environment-creating; i.e., the ability of other components of the geocomplex to enter into the biocoenotic relationships necessary for the proper functioning of the analyzed population

The third point relates to the buffering potential, which manifests itself in the ability to reduce the amplitude of unfavorable external impacts. Different populations usually use the various potentials of the different geocomplexes (patches) within a landscape. Their utilization is capable of being diversified over time, and at the same time is not always optimal. Spatial analysis of differences in the potential of geocomplexes (including the identification of leading functions and those which are of secondary or lesser importance) and analysis of the life requirements of a population represent mutually augmentative studies that are, metaphorically speaking, two sides of the same coin. Thus, the principle of the differentiated potential of the geocomplex is clearly of basic significance in the construction of more realistic models of patches and corridors and their use by organisms.

The principle of the delimitation and bioindicative assessment of the geocomplex on the basis of the vegetation cover

According to the classical definition, indication is a process in which quantitative and/or qualitative characteristics of a single object, or one feature therein, define the state of another object or other features. The theoretical basis of indication results from the principle of the hierarchical ordering of geocomponents. The role of vegetation cover as a bioindicator results from its subordination to other less labile geocomponents. These relationships have been shown, *inter alia*, by Kostrowicki (1976). He demonstrated that structural features of vegetation are correlated with more than 70% of the features of other geocomponents.

Phytoindicators may be divided into two groups, which differ in relation to the object indicated. The first group includes indicators that define the general situation of the environment and the directions of the processes taking place. They define (indicate) the so-called "conditional" and "positional" environmental factors. The second group of indicators is used for the precise characterization of the state of selected components, in particular the level of anthropogenic influence. They indicate the so-called "environmental factors having direct impact" (Van Wirdum, 1981; cited in Zonneveld, 1982).

The application of the indicative approach in basic research to the spatial structure of the landscape is not too widespread. The only exception is the identification of the basic elements of the landscape in accordance with the principle of "one phytocoenosis = one ecosystem." It is much more common, however, for this method to be applied in assessment studies.

The principle of the minimization of energy costs

Unlike the principles discussed previously, which relate to structural relationships, this principle concerns the functioning of geosystems. In accordance with it, the flow of matter and information between systems (geocomplexes) proceeds via routes characterized by the smallest outlays of energy. In other words, the network of information channels is constructed in such a way that the energy costs of transfer are the lowest possible. This principle tends to follow from theoretical considerations of geosystem functioning, rather than from empirical research. Nevertheless, it may be particularly important where attempts are made to restore the landscape or its elements.

Final remarks

The above principles are clearly geographical in nature and are not widely referred to in landscape ecology handbooks. Other widely accepted ideas have developed independently in both geography and ecology, such as the principle that "pattern affects process." The principles are, to some extent, like empirical rules. Although their rectitude is supported by many examples, they cannot be recognized as true "laws of nature." Their status is similar to that of the principles of landscape ecology set out in the works of Forman and Godron (1986) and Farina (1998).

References

Allen, T. F. H. and Starr, T. B. (1982). *Hierarchy: Perspectives for Ecological Complexity*. Chicago, IL: University of Chicago Press.

Barthlott, W., Lauer, W., and Placke, A. (1996). Global distribution of species diversity in vascular plants: towards a world map of phytodiversity. *Erdkunde*, 50, 317–327.

Barthlott, W., Biedinger, N., Braun, G., Feig, F., Kier, G., and Mutke, J. (1999). Terminological and methodological aspects of the mapping and analysis of global biodiversity. *Acta Botanica Fennica*, 162, 103–110.

Cousins, S. H. (1993). Hierarchy in ecology: its relevance to landscape ecology and geographic information systems. In *Landscape Ecology and Geographic Information Systems*, ed. R. Haines-Young, D. R. Green, and S. Cousins. New York, NY: Taylor and Francis, pp. 75–86.

Dansereau, P. (1951). The scope of biogeography and its integrative levels. *Review of Canadian Biology*, 10, 8–32.

Farina, A. (1998). *Principles and Methods in Landscape Ecology*. London: Chapman & Hall.

Farina, A. (2000). The cultural landscape as a model for the integration of ecology and economics. *BioScience*, 50, 313–321.

Forman, R. T. T. and Godron, M. (1986). *Landscape Ecology*. New York, NY: Wiley.

Haase, G. (1964). Landschaftsökologische Detailuntersuchung und naturräumliche Gliederung. *Petermanns Geographische Mitteilungen*, 108, 8–30.

Hansson, L. (1979). On the importance of landscape heterogeneity in northern regions for the breeding population densities of homeotherms: a general hypothesis. *Oikos*, 33, 182–189.

Kostrowicki, A. S. (1976). A system-based approach to research concerning the geographical environment. *Geographia Polonica*, 33, 27–37.

Leser, H. and Rodd, H. (1991). Landscape ecology: fundamentals, aims and perspectives. In *Modern Ecology: Basic and Applied Aspects*, ed. G. Esser and O. Overdieck. Amsterdam: Elsevier, pp. 831–844.

MacArthur, R. H. and Wilson, E. O. (1967). *The Theory of Island Biogeography*. Princeton, NJ: Princeton University Press.

Neef, E. (1984). Applied landscape research. *Applied Geography and Development*, 24, 38–58.

Odum, E. P. (1971). *Fundamentals of Ecology*. Philadelphia, PA: Saunders.

Perez-Trejo, F. (1993). Landscape response units: process-based self-organising systems. In *Landscape Ecology and Geographic Information Systems*, ed. R. Haines-Young, D. R. Green, and S. Cousins. New York, NY: Taylor and Francis, pp. 87–98.

Perry, D. A. (1995). Self-organizing systems across scales. *Trends in Evolution and Ecology*, 10, 241–244.

Richling, A. and Solon, J. (1996). *Ekologia Krajobrazu* [*Landscape ecology*], 2nd edn. Warszawa: PWN.

Sochava, V. B. (1978). *Vviedenie v ucenie o geosistemakch* [*Introduction to Geosystem Science*]. Novosibirsk: Nauka.

Solon, J. (2000). Persistence of landscape spatial structure in conditions of change in habitat, land use and actual vegetation: Vistula Valley case study in Central Poland. In *Consequences of Land Use Changes: Advances in Ecological Sciences* 5, ed. U. Mander and R. H. G. Jongman. Southampton; Boston: WIT Press, pp. 163–184.

Tüxen, R. (1956). Die heutige potentielle natürliche Vegetation als Gegenstand der Vegetationskartierung. *Angewandte Pflanzensoziologie*, 13, 5–42.

Whittaker, R. H. (1956). Vegetation of the Great Smoky Mountains. *Ecological Monographs*, 26, 1–80.

Wolfram, S. (1984). Cellular automata as models of complexity. *Nature*, 311, 419–424.

Zonneveld, I. S. (1982). Principles of indication of environment through vegetation. In *Monitoring of Air Pollutants by Plants: Methods and Problems*, ed. L. Steubing and H. -J. Jager. The Hague: Junk, pp. 3–17.

PART II

Theory, experiments, and models in landscape ecology

3

Theory in landscape ecology

Over the past decade, landscape ecology has seen a period of remarkable progress. Remote imagery has provided new access to spatial data. Geographic information systems (GIS) have facilitated the handling, analysis, and display of spatial data. New theory has provided the means to quantify pattern (O'Neill *et al.*, 1988a), test hypotheses against random expectations (Gardner *et al.*, 1987), and come to grips with complexity (Milne, 1991) and scale (Turner *et al.*, 1993). The stage seems set for breakthroughs in the new millennium. Nowhere in the field of ecology is there greater promise, nowhere are there more exciting challenges.

This paper has a simple outline. The following sections review four areas of theory that have been applied to spatial effects in ecology. Each theory is then examined to identify the key advances that will be needed to apply the theory to our understanding of landscape dynamics. The intent is to propose an explicit list of major challenges for landscape theory.

Hierarchy theory and landscape scale

The concept of spatial hierarchy has already proven its value. Hierarchy theory (Allen and Starr, 1982; O'Neill *et al.*, 1986) states that ecosystem processes are organized into discrete scales of interaction. The scaled temporal dynamics, in turn, impose discrete spatial scales on the landscape. O'Neill *et al.* (1991) examined vegetation transects from four ecosystems and established that multiple scales of pattern actually existed in the field. Holling (1992) showed that peaks in the frequency distributions of vertebrate body weights corresponded to distinct scales of pattern in the landscape.

The spatial hierarchy on the landscape holds great promise for explaining ecological phenomena. Kotliar and Wiens (1990) pointed out that an insect

Issues and Perspectives in Landscape Ecology, ed. John A. Wiens and Michael R. Moss. Published by Cambridge University Press. © Cambridge University Press 2005.

uses one set of criteria to locate a patch, a second set to choose a tree, and yet a third to select an individual leaf. Wallace *et al.* (1995) showed that large ungulates forage randomly within a patch. However, the grazers use a completely different set of sensory clues as they move from one patch to another.

Application of spatial hierarchy theory is currently limited by statistical methods. The available methods have been summarized by Turner *et al.* (1991). In most cases, such as spatial autocorrelation, the technique is designed to detect a single scale of pattern. Trying to extend these methods to detect multiple scales leads to a number of problems. A significant challenge exists, therefore, for landscape theoreticians to develop statistical methods specifically designed to quantify multiple scales of pattern.

Percolation theory and hypothesis testing

Percolation theory deals with the connectance properties of a random landscape (Gardner *et al.*, 1989). If the landscape is considered as a square grid with units of habitat randomly scattered, the habitat tends to coalesce into a single continuous unit if habitat exceeds 59% of the grid. The theory has been used to study epidemics (O'Neill *et al.*, 1992a), to determine the scale at which an organism must operate to reach all resources (O'Neill *et al.*, 1988b), and to predict the spread of disturbances (Turner *et al.*, 1989).

The theory has been expanded to deal with connectance on hierarchically structured landscapes (O'Neill *et al.*, 1992b). Lavorel *et al.* (1994) have considered the dispersal strategies of annual plants competing on a random landscape. Further developments have also occurred in lacunarity theory (Plotnick *et al.*, 1993), which considers the properties of gaps between patches on the landscape.

But while theoretical developments have been fruitful, the real power of the theory has yet to be exercised. A major goal of landscape ecology is to understand the influence of spatial pattern on ecological processes (Urban *et al.*, 1987). Percolation theory permits one to develop a theoretical expectation of the process on a random landscape, that is, without spatial pattern. Deviations from this random expectation are then due explicitly to pattern (Gardner and O'Neill, 1991). Field data can be tested against the quantitative prediction and statistically significant differences can be attributed to patterning. The theory, therefore, holds enormous promise for the statistical testing of hypotheses on the effect of spatial patterning on ecological processes. This application of percolation theory represents another important challenge for both theoreticians and empirical researchers.

Spatial population theory

Ecologists have long considered the impact of spatial heterogeneity on population dynamics and stability. Lack (1942) noted fewer bird species on remote British islands and Watt (1947) pointed out that patches were fundamental to understanding community structure. Huffaker (1958) performed classic experiments showing that the stability of mite populations depended on the spatial configuration of oranges on a laboratory table.

In one body of theory, MacArthur and Wilson (1963) considered biodiversity on oceanic islands. Immigration was a function of distance to a source community and extinction was a function of island size. Although the theory has been criticized for its assumption of equilibrium (Barbour and Brown, 1974), considerable empirical data (Saunders *et al.*, 1991) have confirmed its general properties. The similarities between oceanic islands and landscape patches deserve more investigation.

In mathematical ecology, Levins (1970) proved that an unstable population could persist in a patchy environment. The development of the mathematical theory known as metapopulation theory was actively pursued by Hanski (1983) and is reviewed in Levin (1976) and Hanski and Gilpin (1997). Additional work has dealt with dispersion as a diffusion process (Andow *et al.*, 1990) and with applications of the physics of interacting particles (Durrett and Levin, 1994).

The theories developed by population ecologists have obvious applications to landscape ecology. Yet very little has been done to apply island biogeography or metapopulation theory to landscape problems. I regard this as being an important challenge and a wide-open opportunity to advance our understanding of populations operating on patchy landscapes.

Economic geography

Physical location and transportation costs often determine the profitability of an economic activity. In turn, that economic activity is the primary determiner of landscape pattern and change. So it is surprising that landscape ecology has not taken advantage of the well-developed theory of economic geography (Thoman *et al.*, 1962; Healey and Ilbery, 1990). Applicable areas include central place theory (e.g., Berry and Pred, 1961)., location theory (e.g., Friedrich, 1929; Hall, 1966), and market area analysis (e.g., Losch, 1954). Location theory, for example, considers the value of various products and the cost of transporting them to a central market (Jones and O'Neill, 1993, 1994). The theory then predicts which product will be grown close to the market and which can be profitably grown at greater distances (Jones and O'Neill, 1995).

The theory of economic geography has two obvious applications in landscape ecology. First, it can be used to drive models of land-use change, such as those used to predict deforestation in Brazil (Southworth *et al.*, 1991; Dale *et al.*, 1993). Second, consumers must use very much the same principles to optimize their use of resources on the landscape. Applications are particularly feasible because of the availability of excellent and detailed descriptions of the methodology (e.g., Isard, 1960). Once again, this area seems to hold the potential for real breakthroughs in landscape theory.

Conclusions

These four areas seem to hold the potential for major breakthroughs in our understanding of landscapes. I have made no attempt to be comprehensive or to identify all possible areas of research. These are simply areas where I personally can perceive the potential for breakthroughs. One thing seems clear: landscape theory is a wide-open field with enormous potential. It is certainly where I would be working if I were 27 again!

Acknowledgments

This research is supported by the US Environmental Protection Agency under Interagency Agreement 42WI066010.

References

Allen, T. F. H. and Starr, T. B. (1982). *Hierarchy: Perspectives for Ecological Complexity*. Chicago, IL: University of Chicago Press.

Andow, D. A., Kareiva, P. M., Levin, S. A., and Okubo, A. (1990). Spread of invading organisms. *Landscape Ecology*, 4, 177–188.

Barbour, C. D. and Brown, J. H. (1974). Fish species diversity in lakes. *American Naturalist*, 108, 473–478.

Berry, B. J. L. and Pred, A. (1961). *Central Place Studies: a Bibliography*. Philadelphia, PA: Regional Studies Research Institute, University of Pennsylvania.

Dale, V. H., O'Neill, R. V., Pedlowski, M., and Southworth, F. (1993). Causes and effects of land use change in central Rondonia, Brazil. *Photogrammetric Engineering and Remote Sensing*, 59, 997–1005.

Durrett, R. and Levin, S. A. (1994). Stochastic spatial models: a user's guide to ecological applications. *Philosophical Transactions of the Royal Society of London B*, 343, 329–350.

Friedrich, C. J. (1929). *Alfred Weber's Theory of the Location of Industries*. Chicago, IL: University of Chicago Press.

Gardner, R. H., Milne, B. T., Turner, M. G., and O'Neill, R. V. (1987). Neutral models for the analysis of broad-scale landscape pattern. *Landscape Ecology*, 1, 19–28.

Gardner, R. H., O'Neill, R. V., Turner, M. G., and Dale, V. H. (1989). Quantifying scale dependent effects with simple percolation models. *Landscape Ecology*, 3, 217–227.

Gardner, R. H. and O'Neill, R. V. (1991). Pattern, process and predictability: the use of neutral models for landscape analysis. In *Quantitative Methods in Landscape Ecology*, ed. M. G. Turner and R. H. Gardner. New York, NY: Springer, pp. 289–307.

Hall, P. (ed.) (1966). *Von Thunen's Isolated State.* Oxford: Pergamon Press.

Hanski, I. (1983). Coexistence of competitors in patchy environments. *Ecology*, 64, 493–500.

Hanski, I. and Gilpin, M. E. (eds.) (1997). *Metapopulation Biology: Ecology, Genetics and Evolution.* San Diego, CA: Academic Press.

Healey, M. J. and Ilbery, B. W. (1990). *Location and Change: Perspectives on Economic Geography.* Oxford: Oxford University Press.

Holling, C. S. (1992). Cross-scale morphology, geometry, and dynamics of ecosystems. *Ecological Monographs*, 62, 447–502.

Huffaker, C. B. (1958). Experimental studies on predation: dispersion factors and predator–prey oscillations. *Hilgardia*, 27, 343–383.

Isard, W. (1960). *Methods of Regional Analysis: an Introduction to Regional Science.* Cambridge, MA: MIT Press.

Jones, D. W. and O'Neill, R. V. (1993). Human–environmental influences and interactions in shifting agriculture when farmers form expectations rationally. *Environment and Planning A*, 25, 121–136.

Jones, D. W. and O'Neill, R. V. (1994). Development policies, rural land use, and tropical deforestation. *Regional Science and Urban Economics*, 24, 753–771.

Jones, D. W. and O'Neill, R. V. (1995). Development policies, urban unemployment and deforestation: the role of infrastructure and tax policy in a 2-sector model. *Journal of Regional Science*, 35, 135–153.

Kotliar, N. B. and Wiens, J. A. (1990). Multiple scales of patchiness and patch structure: a hierarchical framework for the study of heterogeneity. *Oikos*, 59, 253–260.

Lack, D. (1942). Ecological features of the bird fauna of British small islands. *Journal of Animal Ecology*, 11, 9–36.

Lavorel, S., Gardner, R. H., O'Neill, R. V., and Burch, J. B. (1994). Spatiotemporal dispersal strategies and annual plant-species coexistence in a structured landscape. *Oikos*, 71, 75–88.

Levin, S. A. (1976). Population dynamic models in heterogeneous environments. *Annual Review of Ecology and Systematics*, 7, 287–310.

Levins, R. (1970). Extinctions. In *Some Mathematical Questions in Biology: Lectures on Mathematics in the Life Sciences.* Providence, RI: American Mathematical Society, pp. 77–107.

Losch, A. (1954). *The Economics of Location.* New Haven, CT: Yale University Press.

MacArthur, R. H. and Wilson, E. O. (1963). An equilibrium theory of insular zoogeography. *Evolution*, 17, 373–387.

Milne, B. T. (1991). Lessons from applying fractal models to landscape patterns. In *Quantitative Methods in Landscape Ecology*, ed. M. G. Turner and R. H. Gardner. New York, NY: Springer, pp. 199–235.

O'Neill, R. V., DeAngelis, D. L., Waide, J. B., and Allen, T. F. H. (1986). *A Hierarchical Concept of Ecosystems.* Princeton, NJ: Princeton University Press.

O'Neill, R. V., Krummel, J. R., Gardner, R. H., *et al.* (1988a). Indices of landscape pattern. *Landscape Ecology*, 1, 153–162.

O'Neill, R. V., Milne, B. T., Turner, M. G., and Gardner, R. H. (1988b). Resource utilization scales and landscape pattern. *Landscape Ecology* 2, 63–69.

O'Neill, R. V., Turner, S. J., Cullinan, V. I., *et al.* (1991). Multiple landscape scales: an intersite comparison. *Landscape Ecology*, 5, 137–144.

O'Neill, R. V., Gardner, R. H., Turner, M. G., and Romme, W. H. (1992a). Epidemiology theory and disturbance spread on landscapes. *Landscape Ecology*, 7, 19–26.

O'Neill, R. V., Gardner, R. H., and Turner, M. G. (1992b). A hierarchical neutral model for landscape analysis. *Landscape Ecology*, 7, 55–61.

Plotnick, R. E., Gardner, R. H., and O'Neill, R. V. (1993). Lacunarity indices as measures of landscape texture. *Landscape Ecology*, 8, 201–212.

Saunders, D., Hobbs, R. J., and Margules, C. R. (1991). Biological consequences of ecosystem fragmentation: a review. *Conservation Biology*, 5, 18–32.

Southworth, F., Dale, V. H., and O'Neill, R. V. (1991). Contrasting patterns of land use in Rondonia, Brazil: simulating the effects on carbon release. *International Social Science Journal*, 43, 681–698.

Thoman, R. S., Conkling, E. C., and Yeates, M. H. (1962). *The Geography of Economic Activity.* New York, NY: McGraw-Hill.

Turner, M. G., Gardner, R. H., Dale, V. H., and O'Neill, R. V. (1989). Predicting the spread of disturbances across heterogeneous landscapes. *Oikos*, 55, 121–129.

Turner, M. G., Romme, W. H., Gardner, R. H., O'Neill, R. V., and Kratz, T. K. (1993). A revised concept of landscape equilibrium: disturbance and stability on scaled landscapes. *Landscape Ecology*, 8, 213–227.

Turner, S. J., O'Neill, R. V., Conley, W., Conley, M. R., and Humphries, H. C. (1991). Pattern and scale: statistics for landscape ecology. In *Quantitative Methods in Landscape Ecology*, ed. M. G. Turner and R. H. Gardner. New York, NY: Springer, pp. 17–49.

Urban, D., O'Neill, R. V., and Shugart, H. H. (1987). Landscape ecology. *BioScience*, 37, 119–127.

Wallace, L. L., Turner, M. G., Romme, W. H., O'Neill, R. V., and Wu, Y. (1995). Scale of heterogeneity of forage production and winter foraging by elk and bison. *Landscape Ecology*, 10, 75–83.

Watt, A. S. (1947). Pattern and process in the plant community. *Journal of Ecology*, 35, 1–22.

ANTHONY W. KING

4

Hierarchy theory and the landscape . . . level? or, Words do matter

The ill and unfit choice of words wonderfully obstructs the understanding

Francis Bacon

The term "level" is often used in association with "landscape," as in "landscape level." What is the, or a, landscape level? Is the landscape a level in a landscape hierarchy? And how do the answers to these questions impact the use of hierarchy theory to investigate and understand landscapes? I will attempt to answer these questions in this essay. Even if I am unable to satisfy you with definitive answers, I will hopefully stimulate your thinking about these topics. In the end I hope to have at least sensitized you to the need for care in choosing to use the words "landscape level."

First, "landscape level" is not synonymous with "landscape scale." Too frequently, "landscape level" is used as if it were interchangeable with "landscape scale." This usage implies (or asserts) a synonymy between "level" and "scale" that does not exist. Scale refers to the physical spatial and temporal dimensions of an object or event, its size or duration. Scale also involves units of measure. The spatial or temporal properties of an object or event are characterized by measurement on some quantitative scale. As we shall see below, "level" refers to a "level of organization" within a hierarchically organized system, and the level of organization is quantified by a rank ordering relative to other levels in the system. A level of organization is not defined by its physical dimensions. A particular substantiation or embodiment of a level of organization may be characterized by its scale (e.g., its size), but that does not mean that scale and level are the same thing. Individual mites and individual blue whales can both be understood as examples of the individual level of organization in a biological hierarchy. The scales of these individuals are, however, quite different. Same level of organization, much different scales – scale and level are simply not the same thing. One does not

Issues and Perspectives in Landscape Ecology, ed. John A. Wiens and Michael R. Moss. Published by Cambridge University Press.
© Cambridge University Press 2005.

measure the "levelness" of an object or event. One can, and does, however, measure the scale of an object or event.

In the case of landscapes, "landscape scale" typically refers to the areal extent, or more simply, the area, of the landscape. This physical characterization of a landscape's spatial (length) dimension is reported in units of square meters, square kilometers, or hectares. It is conceptually correct to talk about the scale of a landscape on a dimension of time (e.g., the time [in units of years] it takes for a landscape pattern to emerge and reach some steady state, or the frequency at which the landscape pattern changes). But this usage is not commonplace and normally the term "landscape scale" is correctly (albeit incompletely) synonymous with "landscape area."

It is important to note that there is no scale (e.g., area) that defines the existence of a landscape. There is no particular scale inherent in the concept of a landscape, only that it has a spatial (length) dimension or scale. There is no threshold value of area, no scale, above which a spatial extent is a landscape and below which it is not a landscape. A landscape, an area, with units of 10 square meters is as legitimately a landscape as an area with units of 10 thousand square kilometers. By convention or common usage it may be "understood" that "the" landscape scale refers to large areas more appropriately measured with units of hectares or square kilometers rather than square meters, but conventional or colloquial usage should not be confused with conceptual definitions. The individual level of organization in the biological hierarchy is not defined by scale; remember the example of the mites and blue whales. The individual level of organization is understood to span a large range of scale (e.g., physical dimensions). The same understanding applies to landscapes if the landscape level is understood to be a level of ecological organization. There is no "*the* landscape scale." The truth of this statement is apparent in the substitution of "area" for "scale." "*The* landscape area" doesn't have the resonance of "*the* landscape scale," but if there is no "*the* landscape area," there is no "*the* landscape scale." The landscape scale does not exist as some conceptual thing. The landscape scale, i.e., the scale of the landscape, is something that is measured on a particular landscape. And it is not the same thing as the landscape level.

So, the "landscape level" is not the "landscape scale." I've not yet defined what the "landscape level" *is*, but hopefully I've convinced you that the landscape level *is not* the landscape scale. Still not convinced? Try another word substitution. Substitute "area" for "level" so that "landscape level" becomes "landscape area." Feel the conceptual shift? If a particular reference to "landscape level" can be understood to mean "landscape area," the user is making the error of synonymizing scale and level, and "landscape level" should be translated to "landscape scale," which itself should be interpreted as shorthand for the "scale of the landscape(s) under consideration." What, then, is the "landscape level" if it is not (and it is not) the same thing as "landscape scale"?

"Landscape level" refers implicitly or explicitly to the landscape as a level of organization in a hierarchically organized ecological system. It is often assumed, again either implicitly or explicitly, that the landscape is a level in an ecological extrapolation of the traditional biological hierarchy (cells, tissues, organs, systems, individuals) such that interacting individuals are organized as populations, populations as communities, communities as ecosystems, and ecosystems as landscapes. Some would have landscapes organized as biomes and biomes combined to form the biosphere. Forman and Godron (1986 11) define the landscape as "a heterogeneous land area composed of a cluster of interacting ecosystems that is repeated in similar form throughout." The landscape as a higher level of organization composed of lower-level ecosystems is clearly implied. Extrapolation of the traditional biological hierarchy to encompass ecological disciplines is highly suspect. Elsewhere, I and others have called for careful interpretation of this purported ecological hierarchy, if not its outright abandonment. Consequently, it is appropriate to ask if there is in fact a "landscape level." Is the assumption that the landscape is a level of hierarchical organization warranted?

Much has been written about the application of hierarchy theory to ecological systems in general and landscapes in particular, following the seminal work of Allen and Starr (1982). I refer you to the references in King (1997). For the present purpose, level refers to level of organization in a hierarchically organized system. Differences in interaction strength and frequency among the components of a middle-number system can lead to the ordering of the system into a hierarchy of levels of organization. A hierarchical system is a system of ordered systems within systems. Members of the system at one level L in the hierarchy are composed of and exist as a consequence of interactions among system elements at the next lower level, $L - 1$. Each of these component system elements is itself a hierarchically organized system. At the same time, member systems of level L are themselves component parts of a level $L + 1$ system. Higher-level systems operate at slower rates than lower-level, and in nested hierarchical systems lower-level entities are physically part of higher levels and consequently are of smaller scale (i.e., spatial extent). Key to the concept of hierarchically organized systems is the constitutive relationship between system members at one level that determines – indeed creates – the systems of the next higher level. In a hierarchically organized system, the elements at one level emerge as a consequence of the interactions and relationships among elements of the next lower level. This emergent behavior is a fundamental property of hierarchically organized systems. Change the interactions and relationships between components and the higher-level properties will be altered; the higher-level system may even cease to exist, even if all the lower-level components remain. Thus, the interactions among system components and this constitutive relationship are the appropriate foci for consideration of hierarchical systems, rather than a cataloging or static description of component parts. The emergent properties of the three-dimensional configuration (secondary structure) of proteins is one of the best biological examples of this constitutive relationship so key to hierarchical organization. The properties of the protein at the level of the secondary structure emerge from the relationships and linkages among amino acids at the lower level organization of the polypeptide chain. Alter these linkages and the function of the protein changes, even though the parts – the amino acid composition of the chain – remain the same.

Jumping from proteins to landscapes, the question of interactions among landscape components becomes critical. If landscapes are composed of interacting ecosystems, what material or information is being exchanged in these interactions that links the components together in a constitutive relationship responsible for the emergent properties of the higher-level landscape? If landscapes are composed of patches, what material or information is being exchanged between patches that links them in a constitutive relationship from which the properties of the landscape level emerge? Are the interactions mediated by the movement of individual organisms among patches, or by the flow of water across the landscape? A change in criteria or the "currency" of the interactions can, and usually will, reveal a different system, a different hierarchy, operating within the same spatial extent. It is not enough to talk about the "landscape level." The reference must be to the "landscape level" of the hierarchy defined by specific interactions or criteria.

The physical superpositioning of systems within systems characteristic of nested hierarchical systems is a necessary but not sufficient condition for the existence of a higher level of organization. Superpositioning is shared with Russian dolls or nested Chinese boxes, where a box contains a smaller box that itself contains a smaller box, and so on. However, because these boxes are not interacting as part of a system to generate the next box in the ordered set, the boxes do not represent a hierarchical system. The relationship can be described as a hierarchical ordering, but it does not represent a hierarchically organized system. Similarly, the Linnaean system of taxonomic classification can be characterized as hierarchical, but the taxonomic groups do not interact to generate a next level of organization. Consequently, hierarchical ordering of patches within patches in a landscape is not sufficient evidence of hierarchical system organization for the landscape or a "landscape level." If the "landscape level" is anything more than a level in a taxonomy of landscape elements, it must be shown that higher-order patches and the landscape emerge as a consequence of a constitutive relationship among lower-order patches.

The importance of the constitutive relationship for hierarchically organized systems suggests a test for the existence of a "landscape level." Interactions among the lower-level components of a posited "landscape level" are most likely related to the spatial pattern of these components. Elements (e.g., patches) in proximity to one another are likely to have stronger and more frequent interactions than elements separated by great distances or by barriers to the flow of materials or information. Thus, if the landscape is a level of organization, a change in spatial pattern would be expected to result in a change in the holistic aggregate properties of the landscape. Failure to observe a change in "landscape level" properties with a change in spatial pattern would be evidence that the landscape was not a "level," but simply an areal extent over which observations were being made. The landscape is simply the stage on which the dynamics of ecological systems are played out. Note that this criterion for the existence of a "landscape level" is in harmony with the view of landscape ecology as the science of understanding how spatial pattern affects ecological function.

It should also be noted that if the "landscape level" is a level of organization in a hierarchically organized, spatially distributed system, the choice of scale of observation of the landscape cannot be arbitrary. The spatial extent, the area, of the observations must be large enough to encompass the entirety of this holistic thing which is the landscape and large enough to capture the interactions from which the landscape-level properties emerge. You cannot understand an individual organism as a level of organization by observing only half of the volume it occupies. Similarly, you cannot understand a landscape as a level of organization by observing only part of the area it occupies. Moreover, if you wish to do more than simply observe the aggregate holistic properties of the landscape level, if you wish to understand how those properties are related to the landscape components, the grain (resolution) of the observation must be chosen so as to resolve the components of the system at the level just below that of the landscape. If the landscape is a "landscape level," arbitrarily identifying the extent of a remote sensing scene or the boundaries of a land management unit as the landscape is inappropriate. Effort must be made to identify the intrinsic scales at which the landscape and its component parts operate.

What is the "landscape level"? If by "landscape level" we mean a level in a hierarchically organized system, hierarchy theory very clearly lays out the fundamental nature and properties of a landscape level. These properties cannot be assumed by naive or thoughtless extrapolation from the traditional biological hierarchy to the landscape. Nor can they be assumed from evidence of a hierarchical ordering of patches within patches on the landscape. This necessary but not sufficient property must be combined with evidence of interactions among patches (or other landscape elements) that lead to emergent, holistic, aggregate properties at the "landscape level." A landscape, an areal extent, may or may not represent a level of organization, with all that implies about holistic emergent properties and relationships with higher and lower levels of organization. It is inappropriate to invoke hierarchy theory to "explain" or justify an assumed landscape level. Hierarchical organization and a landscape level cannot be assumed or imposed arbitrarily a priori. They must be extracted from an analysis of observed data. It is in the provision of objective methods for extracting levels of explanation from observations on a spatially distributed system, or for testing the existence of a hypothesized "landscape level," that hierarchy theory contributes to the science of landscape ecology.

If the "landscape level" is not the "landscape scale" and a "landscape level" of hierarchical organization cannot be assumed to exist a priori, to what, if anything, does the frequent use of "landscape level" actually and correctly refer? I agree with R. V. O'Neill and T. F. H. Allen (Allen, 1998) that all too often the term "level" is gratuitously tacked on to the term "landscape" when "landscape" alone would suffice. When referring simply to an area under investigation, it is sufficient, and most appropriate, to limit oneself to the term "landscape." It is neither necessary nor appropriate to refer to the "forest level" when identifying a forest, or forests in general, as the subject of study. Neither is it appropriate to use the term "landscape level" in this sense. I've already discussed the error of using "landscape level" when one really means "landscape scale" as in the scale (e.g., area) of a landscape. And I've argued that "landscape level" should not be used to refer to a level of hierarchical organization until the existence of such a level has been demonstrated.

Adherence to these guidelines will eliminate many of the inappropriate uses of the term "landscape level." I believe, however, that the term "landscape level" is frequently used when the intent is primarily to communicate that the author is adopting a landscape perspective on an ecological problem. The landscape perspective involves consideration of ecological processes as they are played out in heterogeneous space and attention to how these processes are influenced by spatial pattern. In this circumstance, it is more appropriate to note, for example, that a study "addresses population dynamics from a spatial or landscape perspective" rather than referring to "population dynamics at the landscape level."

Gratuitous or thoughtless use of the term "level" in association with "landscape" should be avoided. At best, it is unnecessary; at worst, it implies the existence of a hierarchical organization and landscape properties that may or may not exist. The latter suggests, perhaps inappropriately, that hierarchy theory can be used to explain the landscape, which in turn can lead to undisciplined invocations of hierarchy theory and inappropriate "tests" of the theory. Both landscape ecology and ecological hierarchy theory deserve better. Tim Allen has argued that the landscape "level" is dead, and should be laid to rest (Allen, 1998). I wouldn't go that far, but I would reserve the use of the term for situations in which hierarchical organization and a "landscape level" have been demonstrated. Otherwise we run the risk of falling prey to Francis Bacon's Idols of the Market-place, where our "ill and unfit choice of words wonderfully obstructs the understanding."

References

Allen, T. F. H. (1998). The landscape "level" is dead: persuading the family to take it off the respirator. In *Ecological Scale*, ed. D. L. Peterson and V. T. Parker. New York, NY: Columbia University Press, pp. 35–54.

Allen, T. F. H. and Starr, T. B. (1982). *Hierarchy: Perspectives for Ecological Complexity*. Chicago, IL: University of Chicago Press.

Forman, R. T. T. and Godron, M. (1986). *Landscape Ecology*. New York, NY: Wiley.

King, A. W. (1997). Hierarchy theory: a guide to system structure for wildlife biologists. In *Wildlife and Landscape Ecology: Effects of Pattern and Scale*, ed. J. A. Bissonette. New York, NY: Springer, pp. 185–212.

5

Equilibrium versus non-equilibrium landscapes

Landscapes have a spatial domain that can be relatively large or small with respect to their disturbance regime. The ratio of typical disturbance size and landscape spatial extent characterizes the overall landscape behavior as well as the relative predictability of this behavior. Large-scale environmental change, human land-use changes, and natural or human-induced changes in the climate can all alter the spatial and temporal domain of the disturbance, and thus change the degree to which one can predict a landscape's dynamic behavior.

Conceptual considerations

When disturbances are sufficiently small or frequent, they are incorporated into the environment of the ecosystem; when they are sufficiently large and infrequent, they are catastrophic (Fig. 5.1A). There is an intermediate scale of extent and occurrence at which disturbance enforces a mosaic pattern to the ecological landscape. In this case, the landscape pattern is a mosaic of patches – each patch with an internal homogeneity of recent disturbance history different from the surrounding patches.[1]

The mosaic landscape is a statistical assemblage of patches. As in any sampled system, when the number of such patches is small, the variability is relatively large with related increased unpredictability (Fig. 5.1B). If the number of patches making up a landscape is large, the landscape dynamics will become more predictable. Climate change and human land-use changes tend to increase the size and synchronization of disturbances and make landscape dynamics less predictable (Fig. 5.1B).

---

[1] The comments made in this essay with regard to spatial extent of disturbances can also be applied to the frequencies of occurrence of disturbance. Infrequent disturbances are catastrophic; often-recurring disturbances are considered part of the "normal environment" of the ecosystem.

36   *Issues and Perspectives in Landscape Ecology*, ed. John A. Wiens and Michael R. Moss. Published by Cambridge University Press.

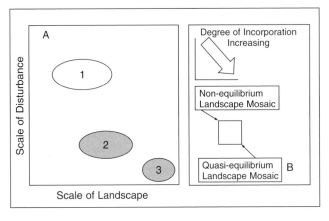

FIGURE 5.1
Landscape and disturbance scales. (A) The relationship between the size range of disturbances and of the landscapes on which they operate can be used to categorize landscape dynamic behavior. (1) indicates a disturbance regime whose spatial scale extent is so large that it could be termed a catastrophe. (2) indicates a disturbance regime whose spatial scale is smaller and is a disturbance in the usual sense of the word. (3) indicates a disturbance regime whose spatial scale is so small with respect to the scale of the landscape that it would normally be considered an internal landscape process. (B) Quasi-equilibrium landscapes are much larger than the disturbances that drive them and the average behavior of these landscapes appears to be relatively more predictable. When the disturbance scale is relatively large with respect to a given landscape system, the resultant landscape is effectively a non-equilibrium system and is predictable only when the disturbance history is known. The relatively smaller a disturbance, the greater the degree of incorporation into the functioning of the ecosystem.

The characterization of a forested landscape as a dynamic mosaic of changing patches was well expressed by Bormann and Likens (1979) in what they call the "shifting mosaic steady-state concept of ecosystem dynamics." This is an old concept in ecology (Aubréville, 1933, 1938; Watt, 1947; Whittaker, 1953; Whittaker and Levin, 1977). In a landscape composed of many patches, the proportion of patches in a given successional state should be relatively constant, and the resulting landscape should contain a mixture of patches of different successional ages – a quasi-equilibrium landscape (Shugart, 1998). In small landscapes (or landscapes composed of relatively few patches), the stabilizing aspect of averaging large numbers is lost and the dynamics of the landscape and the proportion of patches in differing states making up the landscape also becomes more subject to chance variation. If a landscape is small, it takes on many of the attributes of the dynamically changing mosaic patches that make it up – an effectively non-equilibrium landscape (Shugart, 1998).

Examples of different kinds of landscapes

In Fig. 5.2, landscape area is plotted along the horizontal axis; typical disturbance area for each landscape type is plotted along the vertical axis. The 1-to-50 ratio of disturbance area to landscape area is shown as a line. The 1/50 ratio was derived (see Shugart and West, 1981) from using individual-based tree models (Shugart, 1998) to determine the number of samples of simulated plots needed to be averaged to obtain a statistically reliable estimate of landscape biomass. About 50 plots, on average, tend to produce a fairly predictable landscape-level biomass response and can be used as an arbitrary delineation between quasi-equilibrium and effectively non-equilibrium landscapes. Please note that the comments that follow would hold if this ratio were 1/10 or 1/200.

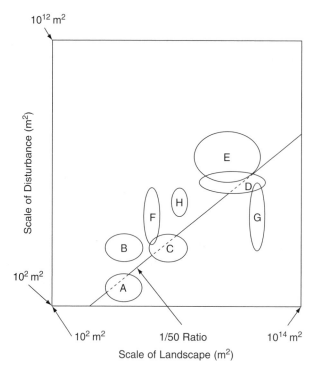

FIGURE 5.2
Examples of quasi-equilibrium and effectively non-equilibrium landscapes. (A) Tree fall size versus size of watershed of first-order streams in the Appalachian region of the USA. (B) Wildfire size versus size of watershed of first-order streams in the Appalachian region of the USA. (C) Wildfire size versus size of national parks in the Appalachian region of the USA. (D) Wildfire size versus spatial extent of the species ranges for commercial Australian *Eucalyptus* species. (E) Size of hurricanes versus spatial area of islands in Caribbean. (F) Size of wildfires in Siberia versus size of a forest stand. (G) Size of wildfires in Siberia versus land area of Siberia. (H) Size of floods versus size of floodplain forests.

For example, in Australia, the amount of land burned each year by fires approaches the size of the actual species ranges of a large number of commercial tree species (Fig. 5.2D). Entire species populations do not have stable age distributions over the entire continent. Some over-represented tree ages are of individuals regenerated in a particular fire and not subsequently destroyed by later fires. *Eucalyptus delegatensis* tree populations in Australia were disturbed in a tremendous set of forest fires in 1939 that burned over the species' range. For this reason, there are fewer than expected trees over 60 years of age. A large number of trees regenerated following the 1939 fire and this cohort is over-represented continentally. There have been other fires since 1939 (notably in 1984) that also created large mortality events followed by large birth events. Thus, for *Eucalyptus delegatensis* throughout southeastern Australia, most of the trees are of only a few age classes. This situation has important consequences. One of these is that several species of animals that require old *Eucalyptus delegatensis* trees as habitat are now considered endangered species. Many of the Australian forests dominated by *Eucalyptus* species are effectively non-equilibrium landscapes with respect to their biomass dynamics.

If the fall of a tree is the disturbance of interest (gap-scale disturbances), then watersheds of first-order streams in the Appalachian Mountains (Fig. 5.2A) would be quasi-equilibrium landscapes. However, if Appalachian wildfires are the focal disturbance (Fig. 5.2B), these same watersheds are too small, relatively, and the dynamics of their biomass would be unpredictable without knowing the fire history (as for an effectively non-equilibrium landscape). Indeed, only in the largest parks in the Appalachian region of the USA (Fig. 5.2C), are the landscapes large enough to average away the effects on biomass dynamics of the disturbance from typical-sized forest fires. Similarly, forest fires in Russia are large enough to make Siberian forest stands effectively non-equilibrium landscapes (Fig. 5.2F), but Siberia as a whole may be large enough to average away these variations and be a quasi-equilibrium landscape (Fig. 5.2G)

In some cases, entire biotas may inhabit effectively non-equilibrium landscapes. One continental-scale example has already been discussed for *Eucalyptus* forest biomass dynamics under the Australian fire disturbance regime (Fig. 5.2D) and another for Siberian forests (Fig. 5.2G). As a further example, the hurricanes that disturb West Indian forests are large when compared to the size of the islands in the Caribbean (Fig. 5.2E). The Caribbean islands are small with respect to the spatial scale of a major climatological feature that disturbs them; for this reason, they may function as effectively non-equilibrium landscapes. A similar example would be the spatial extent of floodplain forests and the spatial extent of floods (Fig. 5.2H) in large rivers.

Consequences

The mosaic dynamics of terrestrial ecosystems are particularly well developed as a theoretical concept in forest ecology. Some of this development is due to the progress made in practical forestry over the past two centuries. The size of mature trees and the damage done by their fall are also at a scale that is naturally observed by humans. In forests, the local influence of a large tree on its associated microenvironment is sufficient to produce a considerable impact on the environment when the tree dies. Tree birth, growth, and death cycles in the gaps left in the canopy of a forest after a large tree falls are processes that can produce a mosaic character to a forest independent of external factors. This tendency for forests to generate a canopy-tree-scale mosaic interacts with external factors. This interaction confers advantages or disadvantages to trees of different species at different stages in their life cycle.

For equilibrium landscapes, the mosaic dynamics underlie the expected pattern of biomass dynamics during recovery from disturbance. There are significant differences in the expected biomass dynamics in landscape ecosystems assumed to be homogeneous and in a mosaic landscape. A homogeneous or "metabolic" view of biomass dynamics of landscapes leads one to expect the net ecosystem productivity to balance net ecosystem losses. Hence, the biomass dynamics of landscapes should rise monotonically to equilibrium. In large mosaic landscapes, however, the expected biomass dynamics involve multiple local balances of production and losses and are also products of the synchrony of the changes in the patches that make up the landscape. One expects the biomass dynamics to overshoot the eventual long-term landscape biomass (Bormann and Likens, 1979; Shugart, 1998). This expected pattern can be modified by compositional or successional change during the landscape transient response.

Along a similar vein, in a landscape that behaves as a shifting mosaic of habitats, species-diversity patterns observed by community ecologists can arise as a consequence of seemingly simple models relating the species carrying capacity to habitat availability on the mosaic landscape. One of these is the species–area curve – an important relationship in the development of the theory of island biogeography (Shugart, 1998).

It is difficult to effectively manage non-equilibrium landscapes. Landscapes that are small with respect to the forces that disturb them can be expected to have an erratic dynamic behavior. Such systems are difficult to manage toward a goal of constancy because they are regularly disequilibriated by disturbance events. Busing (1991) points out that to manage a landscape for a particular habitat type (or for a particular species that uses one of the several habitat types that occur on a dynamic mosaic) requires a landscape area much greater than the biomass-based 50/1 ratio of landscape size to disturbance size used in Fig. 5.2. Habitat dynamics on small landscapes increase the extirpation rate of resident species. These considerations point to the need for very large land areas for nature reserves or parks that are intended to preserve habitat and biotic diversity. The manager of a natural landscape needs the capability to project the future response of the landscape to the particular regime of disturbances and habitat types as a prerequisite to rational management

References

Aubréville, A. (1933). La forêt de la Côte d'Ivoire. *Bulletin du Comité des Etudes Historiques et Scientifiques de l'Afrique Occidentale Française*, 15, 205–261.

Aubréville, A. (1938). La forêt colonaile: les forêts de l'Afrique occidentale française. *Annales Academie Sciences Colonaile*, 9, 1–245. Translated by S. R. Eyre. (1991). Regeneration patterns in the closed forest of Ivory Coast. In *World Vegetation Types*, ed. S. R. Eyre. London: Macmillan, pp. 41–55.

Bormann, F. H. and Likens, G. E. (1979). *Pattern and Process in a Forested Ecosystem*. New York, NY: Springer.

Busing, R. T. (1991). A spatial model of forest dynamics. *Vegetatio*, 92, 167–179.

Shugart, H. H. (1998). *Terrestrial Ecosystems in Changing Environments*. Cambridge: Cambridge University Press.

Shugart, H. H. and West, D. C. (1981). Long-term dynamics of forest ecosystems. *American Scientist*, 69, 647–652.

Watt, A. S. (1947). Pattern and process in the plant community. *Journal of Ecology*, 35, 1–22.

Whittaker, R. H. (1953). A consideration of climax theory: the climax as a population and a pattern. *Ecological Monographs*, 23, 41–78.

Whittaker, R. H. and Levin, S. A. (1977). The role of mosaic phenomena in natural communities. *Theoretical Population Biology*, 12, 117–139.

6

Disturbances and landscapes: the little things count

Disturbances are events that significantly change patterns in the structure and function of landscape systems (Forman, 1995). These events and changes may be small to large, minor to catastrophic, natural to anthropogenic, and short-term to long-lasting. It is almost trite to say that disturbances are a ubiquitous component of all landscapes. Volumes and reviews have been written on landscape disturbances and responses (e.g., Pickett and White, 1985; Turner, 1987; Rundel *et al.*, 1998; Gunderson, 2000), and some aspect of disturbance permeates most of the other papers in this volume.

Rather than attempt another general review of disturbance impacts on landscapes, which in a short paper could only be superficial, my aim here is to present a special perspective, one focused on a framework for how disturbances impact on small landscape structures (vegetation patches) and, consequently, on vital processes that occur at this fine scale. I will illustrate the way these impacts flow on to affect two landscape functions: conserving resources and maintaining diversity. It is these impacts and functions that are of growing interest to ecologists (e.g., McIntyre and Lavorel, 1994, 2001) and of critical importance to a wide spectrum of land managers, from ranchers with economic production goals to park rangers with biodiversity conservation goals (Freudenberger *et al.*, 1997). I hope to convince you, with two examples, that understanding the effect of disturbances on basic landscape functions at a fine scale can lead to principles with much broader implications for both landscape preservation and restoration.

Small landscape structures and their functions

As a patchy mosaic of interconnected and interacting ecosystem units, the structural attributes of a landscape can be defined over a range of scales, from local to global (Forman, 1995). I will restrict my attention to local

Issues and Perspectives in Landscape Ecology, ed. John A. Wiens and Michael R. Moss. Published by Cambridge University Press.
© Cambridge University Press 2005.

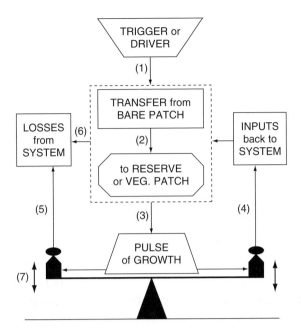

FIGURE 6.1

A trigger–transfer–reserve–pulse framework for how arid and semiarid landscapes are structured to function in time and space to conserve resources and maintain habitats (adapted from Ludwig and Tongway, 1997, 2000). In this framework, examples of key events or processes include: (1) a rain–wind storm that triggers or drives a runoff–erosion event, that (2) transfers resources such as water and soil particles from a source (bare patch) to a sink (vegetation patch) that traps these resources, which in turn (3) initiates a pulse of vegetation growth; products from this pulse of growth can serve as (4) inputs back to the landscape system to maintain or increase its patch structures and functions or, if not, these products may be consumed by fire or livestock and, hence, (5) lost from the landscape system; (6) resources can also be lost from this system in runoff–erosion events if vegetation patches fail to capture and retain these resources within the landscape system, or if these patches are degraded by disturbances such as grazing or fire; and (7) the landscape system will maintain a balance if fluctuating inputs and losses are equal over time and space.

landscapes (e.g., hillslopes) where biotic, resource-rich patches (small patches of dense vegetation and fertile soils) occur within a matrix of bare, poor soils. These two-phase mosaics occur in arid and semiarid landscapes around the world (d'Herbies *et al.*, 2001), where a patchy vegetation structure is maintained by fine-scale source-to-sink processes (Seghieri and Galle, 1999; Tongway and Ludwig, 2001). The bare or open patches within these two-phase mosaic landscapes are the source of materials transferred into sinks as driven (triggered) by water and wind processes (Fig. 6.1). Sinks are those vegetation patches that form surface obstructions to these water- and wind-driven flows – processes that build and maintain patch structures. This local redistribution of resources from source to sink has been termed the "reversed Robin Hood" phenomenon (Tongway and Ludwig, 1997), where vital materials are "robbed from the poor to give to the rich" (i.e., taken from the resource-poor part of the landscape matrix and given to fertile or rich patches).

It is these small patch structures and fine-scale source-to-sink processes which convey two important functions to arid and semiarid landscapes: (1) the capture and concentration of scarce resources such as rainwater, soil nutrients, and litter; and (2) the conservation of a high diversity of organisms. Many such landscapes around the world are strongly patchy at scales of less than 100 m, for example, banded vegetation occurring on ancient, gentle topographies with nutrient-poor, medium-textured soils, and in climates with low and unpredictable rains (Tongway and Ludwig, 2001). In these landscapes, the conservation of limited water and nutrient resources is obviously an important function, especially on lands used by humans for subsistence livestock grazing (e.g., Rietkerk *et al.*, 1997). Small patches within such landscapes also provide habitats for many species (e.g., Wiens, 1997), and during droughts some patches are extremely important as refugia (e.g., Wardell-Johnson and Horwitz, 1996).

What scale really matters to these functions?

Of course, the answer to this question is that all landscape scales are important, from micro to macro, because function cannot be divorced from the material or organism of interest (see Wiens, 1997). However, I think it is fair to say that landscape ecology has had a tendency to emphasize macro scales, for example, the clearing of woodlands and forests on watersheds or the filling of estuaries by urban developments (Forman, 1995). The appeal of working at the macro scale is that these landscape changes can be detected and documented by satellite imagery (e.g., Roderick *et al.*, 1999), providing colorful and interesting maps and digital data for a myriad of spatial metrics and models. However, for two critical landscape functions, conserving water and nutrient resources and maintaining biodiversity, the importance of micro or fine-scale patterns and processes is now emerging (Wiens, 1997; Ludwig *et al.*, 2000a). For example, small water- and nutrient-enriched patches, such as perennial grass clumps, log mounds, shrub hummocks, and tree "islands," are critical for a multitude of species such as ants, termites, beetles, grasshoppers, lizards, and small mammals that inhabit undisturbed and disturbed deserts, grasslands, and savannas (e.g., McIntyre and Lavorel, 1994; With, 1994; Wiens *et al.*, 1995; Ludwig *et al.*, 2000b).

As noted earlier, but worthy of repeating, small landscape patches also form important surface obstructions that function to capture water and soil nutrients being carried in runoff, and to trap litter and soil particles being blown about in winds (Tongway and Ludwig, 1997). Water and nutrients captured and stored in these vegetation patches can trigger pulses of plant, animal, and microbial growth (Fig. 6.1). These biotic activities serve as positive feedbacks to build and enrich patches, maintaining them as habitats and priming them to function again as obstructions with the next runoff or wind erosion event. Without this function, soils excessively erode and are lost from uplands to choke lowlands, creeks, and rivers with rich sediment loads, upsetting or shifting the balance of these ecosystems (Bunn et al., 1999). Flow-on effects can even have long-term, large-scale impacts on out-flow estuaries and offshore barrier islands and reefs (Cavanagh et al., 1999).

Tales from two continents

Two examples will be used to illustrate the importance of disturbance on micro-scale matrix-patch patterns for the two landscape functions being treated here, resource conservation and habitat biodiversity maintenance. Over more than a century, disturbances from extensive cattle ranching and overgrazing of landscapes in the southwestern United States has caused major shifts in vegetation over large areas (Dick-Peddie, 1993; Van Auken, 2000). One shift has been a change from the fine-scale patchiness observed in desert grasslands to the coarser-scale patterns evident in desert shrub dunelands, a process termed desertification (Schlesinger et al., 1990). Although causes of this desertification are widely debated (e.g., Grover and Musick, 1990), it is most probable that cattle grazing reduced the ground cover of grass patches (tussocks and clumps), thereby reducing competition and favoring shrubs (Van Auken, 2000). Wind and water-driven processes favored the formation of a larger-scale patch-matrix pattern of shrub-dune "resource islands" within a matrix of bare, inter-shrub spaces (Reynolds et al., 1999). Autogenic shrub effects and source-to-sink landscape processes now maintain this coarser, patchy landscape.

In these landscapes, the rich diversity of plants and animals that typically inhabits desert grasslands (e.g., Burgess, 1995) has now changed to a different suite of fewer species in the shrub dunelands, although interestingly the above-ground productivity of these dunelands does not appear to have significantly changed from that of the grassland (Huenneke, 1996). This suggests that water and nutrient resources are still being effectively captured by the dune landscape, only the scale or pattern of the distribution of these resources and production has become coarser.

In the tropical savannas of northern Australia, disturbances by cattle near artificial watering points has also caused a change in fine-scale patch structures (Ludwig *et al.*, 1999). Perennial grass tussocks and clumps have been lost to form a more open and bare matrix-patch pattern. This loss of landscape patches near water has reduced the potential for the local landscape system to capture resources, resulting in a loss in diversity of both plants and grasshoppers, the latter requiring the habitats provided by the now missing grass patches. Fires in these grazed landscapes also have impacts on birds and reptiles (Woinarski *et al.*, 1999).

In many of these savanna landscapes, soil surfaces have been exposed to runoff and wind processes, creating significant soil erosion features such as bare soil "scalds," rills and gullies that require restoration (Tongway and Ludwig, 2002a). Soils have been stripped from these landscapes, ending up out of the system, down in creeks and rivers (Bunn *et al.*, 1999; Prosser *et al.*, 2001). This soil erosion can lead to extensive desertification that is difficult to combat (Tongway and Ludwig, 2002b). The basic restoration principle is to rebuild fine-scale patches in the landscape, thereby re-establishing the role of such patches as obstructions to trap and regulate resources (Tongway and Ludwig, 1996).

Disturbances and continua of landscape function

How well a landscape functions to conserve resources and maintain biodiversity can be viewed as a continuum (Fig. 6.2A). Conceptually, landscapes may be termed "fully functional" when they conserve resources to maintain rich and diverse environments that provide many habitats suitable for a rich diversity of species. At the other end of the continuum, a landscape may be totally dysfunctional, where all resources "leak" from the system resulting in a landscape with poor resources and no habitats suitable for species. Of course, the landscapes we observe fall between these two extremes. Comparing different landscapes in terms of their degree of functionality has proven useful (see examples in Tongway and Ludwig, 1997). However, there is a need to improve the methods used to position landscapes along such a continuum, either by indirectly identifying indicators of functionality or by directly using simple measures of resource and habitat attributes (Ludwig and Tongway, 1993).

The concept of ecosystem stability can also be applied to how disturbances relate to this continuum of landscape functionality. In ecological systems, stability has been defined using terms such as resilience and persistence (Holling, 1973; Gunderson, 2000). Persistence refers to how far a system moves away from its dynamic equilibrium or steady state when disturbed without changing into a different state (D. Ludwig *et al.*, 1996). Resilience refers to how quickly this perturbed system will return to its steady state once this disturbance is removed.

(A) Continuum of Landscape Functionality

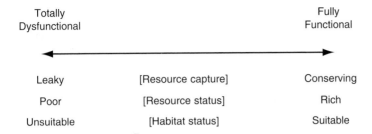

(B) Disturbance and Landscape Persistence

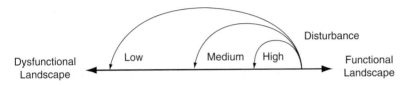

(C) Disturbance and Landscape Resilience

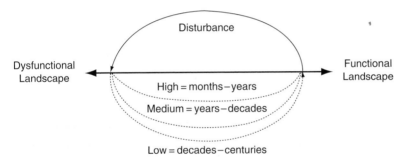

FIGURE 6.2
Landscape functionality as: (A) a continuum from functional to dysfunctional, and in relation to low, medium, and high levels of (B) persistence and (C) resilience to disturbance.

Using these definitions, a landscape has low persistence if a disturbance causes a highly functional ecosystem to shift well away from this state to become dysfunctional (Fig. 6.2B). A landscape with high persistence will only slightly shift down the continuum under the impact of the same disturbance. Highly resilient landscapes will rapidly recover, say in a matter of months or a few years, to a displacement down the continuum caused by a disturbance (Fig. 6.2C). Landscapes with low resilience may take centuries to recover from this same disturbance.

This rather simplistic and equilibrium-based concept of system stability has undergone a significant paradigm shift in recent times (Gunderson, 2000). Resilient ecosystems are now assumed to be complex and to have an adaptive capacity, where the components of the system adapt to disturbances, causing them to reorganize. Humans should now be considered an integral part of any ecosystem, which at times may appear to behave in chaotic and unpredictable ways because we are looking from within the system (see Pahl-Wostl, 1995). I feel these important conceptual and theoretical developments need to be extended to how we view fine-scale landscape functions.

Implications for landscape preservation and restoration

The basic theme of this paper can be stated as a simple first principle:

- Disturbances affect how well landscapes function to conserve resources and maintain biodiversity by degrading fine-scale patch structures and habitats, accelerating landscape processes such as water- and wind-driven erosion (little things count).

This leads to a second principle, applicable when the goal of land management is to preserve patch structures, resources, habitats, and species diversity within a landscape:

- It is far more effective ecologically and efficient economically to prevent landscape degradation by managing levels of disturbance than it is to attempt to rehabilitate a landscape after it has been degraded.

To apply this principle, the land manager must have a firm grasp of management goals, Otherwise, the levels of acceptable disturbance and degradation remain fuzzy or unknown (McIntyre and Hobbs, 1999). To make wise judgments about any landscape degradation, and to manage any disturbances, land managers must have effective monitoring systems in place (Tongway and Hindley, 2000). A high priority should be given to identifying indicators of landscape functionality and building these into monitoring procedures (Ludwig and Tongway, 1993).

A third principle applies when dealing with landscapes that have already been degraded, relative to one's management goals:

- Rehabilitate landscapes by repairing fine-scale patch structures first, then vegetation, soil fertility, habitat complexity, and biodiversity will follow.

This third principle has been successfully applied to degraded rangelands in Australia (Ludwig and Tongway, 1996; Tongway and Ludwig, 1996; Noble *et al.*, 1997). Small patches were constructed on a bare, degraded slope. These patches consisted of piles of tree and shrub branches, which were strategically positioned along slope contours to form obstructions to trap water and sediments running off from upslope. Within three years, soil fertility, infiltration rates, and soil biota increased significantly and perennial plants had established within the small patches, along with many invertebrates such as ants and termites. Although techniques such as contour banking and reseeding have been applied to rangeland rehabilitation and mine-site reclamation, these applications have often failed (Tongway and Ludwig, 1996). These failures are usually caused by a lack of understanding of this third landscape ecology principle: first rebuild fine-scale patch structures, then landscape source-to-sink processes will be set in motion to conserve resources and to build habitats and biodiversity, creating positive feedback systems. In the future, I believe improvements in the successful restoration, rehabilitation, or reclamation of degraded landscapes will be achieved by applying this principle.

Acknowledgments

This paper could not have been written without the years of stimulating research and discussions with CSIRO colleagues such as David Tongway and with Jornada colleagues such as Walt Whitford and Jim Reynolds.

References

Bunn, S. E., Davies, P. M., and Mosisch, T. D. (1999). Ecosystem measures of river health and their response to riparian and catchment degradation. *Freshwater Biology*, 41, 333–345.

Burgess, T. L. (1995). Desert grassland, mixed shrub savanna, shrub steppe or semidesert scrub? The dilemma of coexisting growth forms. In *The Desert Grasslands*, ed. M. P. McClaran and T. R. Van Devender. Tucson, AZ: University of Arizona Press, pp. 31–67.

Cavanagh, J. E., Burns, K. A., Brunskill, G. J., and Coventry, R. J. (1999). Organochlorine pesticide residues in soils and sediments of the Herbert and Burdekin river regions, North Queensland: implication for contamination of the Great Barrier Reef. *Marine Pollution Bulletin*, 39, 367–375.

d'Herbies, J.-M., Valentin, C., Tongway, D. J., and Leprun, J.-C. (2001). Banded vegetation patterns and related structures. In *Banded Vegetation Patterning in Arid and Semiarid Environments: Ecological Processes and Consequences for Management*, ed. D. J. Tongway, C. Valentin, and J. Seghieri. New York, NY: Springer, pp. 1–19.

Dick-Peddie, W. A. (1993). *New Mexico Vegetation: Past, Present and Future*. Albuquerque, NM: University of New Mexico Press.

Forman, R. T. T. (1995). *Land Mosaics: the Ecology of Landscapes and Regions*. Cambridge: Cambridge University Press.

Freudenberger, D., Noble, J., and Hodgkinson, K. (1997). Management for production and conservation goals in rangelands. In *Landscape Ecology, Function and Management: Principles from Australia's Rangelands*, ed. J.A. Ludwig, D. Tongway, D. Freudenberger, J. Noble, and K. Hodgkinson. Melbourne: CSIRO, pp. 93–106.

Grover, H. D. and Musick, H. B. (1990). Shrubland encroachment in southern New

Mexico, U.S.A.: an analysis of desertification processes in the American Southwest. *Climate Change*, 17, 305–330.

Gunderson, L. H. (2000). Ecological resilience: in theory and application. *Annual Review of Ecology and Systematics*, 31, 425–439.

Holling, C. S. (1973). Resilience and stability of ecological systems. *Annual Review of Ecology and Systematics*, 4, 1–23.

Huenneke, L. F. (1996). Shrublands and grasslands of the Jornada long-term ecological research site: desertification and plant community structure in the northern Chihuahuan Desert. In *Proceedings: Shrubland Ecosystem Dynamics in a Changing Environment*, ed. J. R. Barrow, E. D. McArthur, R. E. Sosebee, and R. J. Tausch. USDA Forest Service General Technical Report INT-GTR-338. Ogden, UT: USDA, pp. 48–50.

Ludwig, D., Walker, B., and Holling, C. S. (1996). Sustainability, stability and resilience. *Conservation Ecology*, 1, 1–27.

Ludwig, J. A. and Tongway, D. J. (1993). Monitoring the condition of Australian arid lands: linked plant–soil indicators. In *Ecological Indicators, Vol.* 1, ed. D. H. McKenzie, D. E. Hyatt, and V. J. McDonald. Essex: Elsevier, pp. 765–772.

Ludwig, J. A., and Tongway, D. J. (1996). Rehabilitation of semiarid landscapes in Australia. II. Restoring vegetation patches. *Restoration Ecology*, 4, 398–406.

Ludwig, J. A., and Tongway, D. J. (1997). A landscape approach to rangeland ecology. In *Landscape Ecology, Function and Management: Principles from Australia's Rangelands*, eds. J. A. Ludwig, D. Tongway, D. Freudenberger, J. Noble, and K. Hodgkinson. Melbourne: CSIRO, pp. 1–12.

Ludwig, J. A., and Tongway, D. J. (2000). Viewing rangelands as landscape systems. In *Rangeland Desertification*, ed. O. Arnalds and S. Archer. Dordrecht: Kluwer, pp. 39–52.

Ludwig, J. A., Eager, R. W., Williams, R. J., and Lowe, L. M. (1999). Declines in vegetation patches, plant diversity, and grasshopper diversity near cattle watering-points in the Victoria River District, northern Australia. *Rangeland Journal*, 21, 135–149.

Ludwig, J. A., Wiens, J. A., and Tongway, D. J. (2000a). A scaling rule for landscape patches and how it applies to conserving soil resources in savannas. *Ecosystems*, 3, 84–97.

Ludwig, J. A., Eager, R. W., Liedloff, A. C., *et al.* (2000b). Clearing and grazing impacts on vegetation patch structures and fauna counts in eucalypt woodland, central Queensland. *Pacific Conservation Biology*, 6, 254–272.

McIntyre, S. and Hobbs, R. (1999). A framework for conceptualising human effects on landscapes and its relevance to management and research models. *Conservation Biology*, 13, 1282–1292.

McIntyre, S. and Lavorel, S. (1994). Predicting richness of native, rare, and exotic plants in response to habitat and disturbance variables across a variegated landscape. *Conservation Biology*, 8, 521–531.

McIntyre, S. and Lavorel, S. (2001) Livestock grazing in subtropical pastures: steps in the analysis of attribute response and plant functional types. *Journal of Ecology*, 89, 209–226.

Noble, J., MacLeod, N., and Griffin, G. (1997). The rehabilitation of landscape function in rangelands. In *Landscape Ecology, Function and Management: principles from Australia's rangelands*, ed. J. A. Ludwig, D. Tongway, D. Freudenberger, J. Noble, and K. Hodgkinson. Melbourne: CSIRO, pp. 107–120.

Pahl-Wostl, C. (1995). *The Dynamic Nature of Ecosystems: Chaos and Order Entwined*. New York, NY: Wiley.

Pickett, S. T. A. and White, P. S. (eds.) (1985). *The Ecology of Natural Disturbance and Patch Dynamics*. New York, NY: Academic Press.

Prosser, I. P., Rutherford, I. D., Olley, J. M., Young, W. J., Wallbrink, P. J., and Moran, C. J. (2001). Large-scale patterns of erosion and sediment transport in river networks, with examples from Australia. *Marine and Freshwater Research*, 52, 81–99.

Reynolds, J. F., Virginia, R. A., Kemp, P. R., De Soyza, A. G., and Tremmel, D. C. (1999). Impact of drought on desert shrubs: effects of seasonality and degree of resource island development. *Ecological Monographs*, 69, 69–106.

Rietkerk, M., van den Bosch, F., and van de Koppel, J. (1997). Site-specific properties and irreversible vegetation changes in semi-arid grazing systems. *Oikos*, 80, 241–252.

Roderick, M. L., Noble, I. R., and Cridland, S. W. (1999). Estimating woody and herbaceous vegetation cover from time series satellite observations. *Global Ecology and Biogeography*, 8, 501–508.

Rundel, P. W., Montenegro, G., and Jaksic, F. M. (eds.) (1998). *Landscape Disturbance and Biodiversity in Mediterranean-type Ecosystems.* Berlin: Springer.

Schlesinger, W. H., Reynolds, J. F., Cunningham, G. L., *et al.*(1990). Biological feedbacks in global desertification. *Science*, 247, 1043–1048.

Seghieri, J. and Galle, S. (1999). Runon contribution to a Sahelian two-phase mosaic system: soil water regime and vegetation life cycles *Acta Oecologia*, 20, 209–218.

Tongway, D. J., and Hindley, N. (2000). Assessing and monitoring desertification with soil indicators. In *Rangeland Desertification*, ed. O. Arnalds and S. Archer. Dordrecht: Kluwer, pp. 89–98.

Tongway, D. J. and Ludwig, J. A. (1996). Rehabilitation of semiarid landscapes in Australia. I. Restoring productive soil patches. *Restoration Ecology*, 4, 388–397.

Tongway, D. J., and Ludwig, J. A. (1997). The conservation of water and nutrients within landscapes. In *Landscape Ecology, Function and Management: Principles from Australia's Rangelands*, ed. J. A. Ludwig, D. Tongway, D. Freudenberger, J. Noble, and K. Hodgkinson. Melbourne: CSIRO, pp. 13–22.

Tongway, D. J., and Ludwig, J. A. (2001) Theories on the origins, maintenance, dynamics, and functioning of banded landscapes. In *Banded Vegetation Patterning in Arid and Semiarid Environments: Ecological Processes and Consequences for Management*, ed. D. J. Tongway, C. Valentin, and J. Seghieri. New York, NY: Springer, pp. 20–31.

Tongway, D. J., and Ludwig, J. A. (2002a). Australian semi-arid lands and savannas. In *Handbook of Restoration Ecology, Vol. 2*, ed. M. R. Perrow and A. J. Davy. Cambridge: Cambridge University Press, pp. 486–502.

Tongway, D. J., and Ludwig, J. A. (2002b). Desertification, reversing. In *Encyclopedia of Soil Science*, ed. R. Lai. New York, NY: Marcel Dekker, pp. 343–345.

Turner, M. G. (ed.) (1987). *Landscape Heterogeneity and Disturbance*. New York, NY: Springer.

Van Auken, O. W. (2000). Shrub invasions of North American semiarid grasslands. *Annual Review of Ecology and Systematics*, 31, 197–215.

Wardell-Johnson, G. and Horwitz, P. (1996). Conserving biodiversity and the recognition of heterogeneity in ancient landscapes: a case study from south-western Australia. *Forest Ecology and Management*, 85, 219–238.

Wiens, J. A. (1997). The emerging role of patchiness in conservation biology. In *Enhancing the Ecological Basis of Conservation: Heterogeneity, Ecosystems, and Biodiversity*, ed. S. T. A. Pickett, R. S. Ostfeld, M. Shachak, and G. E. Likens. New York, NY: Chapman and Hall, pp. 93–107.

Wiens, J. A., Crist, T. O., With, K. A., and Milne, B. R. (1995). Fractal patterns of insect movement in microlandscape mosaics. *Ecology*, 76, 663–666.

With, K. A. (1994). Ontogenetic shifts in how grasshoppers interact with landscape structure: an analysis of movement patterns. *Functional Ecology*, 8, 477–485.

Woinarski, J. C. Z., Brock, C., Fisher, A., Milne, D., and Oliver, B. (1999). Response of birds and reptiles to fire regimes on pastoral land in the Victoria River District, Northern Territory. *Rangeland Journal*, 21, 24–38.

7

Scale and an organism-centric focus for studying interspecific interactions in landscapes

Ecologists arguably have been remiss in not developing a formal underpinning for the epistemology of ecology, at least not until the 1980s. At that time, the rather forced imposition of deterministic or heavily constrained stochastic population and community models (see Roughgarden, 1979) drew fire, principally through the emergence of ideas of system "openness" (Wiens, 1984; Gaines and Roughgarden, 1985; Amarasekare, 2000; Hughes et al., 2000; Thrush et al., 2000), non-equilibria (DeAngelis and Waterhouse, 1987; Seastadt and Knapp, 1993) and, especially, "scale" (Wiens et al., 1987; Kotliar and Wiens, 1990; Holling, 1992; Levin, 1992, 2000; Pascual and Levin, 1999). Scales of measurement and observation have tremendous impact on the interpretation of what we think we know about systems and how they operate, which clearly has ramifications for most of the hotly contested areas in community ecology. One such dispute concerns the respective roles of "top-down" (large-scale patterns determine the possibilities for small-scale ones; Whittaker et al., 2001) and "bottom-up" (large-scales are emergent properties of small-scale processes; Wootton, 2001; Ludwig, this volume, Chapter 6) processes in pattern generation in ecological communities (Carpenter et al., 1985).

An increasing number of field studies (e.g., Bowers and Dooley, 1999; Orrock et al., 2000) and simulations (e.g., Bevers and Flather, 1999; Mac Nally, 2000b, 2001) conducted at multiple spatial scales show that outcomes depend upon how the study is constructed and conducted. I focus here on the nature of scaling in studying the interactions of species and suggest a provisional, conceptual framework for judging whether a study has or can be considered to deliver meaningful information about a particular bilateral interaction (e.g., interspecific competition, predator–prey).

Three kinds of problems

While most ecologists probably have an intuitive feel about what they mean by the term "scale," useful general definitions have been harder to come by. Most workers seem comfortable identifying (1) the overall envelope of their study systems in space and time (the ecosystem was studied for the five years 1990–94, and comprised the area bounded by the coordinates...) and (2) the magnitude of the smallest sampling unit with which they probe their study system (0.25 m^2 quadrats were used ...). These are usually known as the extent and grain, respectively, of the study (King, 1991; Morrison and Hall, 2001). These ideas *have* been useful in the sense that they circumscribe the implied relevance of the study (extent) and also the actual spatial and temporal unit about which anything can be said directly (grain). However, these terms are descriptive and provide little help in overcoming the problems associated with identifying appropriate scales.

One distinction that is often missed in relation to the scaling question is the difference between *scaling* problems and *sampling* problems. These are not independent of each other, but they have some characteristics that address different questions. The scaling problem itself is a function of two aspects, which I refer to as (1) the *organism-centric* and (2) the *probing* problems, respectively. The organism-centric problem relates to the scales (how big? how long?) over which ecological processes take place (Petersen and Hastings, 2001). A major aspect of this involves how the participating players perceive, respond to, and move through the world. The probing problem, on the other hand, relates to the ways in which scale influences how ecologists themselves probe and view the world, dictating the nature of experiments, monitoring, and measurement (Mac Nally and Quinn, 1998).

Probing problems interact with organism-centric problems because the use of certain surveying, monitoring, and experimental methods may *artefactually* influence results (Walde and Davies, 1984; Gurevitch *et al.*, 1992; Petersen *et al.*, 1999). For example, caging experiments can confine animals to too small areas (Cooper *et al.*, 1990; Mac Nally, 1997; Petersen and Hastings, 2001), and also may influence ecologically important physical processes (e.g., hydrodynamics) in the vicinity of the cage (Schoener, 1983; Underwood, 1986). Can the ecological observer ever simultaneously construct spatial and temporal probes that are appropriate for all organisms involved in a particular ecological interaction (Mac Nally, 2000b), given that the organisms' individual yardsticks may be very different (Levin, 1992; Solon, this volume, Chapter 2)?

Sampling problems, on the other hand, often are almost purely statistical in nature. How should a program be designed? How many replicates of each treatment? Given observed variation, does the design have sufficient power to detect nominated effect sizes? In sampling, the objects under study can be anything and are represented by numbers – the same methods are used for quadrats and ball bearings. However, it is relatively easy to show that research programs designed with high statistical purity (appropriate randomization, replication, and power) can lead to nonsense results because of inappropriate scaling decisions (Mac Nally, 1997). We must ask: how reliable are tests of ideas and deductions? Are tests *ecologically* critical as distinct from *statistically* critical? What is the quality of the data vis-à-vis the question being posed (Mac Nally and Horrocks, 2002)? Given the explicit ecological focus of this volume, I concentrate almost entirely on the scaling problem and especially the organism-centric problem in an attempt to deal with scales in relation to the ways in which organisms view and respond to their landscapes.

An organism-centric approach

Each individual organism is likely to have an idiosyncratic view of the world as a function of its own attributes and, more importantly, its exposure to environmental variation. This also means that the designation of a "landscape" scale is not to be necessarily pitched at what seems to be a landscape for humans; consideration of the focal organisms themselves makes the term landscape a relative one (King, this volume, Chapter 4; Ims, this volume, Chapter 8).

For simplicity, I impose two restrictions. First, the perception of the organism depends upon just its somatic size and I disregard sensory capabilities (visual, aural), which may greatly increase the effective radius of the perception of some organisms. I use length, but volume or area might be more appropriate in some cases (Petersen and Hastings, 2001). Second, I ignore life history so that conspecific organisms are regarded as being homogeneous, reaching the same maximum length α, over a fixed lifetime δ. This is purely for convenience because of the complications potentially introduced by mortality schedules, differential age- and size-specific growth rates, etc. We can define a characteristic measure, λ, over the lifetime of the organism as just the product of α and δ. I suspect that generally $\lambda = O(\alpha\delta)$, where $O(.)$ denotes "of the order of." Note that λ has dimensions of length $\times$ time and, therefore, is an integrated measure of the spatial extension of the organism throughout its (living) existence.

The units describing size and lifetime might be selected to best suit a description of the organism in question. For example, reasonable maximum lifetimes and lengths for a number of diverse organisms are: *Escherichia coli* – c. 6 h, 0.5 mm; *Thunnus thynnus* – 7 yr, 2 m; *Loxodonta a. africana* – 60 yr,

3 m; and *Sequoiadendron giganteum* – 3000 yr, 20 m, taking crown diameter as the length measure. Common units could be used for all organisms to reflect directly the differences in their characteristic scales; λs in common units (in μm.h) are: *E. coli* – 3, *T. thynnus* – $O(10^{11})$, *L. a. africana* –·1.6 $\times 10^{12}$ and *S. giganteum* – $O(10^{15})$. Most workers will focus on sets of organisms with λs within an order or two of one another (e.g., competitors or predators and their prey). Given empirical functions relating maximum length (α), mass (M), and maximum life-span (δ) (e.g., Peters 1983: $\delta \propto M^{0.15} \propto \alpha^{0.6}$), we generally can expect $\lambda = O(\alpha^{1.6})$.

λ can be pictured as a natural scale against which to gauge the dynamics of the focal organism and the structure and variability of the landscape of that organism. λ, which covers both spatial and temporal aspects of organisms, is more general than measures of just body size that have been used widely (Peters, 1991; Smallwood and Schonewald, 1996; Ziv, 2000). λ can be used to scope the appropriate space-time scales for considering the way in which the landscape looks to the focal organism and how the organism *can* respond to landscape variation. Let E be a measure of the mobility of the organism (expressed in multiples of the characteristic measure λ), which is a function of the total movement of the organism over its lifetime. I refer to this as the *experience* of the organism.

Also, let L be a pertinent measure of landscape variation (e.g., separations of forested blocks) or resource fluctuation in the landscape (e.g., distribution of seeds), also scaled in units of λ. It is critical to clearly understand that L is a measure of *variation* in the landscape in both space and time. We often may think of L in terms of the extent of a study (e.g., 100 km^2 × 3 yr), but we should be interested in the variation of landscape structure pertinent to the organism (e.g., possibly the standard deviation of resource variation or an appropriately defined fractal characterization; Milne, 1991; Palmer, 1992). I assume for simplicity that the relationship between landscape fluctuations and spatial/temporal scale is linear up to a certain distance or time in what follows, but many relationships are possible and have been described (e.g., Schneider, 1994).

A scoping diagram can be constructed that relates E and L and tells us about the perception of the landscape from the perspective of the focal organism. If E and L are relatively similar, then the scaling suggests that the organism can perceive and is able to react to landscape patterns in a "concordant" way. This implies a resonance between the perceptive and potential reactivity of the organism and the scales over which the landscape varies or fluctuates. This is an intuitive assertion sharing logic with optimal foraging/habitat selection theory; if E and L are concordant, then the organism should be best able to exploit landscape characteristics pertinent to its ecological requirements.

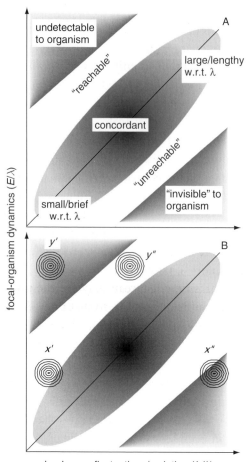

FIGURE 7.1
(A) Scoping diagram relating focal-organism dynamics E (expressed in units of λ) to fluctuations and variation in the landscape L (also expressed in units of λ). See text for description of named planar regions. (B) Scoping diagrams illustrating positions for the one organism relative to two landscape features with very different patterns of variation (x′, x″ or y′, y″).

Note that I present the concordant zone as a "fuzzy" ellipse in Fig. 7.1A, which indicates that the there are no "hard" boundaries as such but the farther from the equality line the less concordant are E and L.

While scaling by λ is not necessary when dealing with one taxon because this involves dividing both E and L by the same constant, it is important when interactions are considered because each taxon has its own characteristic λ and λ becomes the taxon-specific scaling factor that enables placement of each taxon in a common scoping diagram.

The concordant region divides the plane into two halves in which $E > L$ and $L > E$ (Fig. 7.1A). In the former, the organism is capable of perceiving and responding to landscape-scale variability, so that the variability and fluctuations are reachable or potentially exploitable by the organism. Landscape variation is not as well attuned to the organism's capabilities and cannot be exploited as well as in the concordant case. When $E >> L$, the organism is no longer able to identify the landscape-scale variability because it is too fine compared with the organism's spatial and temporal perspective – the landscape appears "flat" to the organism (Fig. 7.1). When $L > E$, the organism is unable to adequately perceive and especially to respond to and exploit landscape-scale variability and fluctuation, and when $L \gg E$, that variation is completely shielded from the capabilities of the organism (Fig. 7.1).

The $E \gg L$ and $L \gg E$ cases may seem similar superficially, but they differ very markedly. A concrete way of distinguishing between them is to consider the distribution of mussels in a bed on a rocky shore. To an oystercatcher (*Haematopus* sp.), discerning variation in nutrient content of potentially consumable mussels when sampled at stride lengths of 20 cm – hence O (km h^{-1}) – would be analogous to the $E \gg L$ situation, and would be even more extreme in flight. The appearance of this same mussel bed to a thaid predatory mollusc, which moves at O(cm h^{-1}), may correspond to $L \sim E$ ("$\sim$" means approximates), while a micro-parasitic crustacean may see the same bed as being a choice between at most a couple of mussels, so that $L \gg E$. So, the $E \gg L$ and $L \gg E$ cases differ because at one extreme the organism smooths over variation due to its large experience, while at the other extreme, the organism cannot experience much of the variation at all.

This scheme is capable of simultaneously representing different elements of landscape variation. For example, some landscape characteristics may change rapidly, such as food-resource distributions (point x' in Fig. 7.1B). In units of λ, such variation may be perceived well by the organism and be exploited effectively; i.e., $L \sim E$. On the other hand, a longer-term landscape change such as a shift in vegetation composition due to climate change (point x'', Fig. 7.1B) may be unperceived by the organism ($L \gg E$).

A case study

I use an example here to illustrate how one can think about scaling different resources with respect to λ. The swift parrot *Lathamus discolor* is a migrant of southeastern Australia; it is considered endangered, with perhaps only 2000 adults alive (Garnett and Crowley, 2000). The birds are about 0.25 m in length and may live for > 20 yr. Thus, $\lambda \sim 5$ m·yr. Migration occurs in autumn when the birds cross Bass Strait from breeding grounds in north-western Tasmania to the mainland, mostly residing in central Victoria for the winter (Mac Nally and Horrocks, 2000). In the overwintering period of the year, movements (= experience) are of the order of 1000 km in about 0.5 yr for 20 yr lifetimes, which is about 2×10^6 λ. Swift parrots in central Victoria appear to depend upon flowering of eucalypts and the availability of lerp (carbohydrate houses secreted on leaves for protection by psyllid bugs). Although difficult to measure directly because of the large areas involved, it seems that flowering in eucalypt forests varies at spatial extents of tens of kilometers over 0.25–0.5 yr (i.e., distances between points with the greatest differences in flowering intensities; Wilson and Bennett, 1999). This represents fluctuations of the order of $1–2 \times 10^3 \; \lambda$, about three orders of magnitude below the mobility scales (and hence experience) of swift parrots. It is more difficult to characterize spatial and temporal variation in lerp production, but it is likely to be of much smaller extent (hundreds of meters) but perhaps of longer duration (c. 1–2 yr), yielding scales of variability of about 50 λ. Thus, $E > L_{\text{flowering}}$ and $E >> L_{\text{lerp}}$. In the scoping plane, the eucalypt-flowering case might be at position y'' in Fig. 7.1B , which is reachable but not concordant, while the lerp condition may be at position y', corresponding to an undetectable scale of landscape variability. In principle, such dimensional arguments might be constructed for most of the landscape characteristics that are pertinent to an organism.

Some provisos

There are several key issues worth considering at this point. First, an organism's dynamics may be so large in space and time that we might have to seriously consider *not* studying some aspects, such as some interspecific interactions. I suggest maximum $E \sim O(10^6 \; \text{m·yr})$ might be a useful heuristic. For example, the average adult individual of the insectivorous bird, the rufous whistler *Pachycephala rufiventris*, in southeastern Australia migrates c. 4000 km yr^{-1} over lifetimes of about 10 yr, yielding $E \sim O \, (10^8 \; \text{m·yr})$. This suggests that competition between rufous whistlers and other insectivorous birds cannot be properly studied, at least by means known and used (and conceived?) by ornithologists up until now. Such studies often have been conducted at a local scale (typically < 50 ha) but competitive impacts and mechanisms of coexistence clearly are operating at continental scales (> 10^6 km^2; see Mac Nally, 2000a). There will be similar lower bounds on E at which it will be effectively impossible to conduct meaningful work *in situ*.

Second, it is necessary to develop theoretical bounds for when "concordant" changes to "reachable" and then to "undetectable," and similarly for the lower half of the plane. For example, do two orders of magnitude difference between L and E place the organism in the undetectable or in the reachable regions? How close to equality do L and E need to be for concordance? If such bounds cannot be constructed in a reasonable theoretical framework, then we will maintain a qualitative picture rather than develop a quantitative description. The latter clearly is to be preferred.

And third, the calculation of E and L as functions of λ is not absolute but refers to a mode of study, especially the time and extent over which the work is done. While E and L will have an "absoluteness" from the organism's perspective, we can rarely if ever determine this because ecologists choose – or are forced – to design sampling or observational programs that apply a possibly artefactual structure on E and L.

In general, the design of a research program will have a bigger effect on L than on E because both the spatial and temporal aspects of the research program will affect L (how big? and how long? for the study) but only the temporal component of E will be much influenced. However, it is possible to modify both E and L in a similar way that might position the study in the concordant zone but in a fashion that may be undesirable. For example, consider a species of nectarivorous bird that routinely moves over very extensive areas feeding from flowers. This may place this species in position x in Fig. 7.2A. A manipulation in which artificial feeders are supplied in an experimental area may cause the birds to move much less than before due to a regular supply of food, contracting E and possibly repositioning the species in the scoping plane to y (Fig. 7.2A). Even though now in the concordant zone, results will probably be artefactual; a more sensible repositioning would be to z (Fig. 7.2A). That is, either the spatial extent of the study or its duration (i.e., increase L) should be expanded to reach the concordant zone. This illustrates what I believe to be a general principle: as far as possible, do not manipulate E (or do so as little as possible) to force the position into the concordant zone because this will most likely lead to scaling artefacts (similarly, therefore, move from x' to z' not to y'). Confinement experiments are a classic case of this phenomenon – organisms may be restricted to a spatial extent far less than they would cover in normal circumstances (i.e. artificially small E; e.g, Schmitz et al., 1997). I consider this in detail elsewhere (Mac Nally, 2000b).

Scoping: interspecific interactions

Many ecologists and conservation biologists are interested in interspecific interactions, and some have questioned whether existing methods for studying interactions are providing relevant inferences, especially because of scaling difficulties (e.g., Frost et al., 1988; Carpenter, 1996). The implications of species-specific locations within the scoping plane are informative. One might start by identifying the principal aspect of landscape variation or fluctuation pertinent to each species. For simplicity, we will for the moment gloss over multiple positions in the scoping plane vis-à-vis different resources or landscape characteristics (see Fig. 7.1B).

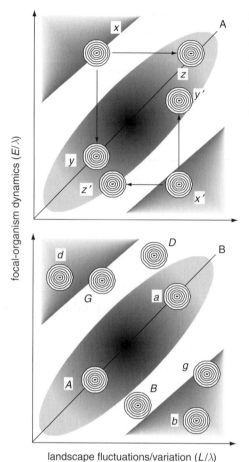

FIGURE 7.2

(A) Illustration of how different study designs can influence the position of an organism in the scoping plane (x, y, and z; x′, y′, and z′). (B) Scoping diagram illustrating some contrasting patterns among pairs of species a and A, b and B, d and D, and g and G.

focal-organism dynamics (E/λ)

landscape fluctuations/variation (L/λ)

I focus on resource competition here because it is the simplest case; it is implicit that both species focus on the same resource. Other interactions are much more complicated. Often, one partner in an interaction (e.g., predator–prey, host–parasite) may be held to contribute strongly to the landscape variation for the other (e.g., the distribution of prey for predators; DeAngelis and Petersen, 2001). Ideally, a general framework should be independent of this limitation, but at this stage, concentrating on resource competition is more clear-cut.

The scoping plane informs us how to regard the experience–landscape relationship of each species in an interaction *under a specified research program.* As described above, the location of species in the plane depends on a nominated program. The program must be developed with a view to how organisms will be positioned in the scoping plane as a function of that program. It also suggests that one should use an iterative procedure where best estimates of the characteristics of organisms and spatial and temporal variability of pertinent landscape characteristics are played off against possible program designs to estimate the location of the organisms in the plane; marrying field experiments with modeling seems a promising avenue (e.g., Cernusca *et al.* 1998, Schmitz 2000). Ideally, any study should aim to place focal organisms in the concordant zone, noting once more that manipulating L is preferable to changing E.

There are four main configurations of two-species interactions (Fig. 7.2B). Note that these scenarios arise from a nominated research program and use best estimates for the organisms and the landscape characteristics involved.

Case 1. This is the desired state. Both species are positioned in the concordant zone (a, A; Fig. 7.2B). This means that the landscape variability/fluctuation to which each species is most responsive conforms well to the species capabilities, scaled by λ (i.e., $E \sim L$ for both organisms). Thus, the researcher's designated program is likely to produce correct inferences in relation to the interaction.

Case 2. If both species lie in the lower region of the plane (b, B; Fig. 7.2B), then the experience, E, of individuals of both species is insufficient to allow them to recognize and to respond to landscape fluctuations and variation L. This may occur when one's program is too coarsely organized to detect appropriate variation in L with respect to E. For example, a study may explore competition among rotifers with sampling extending for 10 m^2 and samples being collected every month. Now, if the rotifer populations cycle within two weeks and individuals experience only 0.01 m^2, then the research program is too coarse to correctly identify the nature of interactions between populations. L needs to be rectified.

Case 3. In the upper part of the scoping plane, Es exceed (possibly greatly) landscape fluctuations L (d, D; Fig. 7.2B). This implies that the research program is unduly constrained in space and time and cannot correctly examine the interaction between the two populations of organisms, especially if one wishes to relate these to characteristics of the landscape. I suspect that much ecological research is conducted within this region of the scoping plane. Confinement experiments and many supply-and-demand research programs on resource competition are spatially limited, while numerous studies of vertebrates and long-lived invertebrates are too short. In most cases, the effect of program design will be to restrict L relative to E mainly due to the limitations of logistics. For example, if one were to look at competition between nomadic birds by using a system of plots covering 1 km², then the program design makes L very small compared with E, and so, the inferences are likely to be unreliable.

Case 4.   The last combination is where one species lies above and the other below the concordance region of the scoping plane (g, G; Fig. 7.2B). This situation implies that there are inverse relationships between E and L for the two organisms: the experience of G (scaled by λ_G) is large compared with landscape fluctuations of the resource, while the inverse is true for g. Given that there is a common resource, what does this mean?

One possibility is that $\lambda_G \gg \lambda_g$, so that the same resource fluctuation (in space and/or time) in the landscape appears small to G but large to g. Moreover, the experience of G relative to g (scaled by the respective λs) is large. Such an interaction may occur between birds and insects competing for nectar (Irwin and Brody, 1998; Lange and Scott, 1999; Navarro, 1999). If the birds are substantially larger and live much longer (hence $\lambda_G \gg \lambda_g$), and move much farther than the insects ($E(\lambda_G) \gg E(\lambda_g)$; e.g., Mac Nally and McGoldrick, 1997), then the G–g scenario may be satisfied. Note how difficult it would be to establish a definitive program to explore this interaction. Bird-scale observational studies (thousands of hectares) would be far too coarse to establish an impact on the insects, while insect-scale studies, which possibly may involve bird-exclusion experiments using netting (Fleming *et al.*, 2001), would not be capable of dealing with the simple option available to the birds of moving 10, 100, or 1000 m to other sources of nectar not included in the experiment.

Another possibility is that G and g differ mainly in mobility. Thus, both taxa would be similar in size and in life-length (i.e., $\lambda_G \sim \lambda_g$). The difference in mobility may correspond to competition between nomadic and sedentary birds, for example. The difficulty in designing an appropriate study in this case is that the competitive effects experienced by the sedentary taxon are very localized (although potentially measurable), while the analogous impacts on the nomads are integrated over possibly vast areas, effectively defying measurement by current methods. Of course, there will be situations in which the scales of study may be small by the ecologist's standards (e.g., 100 m² of rocky shore), so that in principle both the mobile and sedentary competitors might be studied (e.g., Mac Nally, 2000b). However, much ingenuity would be needed to attempt to discern the way in which the competition for the resource is expressed in the two organisms, which amounts to designing a program in which the G and g populations are more nearly co-located in the concordant zone.

To conclude, the scoping plane can be used to: (1) determine whether a particular interaction can be studied by using a particular, or indeed any, research-program design; or (2) refine and plan a research program to attempt to force both interacting species into the concordant zone. At worst, knowledge of how a program design positions potentially interacting populations within the scoping plane can alert one to the possibilities of inferential problems (Morrison and Hall, 2001).

Extensions

I would like to delve more deeply into the other aspects of organism-centric thinking, but these need greater development and detailed analyses are beyond the scope of this essay. As a sampler, the following issues need to be considered thoroughly.

(1) *The importance of the concordant zone.* At present, the assertion that interactions between populations are best studied when each population is in the concordant zone of the plane is just an assertion based on reasonable intuition. If the positions of *A* and *a* in the concordant zone are very different, as depicted in Fig. 7.2B, then this assertion amounts to the populations having different "harmonics" of landscape and experiential variation. It is important to establish first whether the assertion is generally supportable, and second to determine what major differences in position along the axis of the concordant zone might mean when inferring the nature of interactions.

(2) *Reconciliation of responses to multiple resources or aspects of landscape variation.* Different resources may scale quite differently in landscapes (e.g., distributions of flowering by eucalypt trees and the availability of lerp for swift parrots). This represents a general ecological difficulty in the sense that while one may be tempted to focus on resources that might appear most important (perhaps energetically or nutritionally), other resources that are critical for short periods of time (e.g., invertebrates for breeding nectarivorous birds; Paton, 1980) may be neglected. Nevertheless, attempting to design studies to cater for possibly several or many resources or landscape structural elements is challenging.

(3) *Time variation in the significance of alternative resources or aspects of landscape variation.* Similar comments to point (2) apply.

(4) *Ontogenetic changes and individual-specific responses.* In many taxa, larval or juvenile stages have very different ecological requirements to their adult counterparts, necessarily associating them with different suites of trophic interactors (e.g., Delbeek and Williams, 1987). Ontogenetic differences often may have a major influence on planning programs because different life-history stages may have to be considered as

separate entities in dealing with scale issues in ecological research. For example, pelagically dispersed marine larvae and their sedentary adults (e.g., Gaines and Roughgarden, 1985; Hughes *et al.*, 2000) may have to be treated as essentially distinct entities with different λs.

(5) *Impact of stochastic contingencies.* Some ecological factors, such as drought, wildfire, cyclones, etc., will have extents and intensities that are likely to vary dramatically on a case-by-case basis (Hobbs, 1987; Turner and Dale, 1991; Whigham *et al.*, 1991; Richards *et al.*, 1996; Lindbladh *et al.*, 2000). Such events may reconfigure entire landscapes, smoothing or fragmenting resource distributions and landscape features in a myriad of ways (Shugart, this volume, Chapter 5; Ludwig, this volume, Chapter 6). For long-lived organisms, *L*, a measure of landscape-scale variation, may change abruptly through the impact of such factors.

(6) *Point-to-point movement.* Large-scale migrants may effectively operate at much smaller scales over most of their lifetimes, separated by bouts of extensive movement (e.g., neotropical migrant birds; Williams and Webb, 1996; Linder *et al.*, 2000). This may need to be considered by using a series of different *E–L* scoping planes for different phases of the year (or, in some cases, life-history stages).

(7) *Operational estimation of "sufficiently large" sampling or experimental units.* What is the minimum size (space or time) needed for correct inferences (Frost *et al.*, 1988)? Englund (1997) and others interested in predator–prey interactions have begun to address this issue. Englund distinguished between population or global effects and local effects, where the former refer to the overall impacts on dynamics computed for the entire landscape, while local effects are manifestations of patchiness, such as the heterogeneity of distributions of prey or competitors, generated by interactions. Englund (1997) modeled predator–prey systems in a form in which enclosures were "permeable," allowing both predators and prey to move freely. He deduced that enclosures need to be so large that a measure of prey throughput, area-specific migration rate, would have to be < 5% per modeling time-step for local-scale estimates to lie within 10% of global population estimates of predation intensity. While Mac Nally's (2000b) modeling did not support this conclusion, Englund's (1997) approach is laudable and much more thought needs to be given to this area.

(8) *The marriage of data streams: observational and experimental information.* Given that it is difficult to evaluate experimentally all pair-wise interactions in a community because there are $(N/2)(N-1)$ such pairs among *N* taxa (Mac Nally, 2000b), some workers have advocated focusing

experimentally on the probable "strong interactors" to evaluate the main per capita interaction coefficients, and then to use regression approaches to "fill in the gaps" of the other elements of the community matrix. This is the basis of path analysis (Wootton, 1994a, 1994b, 1997; Berlow, 1999; Berlow *et al.*, 1999), a technique for combining measurements from diverse data streams. While thought to be problematic for statistical reasons (Petraitis *et al.*, 1996; Smith *et al.*, 1997), scale considerations require that data derived from alternative means need to be compatible. That is, biases, if they exist, must at least have similar scaling dependencies for combinations of different sources of data to be integrated. In a model system looking at interactions among pairs of grazing species having different mobilities (and hence experiences), I found that in some situations data derived from experimental manipulations (enclosures) may produce results that scale differently to results derived from quadrat-based measurements (Mac Nally, 2001). This is not unexpected given the earlier discussion about manipulating E, which experimental enclosures are designed to do; this should be avoided or at least limited.

Conclusions

One of the defining features of ecology as a discipline is the diversity of the characteristics of organisms with which we deal. A particular research program may be adequate to examine one organism but may be hopelessly inappropriate for investigations of another, similar organism if the latter is more routinely mobile, for example (Mac Nally, 2000b). There is a relativity of the experience of the organism and the nature of landscape-scale variation to each research program. By relating experience and landscape features to the characteristic measure λ of organisms, ecologists can assess more acutely the appropriateness of a proposed or existing program to the inferences that can be derived from the work. Ecologists should take stock of the existing compendia of information to assess the amount of faith that should be attached to published studies. The principal question is: could the workers demonstrate that the research was undertaken in the concordant zone of the scoping plane? If not, then how much faith can we have in the outcomes and inferences (Morrison and Hall, 2001)?

Acknowledgments

I thank John Wiens for kindly extending an invitation to contribute to this volume. I also thank Sam Lake for commenting on an earlier version of this manuscript. Erica Fleishman (Stanford University), as ever, applied the hot needle of inquiry to the manuscript, while members of the Aquatic Laboratory discussion group (Nick Bond, Rhonda Butcher, Gerry Quinn, Andrea Ballinger, Claudette Kellar, Natalie Lloyd) helped clarify certain points in the latest version. The author gratefully acknowledges the support of the Australian Research Council (Grant F19804210).

References

Amarasekare, P. (2000). The geometry of coexistence. *Biological Journal of the Linnean Society*, 71, 1–31.

Berlow, E. L. (1999). Strong effects of weak interactions in ecological communities. *Nature*, 398, 25.

Berlow, E. L., Navarette, S. A., Briggs, C. J., Power, M. E., and Menge, B. A. (1999). Quantifying variation in the strengths of species interactions. *Ecology*, 80, 2206–2224.

Bevers, M., and Flather, C. H. (1999). The distribution and abundance of populations limited at multiple spatial scales. *Journal of Animal Ecology*, 68, 976–987.

Bowers, M. A. and Dooley, J. L. (1999). A controlled, hierarchical study of habitat fragmentation: responses at the individual, patch, and landscape scale. *Landscape Ecology*, 14, 381–389.

Carpenter, S., Kitchell, J., and Hodgson, J. (1985). Cascading trophic interactions and lake productivity. *BioScience* 35, 634–639.

Carpenter, S. R. (1996). Microcosm experiments have limited relevance for community and ecosystem ecology. *Ecology*, 77, 677–680.

Cernusca, A., Bahn, M., Chemini, C., et al.(1998). ECOMONT: a combined approach of field measurements and process-based modelling for assessing effects of land-use changes in mountain landscapes. *Ecological Modelling*, 113, 167–178.

Cooper, S. D., Walde, S. J., and Peckarsky, B. L. (1990). Prey exchange rates and the impact of predators on prey populations in streams. *Ecology*, 71, 1503–1514.

DeAngelis, D. L. and Petersen, J. H. (2001). Importance of the predator's ecological neighborhood in modeling predation on migrating prey. *Oikos*, 94, 315–325.

DeAngelis, D. L. and Waterhouse, J. C. (1987). Equilibrium and nonequilibrium concepts in ecological models. *Ecological Monographs*, 57, 1–21.

Delbeek, J. C. and Williams, D. D. (1987). Food resource partitioning between sympatric populations of brackish water sticklebacks. *Journal of Animal Ecology*, 56, 949–967.

Englund, G. (1997). Importance of spatial scale and prey movements in predator caging experiments. *Ecology*, 78, 2316–2325.

Fleming, T. H., Sahley, C. T., Holland, J. N., Nason, J. D., and Hamrick, J. L. (2001). Sonoran Desert columnar cacti and the evolution of generalized pollination systems. *Ecological Monographs*, 71, 511–530.

Frost, T. M., DeAngelis, D. L., Bartell, S. M., Hall, D. J., and Hurlbert, S. H. (1988). Scale in the design and interpretation of aquatic community research. In *Complex Interactions in Lake Communities*, ed. S. R. Carpenter. New York, NY: Springer, pp. 229–258.

Gaines, S. and Roughgarden, J. (1985). Larval settlement rate: a leading determinant of structure in an ecological community of the marine intertidal zone. *Proceedings of the National Academy of Sciences USA*, 82, 3707–3711.

Garnett, S. T. and Crowley, G. M. (2000). *The Action Plan for Australian Birds*. Canberra: Environment Australia.

Gurevitch, J., Morrow, L. L., Wallace, A., and Walsh, J. S. (1992). A meta-analysis of competition in field experiments. *American Naturalist*, 140, 539–572.

Hobbs, R. J. (1987). Disturbance regimes in remnants of natural vegetation. In *Nature Conservation: the Role of Remnants of Native Vegetation*, ed. D. A. Saunders, G. W. Arnold, A. A. Burbidge, and A. J. M. Hopkins. Sydney: Surrey Beatty, pp. 233–240.

Holling, C. S. (1992). Cross-scale morphology, geometry, and dynamics of ecosystems. *Ecological Monographs*, 62, 447–502.

Hughes, T. P., Baird, A. H., Dinsdale, E. A., *et al.* (2000). Supply-side ecology works both ways: the link between benthic adults, fecundity, and larval recruits. *Ecology*, 81, 2241–2249.

Irwin, R. E. and Brody, A. K. (1998). Nectar robbing in *Ipomopsis aggregata* : effects on pollinator behavior and plant fitness. *Oecologia*, 116, 519–527.

King, A. W. (1991). Translating models across scales in the landscape. In *Quantitative Methods in Landscape Ecology*, ed. M. G. Turner and R. H. Gardner. New York, NY: Springer, pp. 479–517.

Kotliar, N. B. and Wiens, J. A. (1990). Multiple scales of patchiness and patch structure: a hierarchical framework for the study of heterogeneity. *Oikos*, 59, 253–260.

Lange, R. S. and Scott, P. E. (1999). Hummingbird and bee pollination of *Penstemon pseudospectabilis*. *Journal of the Torrey Botanical Society*, 126, 99–106.

Levin, S. A. (1992). The problem of pattern and scale in ecology. *Ecology*, 73, 1943–1967.

Levin, S. A. (2000). Multiple scales and the maintenance of biodiversity. *Ecosystems*, 3, 498–506.

Lindbladh, M., Bradshaw, R., and Holmqvist, B. H. (2000). Pattern and process in south Swedish forests during the last 3000 years, sensed at stand and regional scales. *Journal of Ecology*, 88, 113–128.

Linder, E. T., Villard, M. A., Maurer, B. A., and Schmidt, E. V. (2000). Geographic range structure in North American landbirds: variation with migratory strategy, trophic level, and breeding habitat. *Ecography*, 23, 678–686.

Mac Nally, R. (1997). Scaling artefacts in confinement experiments: a simulation model. *Ecological Modelling*, 99, 229–245.

Mac Nally, R. (2000a). Co-existence of a locally undifferentiated foraging guild: avian snatchers in a southeastern Australian forest. *Austral Ecology*, 25, 69–82.

Mac Nally, R. (2000b). Modelling confinement experiments in community ecology: differential mobility among competitors. *Ecological Modelling*, 129, 65–85.

Mac Nally, R. (2001). Interaction strengths and spatial scale in community ecology: quadrat-sampling and confinement experiments involving animals of different mobilities. *Ecological Modelling*, 144, 139–152.

Mac Nally, R. and Horrocks, G. (2000). Landscape-scale conservation of an endangered migrant: the Swift Parrot *Lathamus discolor* in its winter range. *Biological Conservation*, 92, 335–343.

Mac Nally, R. and Horrocks, G. (2002). Proportional spatial sampling and equal-time sampling of mobile animals: a dilemma for inferring areal dependence. *Austral Ecology*, 27, 405–415.

Mac Nally, R. and McGoldrick, J. M. (1997). Landscape dynamics of bird communities in relation to mass flowering in some eucalypt forests of central Victoria, Australia. *Journal of Avian Biology*, 28, 171–183.

Mac Nally, R. and Quinn, G. P. (1998). Symposium introduction: the general significance of ecological scale. *Australian Journal of Ecology*, 23, 1–7.

Milne, B. T. (1991). Lessons from applying fractal models to landscape patterns. In *Quantitative Methods in Landscape Ecology,* ed. M. G. Turner and R. H. Gardner. New York, NY: Springer, pp. 199–238.

Morrison, M. L. and Hall, L. S. (2001). Standard terminology: toward a common language to advance ecological understanding and application. In *Predicting Species Occurrences: Issues of Accuracy and Scale*, ed. J. M. Scott, P. J. Heglund, and M. L. Morrison. Washington, DC: Island Press, pp. 43–53.

Navarro, L. (1999). Pollination ecology and effect of nectar removal in *Macleania bullata* (Ericaceae). *Biotropica*, 31, 618–625.

Orrock, J. L., Pagels, J. F., McShea, W. J., and Harper, E. K. (2000). Predicting presence and abundance of a small mammal species: the effect of scale and resolution. *Ecological Applications*, 10, 1356–1366.

Palmer, M. W. (1992). The coexistence of species in fractal landscapes. *American Naturalist*, 139, 375–397.

Pascual, M. and Levin, S. A. (1999). From individuals to population densities: searching for the intermediate scale of nontrivial determinism. *Ecology*, 80, 2225–2236.

Paton, D. C. (1980). The importance of manna, honeydew and lerp in the diets of honeyeaters. *Emu*, 80, 213–226.

Peters, R. H. (1983). *The Ecological Implications of Body Size*. Cambridge: Cambridge University Press.

Peters, R. H. (1991). *A Critique for Ecology*. Cambridge: Cambridge University Press.

Petersen, J. E. and Hastings, A. (2001). Dimensional approaches to scaling experimental ecosystems: designing mousetraps to catch elephants. *American Naturalist,* 157, 324–333.

Petersen, J. E., Cornwell, J. C., and Kemp, W. M. (1999). Implicit scaling in the design of experimental aquatic ecosystems. *Oikos,* 85, 3–18.

Petraitis, P. S., Dunham, A. E., and Niewiarowski, P. H. (1996). Inferring multiple causality : the limitations of path analysis. *Functional Ecology,* 10, 421–431.

Richards, C., Johnson, L. B., and Host, G. E. (1996). Landscape-scale influences on stream habitats and biota. *Canadian Journal of Fisheries and Aquatic Sciences,* 53, 295–311.

Roughgarden, J. (1979). *Theory of Population Genetics and Evolutionary Ecology: an Introduction*. New York, NY: Macmillan.

Schmitz, O. J. (2000). Combining field experiments and individual-based modeling to identify the dynamically relevant organizational scale in a field system. *Oikos,* 89, 471–484.

Schmitz, O. J., Beckerman, A. P., and O'Brien, K. M. (1997). Behaviorally induced risk on food-web interactions. *Ecology,* 78, 1388–1399.

Schneider, D. C. (1994). *Quantitative Ecology: Spatial and Temporal Scaling*. San Diego, CA: Academic Press.

Schoener, T. W. (1983). Field experiments on interspecific competition. *American Naturalist,* 122, 240–285.

Seastadt, T. R. and Knapp, A. K. (1993). Consequences of nonequilibrium resource availability across multiple time scales: the transient maxima hypothesis. *American Naturalist,* 141, 621–633.

Smallwood, K. S. and Schonewald, C. (1996). Scaling population density and spatial pattern for terrestrial, mammalian carnivores. *Oecologia,* 105, 329–335.

Smith, F. A., Brown, J. H., and Valone, T. J. (1997). Path analysis : a critical evaluation using long-term experimental data. *American Naturalist,* 149, 29–42.

Thrush, S. F., Hewitt, J. E., Cummings, V. J., Green, M. O., Funnell, G. A., and Wilkinson, M. R. (2000). The generality of field experiments: interactions between local and broad-scale processes. *Ecology,* 81, 399–415.

Turner, M. G. and Dale, V. H. (1991). Modeling landscape disturbance. In *Quantitative Methods in Landscape Ecology*, ed. M. G. Turner and R. H. Gardner. New York, NY: Springer, pp. 323–352.

Underwood, A. (1986). The analysis of competition by field experiments. In *Community Ecology: Pattern and Process*, ed. J. Kikkawa and D. J. Anderson. Melbourne: Blackwell, pp. 240–268.

Walde, S. J. and Davies, R. W. (1984). Invertebrate predation and lotic prey communities: evaluation of in situ enclosure/exclosure experiments. *Ecology,* 65, 1206–1213.

Whigham, D. F., Olmstead, I., Cano, E. C., and Harmon, M. E. (1991). The impact of hurricane Gilbert on trees, litterfall, and woody debris in a dry tropical forest in the northwestern Yucatan peninsula. *Biotropica,* 23, 434–441.

Whittaker, R. J., Willis, K. J., and Field, R. (2001). Scale and species richness: towards a general, hierarchical theory of species diversity. *Journal of Biogeography,* 28, 453–470.

Wiens, J. A. (1984). Resource systems, populations and communities. In *A New Ecology: Novel Approaches to Interactive Systems*, ed. P. W. Price, C. N. Slobodchikoff, and W. S. Gaud. New York, NY: Wiley, pp. 397–436.

Wiens, J. A., Rotenberry, J. T., and Van Horne, B. (1987). Habitat occupancy patterns of North American shrubsteppe birds: the effects of spatial scale. *Oikos,* 48, 132–147.

Williams, T. C. and Webb, T. (1996). Neotropical bird migration during the ice ages: orientation and ecology. *Auk,* 113, 105–118.

Wilson, J. and Bennett, A. F. (1999). Patchiness of a floral resource: flowering of red ironbark *Eucalyptus tricarpa* in a box and ironbark forest. *Victorian Naturalist,* 116, 48–53.

Wootton, J. T. (1994a). The nature and consequences of indirect effects in ecological communities. *Annual Review of Ecology and Systematics,* 25, 443–466.

Wootton, J. T. (1994b). Predicting direct and indirect effects: an integrated approach using experiments and path analysis. *Ecology,* 75, 151–165.

Wootton, J. T. (1997). Estimates and tests of per capita interaction strength : diet, abundance, and impact of intertidally foraging birds. *Ecological Monographs*, 67, 45–64.

Wootton, J. T. (2001). Local interactions predict large-scale pattern in empirically derived cellular automata. *Nature*, 413, 841–844.

Ziv, Y. (2000). On the scaling of habitat specificity with body size. *Ecology*, 81, 2932–2938.

8

The role of experiments in landscape ecology

Why should landscape ecologists conduct experiments?

Experiments play a crucial role in science. They provide the most reliable and efficient means of establishing knowledge. Only proper experiments can establish cause–effect relations between processes and patterns as well as unambiguous links between abstract theory and material nature. Thus, experiments should be a part of scientific enquiries, whenever feasible and ethical.

Landscape ecology, however, is a scientific discipline relatively devoid of experiments. This well known, albeit undesirable, state of affairs is often said to stem from lack of practical feasibility to conduct landscape ecological experiments. True, landscape ecologists are frequently concerned with phenomena covering temporal and spatial scales that are too broad to facilitate an essential ingredient of proper experimental design; that is, replicates of treatment levels are randomized among a sample of experimental units. Clearly, if the extent of the experimental units encompasses region-wide landscapes and the treatments constitute levels of landscape variables such as composition and connectivity, proper experiments may not be feasible. So-called "quasi-experiments" or "natural experiments," which denote single large-scale accidental or intentional perturbations at the landscape level, or "mensurative experiments," referring to any kind of comparison with respect to a focal environmental variable (Hulbert, 1984; McGarigal and Cushman, 2002), provide unique opportunities for informative observations in landscape ecology. However, such approaches do not necessarily give rise to unbiased estimation of effect sizes and confidence intervals. This can only be reliably obtained through proper experiments. To avoid confusion about what kind of inference could be made from empirical studies, the term "experiment" should only be used when all ingredients of proper experiments are present (i.e., randomization, manipulation, replication).

 Issues and Perspectives in Landscape Ecology, ed. John A. Wiens and Michael R. Moss. Published by Cambridge University Press. © Cambridge University Press 2005.

Because of the difficulties in exploring the causal mechanisms underlying regional ecological dynamics, landscape ecology is sometimes claimed to share the constraints of other highly credible sciences dealing with broad-scale phenomena, such as geo- and astrophysics (Hargrove and Pickering, 1992). Given the apparent success of these physical sciences, this comparison, if valid, may seem encouraging. Why shouldn't landscape ecology be conducted without experiments when other disciplines do well without them?

There are several reasons why landscape ecologists should not look to the success of other experiment-poor disciplines to escape the practice of doing experimental work. The main reason regards the dialogue between theory and empirical work. This dialogue is facilitated by a precise theory on one hand and good data on the other. R. A. Fisher, the founder of modern experimental designs and inferential statistics, maintained that progress based on non-experimental data was dependent on a very elaborate and precise theory (Fisher's dictum; see Cox, 1992). But, whereas disciplines addressing broad-scale phenomena in physics have a strong unified theory that facilitates precise predictions (even about yet unobserved phenomena), landscape ecology has no such theoretical basis (Wiens et al., 1993; Wiens, 1995).

Improvement of theory is dependent upon good data. While physical sciences have the means to obtain a large number of precise, non-experimental measurements, observational studies in landscape ecology typically yield estimates of process–pattern relations that are far from precise. Confidence intervals around parameter values are large due to unexplained process variance and measurement errors. Moreover, estimates may be severely biased because of a great deal of uncertainty about what is the correct statistical model. This model uncertainty stems from the choice between a large number of candidate models, a choice that is guided by *post-hoc* statistical criteria (Burnham and Anderson, 1992, 1998) instead of a-priori formulations of causal models based on robust theory.

There is another snag in the analogy between landscape ecology and the "large-scale" physical sciences. In fact, it is not entirely true that the disciplines that some landscape ecologists use as examples of scientific "success without experimentation" are devoid of experiments. It is hard to imagine what would have been the status of geophysics without experiments to establish basic principles (e.g., the laws of thermodynamics), some of which operate on a fine scale. In this context, theory (i.e., mathematical models) provides the link between microscopic mechanisms amenable to experimental explorations and macroscopic phenomena beyond the reach of experiments. Eventual feedback loops between emergent macroscopic processes and their generating mechanisms may also be specified by such models. As yet, there is no such thing as an established set of basic principles for landscape ecology on which a firm predictive theory could build, although we may have hypotheses about what such principles may be (see below).

What kind of experiments should landscape ecologists conduct?

Are landscape ecological experiments at all feasible?

From a conceptual point of view, there are several reasons why experimental approaches should be applied in landscape ecology. Few landscape ecologists would probably disagree on that. However, opinions are more likely to differ with respect to the question of whether experiments addressing issues that are within the realm of landscape ecology are indeed feasible. One may suspect that differing opinions would reflect the variety of views on what landscape ecology really is. The least positive attitude toward experimentation would probably be held by those taking the view that landscape ecology should exclusively deal with ecological phenomena appearing at regional spatial scales and over long time periods (Hargrove and Pickering, 1992), and also that social, cultural, and political issues of the human interface with ecological processes need to be included (Naveh and Lieberman, 1990; Klijn and Vos, 2000). On the other hand, landscape ecologists who believe that questions about how spatial structure interacts with ecological processes, at any spatial and temporal scale (Wiens et al., 1993; Pickett and Cadenasso, 1995), are more likely to accept experiments as a feasible approach. When landscape ecological phenomena are not restricted to broad temporal and spatial scales, experiments should not be more difficult to conduct in landscape ecology than in any other branch of ecology.

Experiments on fundamental landscape ecological mechanisms

Whether experiments can be done in landscape ecology, however, may not be so dependent on which scales are of ultimate interest to landscape ecologists. Of greater importance is whether there are some fundamental ecological mechanisms that underlie landscape ecological phenomena that may be subject to experimental investigations, akin to the microscopic mechanisms underlying physical phenomena.

It has been argued that the movements of organisms within and between landscape elements are fundamental mechanisms underlying most landscape ecological phenomena (Wiens et al., 1993; Ims, 1995; With and Crist, 1996; Lima and Zollner, 1996). Movement processes may be expressed at any scale of resolution as spatial transition probabilities (Turchin, 1998). Consequently, experiments may be designed at manageable scales so as to treat transition probabilities as response variables that are functions of properties of the spatial mosaic being manipulated. Some potentially important spatial features such as patch-boundary characteristics (e.g., sharpness and curvature) may be manipulated in a randomized, replicated fashion at manageable scales in most systems. In light of practical feasibility, it is surprising that so few proper experimental designs have been applied to probe the effects of patch-boundary variables on the movement of individual organisms and the flow of matter between landscape elements. This is especially the case in view of the perceived importance of such processes in landscape ecology (Wiens *et al.*, 1985). Spatial characteristics at the patch scale, such as patch quality, size, and shape, require larger extents of experimental plots but are manageable in terms of manipulations that follow proper experimental designs. Many experiments that consider patch-scale parameters have been conducted over the last decade (for reviews see Debinski and Holt, 2000; McGarigal and Cushman, 2002). Above the patch scale, experimental studies have typically considered inter-patch distance and/or connectivity (Debinski and Holt, 2000), but usually include a small number of patches and a limited range of inter-patch distances. Experiments operating at a scale approaching what we usually term a landscape are still rare (e.g., Lovejoy *et al.*, 1986; Margules, 1992).

From small-scale experiments on mechanisms to inferences about landscape-level phenomena

Although experiments on movement responses to spatial heterogeneity are possible to conduct on fine spatial and temporal scales, landscape ecologists are ultimately interested in predicting the consequences of interactions between movement and spatial structure at larger spatial and temporal scales (Ims, 1995). An important issue is, therefore, whether knowledge about microscopic mechanisms (i.e., movement/spatial-structure interactions) firmly established by experiments can be used to derive predictions about macroscopic, emergent phenomena such as population or community dynamics. It is in this context that theoretical modeling should play a crucial role. Mechanistic models may be used to bridge the gap between fundamental mechanisms at the organismal level and dynamics at higher levels of organization (DeAngelis and Gross, 1992; With and Crist, 1996). Such models may also include feedback loops between the macroscopic emergent properties and the microscopic mechanisms from which these properties are derived (Bascompte and Solé, 1995).

As the gap between mechanisms and predictions in terms of levels of organization and temporal and spatial scales increases, the more likely it is that prediction errors will also increase. For example, a model based on known transition probabilities of individual organisms across patch boundaries conditional on patch and boundary properties is likely to yield larger prediction errors at the metapopulation level/landscape scale than at the population level/patch scale. Research protocols should be established to keep prediction errors in check by validating model predictions against empirical data step by step among levels of organizational and spatial or temporal hierarchies. The most reliable empirical checks in such a step-wise dialogue between theoretical modeling and empirical results, of course, are provided by experimental data. However, experimental testing is usually increasingly difficult as one moves upward in the hierarchy, especially when considering systems in which region-wide landscapes constitute the uppermost level.

Experimental model systems (EMS)

Spatial mosaics large enough to capture the phenomena in which landscape ecologists are ultimately interested do not necessarily need to have region-wide spatial extents and very slow process rates. In fact, spatial mosaics may be constructed (or physically modeled) for the particular purpose of encompassing landscape-level processes at a relatively fine scale, small enough to be amenable to experimental design. In such experimental model systems (Ims and Stenseth, 1989; Wiens et al., 1993; Bowers et al., 1996; Bowers and Doley, 1999), entire (micro)landscapes may be the replicate experimental units, the experimental treatments different levels of landscape heterogeneity (e.g., connectivity and composition), and the response variables landscape-level processes (e.g., source-sink and metapopulation dynamics).

The use of experimental model systems (EMS) has a long tradition in ecology. Early EMS studies in population ecology and community ecology were instrumental in the generation of new ideas and principles (McIntosh, 1985; Kingsland, 1995). Although EMS have been applied to all levels of organization within ecological systems, the practice of building empirical models to experimentally explore the dynamics of ecological systems has not been recognized as a distinct approach in ecology to the same extent as have theoretical models and other empirical approaches. Relatively few ecologists use EMS as a research tool. This situation may be changing, however, as some research teams are presently applying EMS systematically to explore aspects of the dynamics of single and interacting populations (e.g., Constantino et al., 1997; Maron and Harrison, 1998) and ecosystem processes (e.g., Lawton, 1995).

What is the current status of EMS studies in landscape ecology? The first study to establish the fact that spatial heterogeneity may be a key variable in ecological dynamics was laboratory-based EMS (Huffaker, 1958). However, it was not until the late 1980s that the EMS approach again started to play a significant role in probing the relationship between spatial heterogeneity and ecological processes (e.g., Kareiva, 1987; Forney and Gilpin, 1989; Wiens and Milne, 1989). Landscape ecologists applying the EMS approach have, to an increasing degree, brought their systems outdoors so as to include larger spatial dimensions and more realistic features in their model landscapes than can be included in the typical laboratory bottle experiments (Kareiva, 1989). The use of larger experimental plots in the field has also opened the possibility of including model organisms other than arthropods and protists, which for a long time dominated EMS studies. Vertebrates such as small mammals are currently some of the most frequently used organisms in landscape ecological EMS (e.g., Robinson *et al.*, 1992; Harper *et al.*, 1993; Bowers *et al.*, 1996; Johannesen and Ims, 1996; Wolff *et al.*, 1997; Andreassen *et al.*, 1998; Barett and Peles, 1999). Still, there is a great need to include a wider variety of taxonomic groups possessing different life-history characteristics and trophic positions in future EMS studies. A "model organism bias" may severely limit the generality of insights derived from EMS (Burian, 1992).

Modern landscape ecological EMS address processes at many scales and levels of organization. These range from the behavioral decisions of individual organisms moving in fine-scale vegetation mosaics (Wiens *et al.*, 1995), through the demography of single populations in patchy habitats (e.g., Dooley and Bowers, 1998; Boudjemadi *et al.*, 1999; Ims and Andreassen, 1999), predator–prey dynamics (e.g., Kareiva, 1987; Warren, 1996; Burkey, 1997; Ims and Andreassen, 2000), up to the level of species richness and ecosystem processes (Gonzales *et al.*, 1998; Golden and Crist, 1999; Collinge, 2000; Gonzales and Chaneton, 2002). In some EMS, responses at several spatial scales and levels of organization are simultaneously explored (Bowers and Dooley, 1999). EMS of this kind are particularly valuable, as the step-wise protocol of predictions and experimental tests in spatial/organizational hierarchies can be adopted. Establishing reliable knowledge about which processes are most likely to propagate through many levels of organization and spatial scales in spatial mosaics will be crucial for establishing a firmer theoretical basis for landscape ecology. Such knowledge will most likely be derived from multiscale EMS studies in conjunction with theoretical modeling.

Conclusion

Some landscape ecologists express doubts that designed experiments, which necessarily have to be conducted on fine temporal and spatial scales and at a mechanistic level in region-wide landscapes or in an EMS setting, are of much use in landscape ecology. The most pronounced skeptics seem to be those who view landscape ecology as primarily a tool for tackling management problems in region-wide landscapes. Such a view, however, is probably due to the misconception that new knowledge is most significant and relevant if it can be immediately applied to "real problems." The role of landscape ecological experiments in contributing to the establishment of a solid theoretical foundation for an immature scientific discipline is more important than any instant applicability of experimental results to applied problems. Poor theories are likely to yield poor guidelines for experimental designs. Theory will not readily advance without having its basic principles firmly established through the sort of strong empirical inferences only proper experiments can provide. No science is likely to remain viable without sound, well-developed theory. Because theory building and experimentation are intimately intertwined, landscape ecologists need to consider properly designed experiments as a necessary approach within their science in the twenty-first century.

References

Andreassen, H. P., Hertzberg, K., and Ims, R. A. (1998). Space-use responses to habitat fragmentation and connectivity in the root vole *Microtus oeconomus*. *Ecology*, 79, 1223–1235.

Barett, G. W. and Peles, J. D. (1999). *Landscape Ecology of Small Mammals*. New York, NY: Springer.

Bascompte, J. and Solé, R. V. (1995). Rethinking complexity: modeling spatiotemporal dynamics in ecology. *Trends in Ecology and Evolution*, 10, 361–366.

Boudjemadi, K., Lecompte, J., and Clobert, J. (1999). Influence of connectivity on demography and dispersal in two contrasted habitats: an experimental approach. *Journal of Animal Ecology*, 68, 1207–1224.

Bowers, M. A. and Dooley, J. L. (1999). EMS studies at the individual, patch, and landscape scale: designing landscapes to measure scale-specific responses to habitat fragmentation. In *Landscape Ecology of Small Mammals*, ed. G.W. Barrett and J.D. Peles. New York, NY: Springer, pp. 147–174.

Bowers, M. A., Gregario, K., Brame, C. J., Matter, S. F., and Dooley, J. L. (1996). Use of space and habitats by meadow voles at the home range, patch and landscape scales. *Oecologia*, 105, 107–115.

Burian, R. M. (1992). How the choice of experimental organism matters: biological practices and discipline boundaries. *Synthese*, 92, 151–166.

Burkey, T. V. (1997). Metapopulation extinction in fragmented landscapes: using bacteria and protozoa communities as model ecosystems. *American Naturalist*, 150, 568–591.

Burnham, K. P. and Anderson, D. R. (1992). Data-based selection of an appropriate biological model: the key to modern data analysis. In *Wildlife 2001: Populations*, ed. D.R. McCullough and R.H. Barrett. New York, NY: Elsevier, pp. 16–30.

Burnham, K.P., and Anderson, D.R. (1998). *Model Selection and Inference: a Practical Information-theoretic Approach*. New York, NY: Springer.

Collinge, S. K. (2000). Effects of grassland fragmentation on insect species loss, colonization, and movement patterns. *Ecology*, 81, 2211–2226.

Costantino, R. F., Desharnais, R. A., Cushing, J. M., and Dennis, B. (1997). Chaotic dynamics in an insect population. *Science*, 275, 389–391.

Cox, D. R. (1992). Causality: some statistical aspects. *Journal of the Royal Statistical Society*, 155, 291–301.

DeAngelis, D. L., and Gross, L. J. (1992). *Individual-Based Models and Approaches in Ecology: Populations, Communities and Ecosystems*. New York, NY: Chapman and Hall.

Debinski, D. M. and Holt, R. D. (2000). A survey and overview of habitat fragmentation experiments. *Conservation Biology*, 14, 342–355.

Dooley, J. L. and Bowers, M. A. (1998). Demographic responses to habitat fragmentation: experimental tests at the landscape and patch scale. *Ecology*, 79, 969–980.

Forney, K. A. and Gilpin, M. E. (1989). Spatial structure and population extinction: a study with *Drosophila* flies. *Conservation Biology*, 3, 45–51.

Golden, D. M. and Crist, T. O. (1999). Experimental effects of habitat fragmentation on old-field canopy insects: community, guild and species responses. *Oecologia*, 118, 371–380.

Gonzales, A. and Chaneton, E. (2002). Heterotroph species extinction, abundance and biomass dynamics in an experimentally fragmented microecosystem. *Journal of Animal Ecology*, 71, 594–602.

Gonzales, A., Lawton, J. H., Gilbert, F. S., Blackburn, T. M., and Evans-Freke, I. (1998). Metapopulation dynamics abundance and distribution in microecosystems. *Science*, 281, 2045–2047.

Hargrove, W. W. and Pickering, J. (1992). Pseudoreplication: a sine qua non for regional ecology. *Landscape Ecology*, 4, 251–258.

Harper, S. J., Bollinger, E. K., and Barrett, G. W. (1993). Effects of habitat patch shape on population dynamics of meadow voles (*Microtus pennsylvanicus*). *Journal of Mammalogy*, 74, 1045–1055.

Huffaker, C. B. (1958). Experimental studies on predation: dispersion factors and predator–prey oscillations. *Hilgardia*, 27, 343–383.

Hulbert, S. H. (1984). Pseudoreplication and the design of ecological field experiments. *Ecological Monographs*, 54, 187–211.

Ims, R. A. (1995). Movement patterns in relation to landscape structures. In *Mosaic Landscapes and Ecological Processes*, ed. L. Hansson, L. Fahrig, and G. Merriam. New York, NY: Chapman and Hall, pp. 85–109.

Ims, R. A. and Andreassen, H. P. (1999). Effects of experimental habitat fragmentation and connectivity on vole demography. *Journal of Animal Ecology*, 68, 839–852.

Ims, R. A., and Andreassen, H.P. (2000). Spatial synchronization of vole population dynamics by predatory birds. *Nature*, 408, 194–197.

Ims, R. A. and Stenseth, N. C. (1989). Divided the fruitflies fall. *Nature*, 342, 21–22.

Johannesen, E. and Ims, R. A. (1996). Modeling survival rates: habitat fragmentation and destruction in root vole experimental populations. *Ecology*, 77, 1196–1209.

Kareiva, P. (1987). Habitat fragmentation and the stability of predator–prey interactions. *Nature*, 326, 388–390.

Kareiva, P. (1989). Renewing the dialogue between theory and experiments in ecology. In *Perspectives in Ecological Theory*, ed. J. Roughgarden, R. M. May, and S. A. Levin. Princeton, NJ: Princeton University Press, pp. 68–88.

Kingsland, S. L. (1995). *Modeling Nature: Episodes in the History of Population Ecology*. 2nd edn. Chicago, IL: University of Chicago Press.

Klijn, J. and Vos, W. (2000). A new identity for landscape ecology in Europe: a research strategy for the next decade. In *From Landscape Ecology to Landscape Science*, ed. J. A. Klijn and W. Vos. Dordrecht: Kluwer, pp. 149–161.

Lawton, J. H. (1995). Ecological experiments with model systems. *Science*, 269, 328–331.

Lima, S. L. and Zollner, P. A. (1996). Towards a behavioral ecology of ecological landscapes. *Trends in Ecology and Evolution*, 11, 131–135.

Lovejoy, T. E., Bierregaard, R. O, Rylands, A. B. Jr., *et al.* (1986). Edge and other effects of isolation on Amazon forest fragments. In *Conservation Biology: the Science of Scarcity and Diversity*, ed. M. E. Soulé. Sunderland, MA: Sinauer Associates, pp. 257–285.

Margules, C. R. (1992). The Wog Wog habitat fragmentation experiment. *Environmental Conservation*, 19, 316–325.

Maron, J. L. and Harrison, S. (1998). Spatial pattern formation in an insect host–prasitoid system. *Science*, 278, 1619–1621.

McGarigal, K. and Cushman, S. A. (2002). Comparative evaluation of experimental approaches to the study of habitat fragmentation. *Ecological Application*, 12, 335–345.

McIntosh, R. P. (1985). *The Background of Ecology: Concepts and Theory*. Cambridge: Cambridge University Press.

Naveh, Z. and Lieberman, A. S. (1990). *Landscape Ecology: Theory and Application*. New York, NY: Springer.

Pickett, S. T. A. and Cadenasso, M. L. (1995). Landscape ecology: spatial heterogeneity in ecological systems. *Science*, 269, 331–334.

Robinson, G. R., Holt, R. D., Gaines, M. S., *et al.* (1992). Diverse and contrasting effects of habitat fragmentation. *Science*, 257, 524–526.

Turchin, P. (1998). *Quantitative Analysis of Movements*. Sunderland, MA: Sinauer Associates.

Warren, P. H. (1996). Dispersal and destruction in a multiple habitat system: an experimental approach using protist communities. *Oikos*, 77, 317–325.

Wiens, J. A. (1995). Landscape mosaics and ecological theory. In *Mosaic Landscapes and Ecological Processes*, ed. L. Hansson, L. Fahrig, and G. Merriam. London: Chapman and Hall, pp. 1–26.

Wiens, J. A. and Milne, B. (1989). Scaling of landscape in landscape ecology, or landscape ecology from a beetle's perspective. *Landscape Ecology*, 3, 387–397.

Wiens, J. A., Crawford, C. S., and Gosz, J. R. (1985). Boundary dynamics: a conceptual framework for studying landscape ecosystems. *Oikos*, 45, 421–427.

Wiens, J. A., Stenseth, N. C., Van Horne, B., and Ims, R. A. (1993). Ecological mechanisms and landscape ecology. *Oikos*, 66, 369–380.

Wiens, J. A., Crist, T. O., With, K., and Milne, B. T. (1995). Fractal patterns of insect movement in microlandscape mosaics. *Ecology*, 76, 663–666.

With, K. A. and Crist, T. O. (1996). Translating across scales: simulating species distributions as the aggregate response of individuals to heterogeneity. *Ecological Modelling*, 93, 125–137.

Wolff, J. O., Schauber, A., Edge, E. M., and Daniel, W. (1997). Effects of habitat loss and fragmentation on the behavior and demography of gray-tailed voles. *Conservation Biology*, 11, 945–956.

JANA VERBOOM

WIEGER WAMELINK

9

Spatial modeling in landscape ecology

Spatial models, expert knowledge, and data

Bringing together models and data yields more than the sum of both

The Netherlands experienced quite a controversy in January 1999 when an employee of the National Institute of Public Health and the Environment (RIVM) accused his employer, in the media, of relying too much upon unvalidated models instead of empirical data. He argued that the model outcomes were unreliable and that politicians are led to believe that they represent reality, when in fact they represent an artificial universe with no link to real data (Fig. 9.1). He made an interesting point, because models are often used without being calibrated, tested, validated, or analyzed for sensitivity and/or uncertainty. Furthermore, it is usually unclear what part of the model is based upon hard data and where expert knowledge fills in the gaps.

This essay is about models, expert knowledge and data, calibration, validation, and model analysis, and how we can apply these for evaluation or prediction. We argue that all these combined produce a more powerful tool than models, experts, or data do alone. We will not discuss the importance of space, or the merits of spatially explicit versus non-spatial or non-spatially explicit models. This issue has been thoroughly discussed elsewhere (Durrett and Levin, 1994a, 1994b; Wiens, 1997). This essay is a little biased toward spatial population models and vegetation dynamics models, which are our primary fields of interest. Although we offer several critical remarks, we are enthusiastic about the merits of spatial modeling for applying landscape ecological knowledge.

Issues and Perspectives in Landscape Ecology, ed. John A. Wiens and Michael R. Moss. Published by Cambridge University Press.
© Cambridge University Press 2005.

Tom Janssen

FIGURE 9.1
"...and here we are again <u>exactly</u> where we should be, according to my model...!" From newspaper Trouw (January 22, 1999), by permission of Tom Janssen.

Models are necessary for prediction

Correctly used, models are more powerful than crystal balls or experts

The times are long gone when a scientist could work on a problem undisturbed for a decade or longer, analyzing it in all its facets and unraveling all the details and, in the end, perhaps coming up with the perfect solution. With the growing need for applying landscape ecological knowledge, and for insights now, before biodiversity decreases even more, spatial models are increasingly useful for ecological impact assessment. They can apply the integrated knowledge of different disciplines (and experts) in a clear, reproducible way. Models are thus indispensable tools for prediction and ecological impact assessment. The problem is how to deal with incomplete knowledge and model uncertainty. The first point we want to discuss is how different kinds of models can be used for different purposes.

Strategic versus tactical models, or simple versus complex models

Strategic models are simple models useful for gaining insight into the process; tactical models are complex models useful for practical purposes such as prediction

"Make everything as simple as possible, but not simpler"

Einstein

Strategic models (*sensu* May, 1973) are general, simple, and parameter-sparse. A strategic model is based upon the most crucial underlying processes of the system under study, stripping reality to its bare essentials. Although unrealistic for any specific situation, and hence unsuitable for exact predictions, it leads to general insight. Strategic models are therefore of great value. For example, the metapopulation model derived by Levins (1970) includes only the processes of colonization and extinction. Two parameters describe the dynamics of the fraction of patches occupied in a world with an infinite number of equally sized and equally connected patches. In spite of its simplicity, this model provides general insight into metapopulation behavior and serves as a reference or limit case for more complex metapopulation models. It should be the starting point of all metapopulation modeling exercises. The spatially explicit counterpart of the Levins model is the contact process.

Tactical models (*sensu* May, 1973), on the other hand, are specific, complex, detailed, and have many parameters. If input processes are well understood qualitatively and input parameters are well known quantitatively, the models are realistic and suitable for exact predictions. Tactical models, however, do not lead to general insight. There are many examples of complex spatial models: e.g., the models used to forecast the weather and the "Across Trophic Level System Simulation" (ATLSS: DeAngelis *et al.*, 1998). Results of tactical models should be compared to the framework provided by strategic models as a first test: are results in accordance?

In this field of tension between simple and complex models, one has to compromise. A model should have just enough realism and accuracy for its purpose, yet the results should be generalizable. As no model can ever embody the full truth, any specific problem can be tackled through a series of models, ranging from simple models, which provide a better route to understanding, to complex models, which yield more specific results. Furthermore, it is important to work in close connection to empirical research: it only makes sense to include those parameters in the model of which we have or can obtain reasonable estimates, now or in the near future! Although the division described above looks very strict, in practice strategic models are not always simple and tactical models not always complex. An example of the first is the model

NUCOM (Oene *et al.* 1999), which is quite complex and used for understanding ecological and spatial processes in forest succession.

Mechanistic versus descriptive/statistic models

The trouble with descriptive models is that the relations are not necessarily causal; the trouble with mechanistic models is that they may miss an essential process

We can distinguish mechanistic, process-based, causal models from descriptive, static models that are often based upon a statistical relation found in a data set. Both classes have their merits in landscape ecology. Regression techniques are employed to detect relationships in empirical data sets. These relations then can be used for making predictions. Regression models, however, are purely descriptive, and the equations do not necessarily represent causal effects. For example, in the Netherlands, stork numbers and human birth rate are nicely correlated, but one should not apply this relation for predictive purposes. Descriptive models should therefore be applied with caution, especially when extrapolating outside the range of values of the specific situation on which the model was tuned. Moreover, the model may not be valid in another time or for another location.

Mechanistic models are based upon the underlying causal mechanisms or processes of a system. The challenge is to strip the complex, everyday reality of all the details, leaving only the key processes that matter. Such models can be used for impact assessment by modifying the input parameter values and surveying the change in the relevant model output, i.e., "turning the knobs" (Verboom, 1996). However, there are some problems with the use of mechanistic models. First, they are always a simplification of reality (do they capture all the essential causal mechanisms?) and second, parameterization, calibration, and validation are difficult. Resolution of the former problem depends to a great extent upon the level of expert knowledge available. The latter problem will be discussed below.

Chaos and stochasticity

Chaos is a surrogate of stochasticity in spatial population models

Empirical data often show huge fluctuations, occurring in space and time. There are four options for dealing with these fluctuations. First, in the case of predictable, externally driven fluctuations, one may unravel the mechanism that causes the fluctuations and include it in the model. For example, seasonality and latitude effects can be modeled this way. Second, in the case of unpredictable, externally driven fluctuations, one may add environmental noise to the input parameters. For example, a random noise may be added to the population growth rate, representing a fluctuating environment with good and bad years. Third, if one expects predictable, internally driven fluctuations, one may add feedback mechanisms that cause the system to behave chaotically. For example, strong density dependence with a time lag may cause chaotic fluctuations. And, fourth, random sampling effects caused by small numbers may cause fluctuations. In this case, adding demographic stochasticity to the model is the best solution. For example, genetic drift may occur in small and isolated populations.

Unfortunately, we often do not know the cause of fluctuations and, thus, which option to choose. Over the past decade there has been a strong interest in chaos theory among scientists, especially mathematicians, who like deterministic, strategic models that can be analyzed (semi-) analytically. It is our opinion, however, that fluctuations in empirical data sets are superimposed externally by a fluctuating environment or by small numbers, rather than internally by complex feedback mechanisms, especially when it concerns spatial population dynamics. Therefore, the models should have environmental noise and possibly demographic stochasticity added, not chaos-causing feedback mechanisms.

Model parameterization, calibration, and validation

Complex spatial models cannot be validated; calibration may result in the right results on the wrong grounds

"Give me five parameters and I will draw you an elephant; six, and I will have him wave his trunk"

Euler

This quotation (in Mollison, 1986) illustrates the first pitfall of model parameterization and calibration. Without restrictions, a complex model can be fitted to any data set, sometimes resulting in a remarkably good fit. However, the good "result" can very well be derived on the wrong grounds if the parameter values or, even worse, the model assumptions are wrong. Fortunately, there are usually some restrictions for the parameter values from expert knowledge or published field data, which indicate the range within which the parameter value is most likely to lie. With spatial population models, the results are often compared to patterns of presence and absence or to time series of patterns showing turnover and indicating occurrence probability. These data sets tend to be larger than the number of model parameters, making a unique calibration possible, at least in theory.

Unfortunately, in practice, several different combinations of parameter values can yield the same fit to the data (see, for example, ter Braak *et al.* 1997).

A second pitfall arises from stochastic or deterministic fluctuations at different space and time scales in the real world. We should be aware of the fact that models tend to extrapolate "trends" in data. These trends may be real or artefacts. An example of a real trend is the decline of a species in a region due to habitat loss. An example of an artefact is local extinction in a metapopulation: changes in time or space may occur in a small sample while the overall situation is stable. There is usually no way of telling whether an observed trend is real or not. However, the reverse may also occur: there is a trend but the data do not show it. For example, changes in the response variable may lag behind changes in the landscape, as in the hypothesized extinction debt. In summary, what goes into the model and what comes out are often linked in a fuzzy way and chance events and sampling errors in small data sets may have large and unwanted effects upon the outcome.

Model validation is often impossible because there are simply not enough data and no time series long enough. We realize that there is quite a difference between different types of models. For example, spatial vegetation data are often more readily available than spatial animal population data; animal movement data are especially hard to find. For example, testing a predicted MVP size (MVP stands for minimum viable population, defined as the population size with an extinction probability of 5% in 100 years) would mean waiting 100 years with, say, 100 independent replicas (populations of size MVP at year 0). On the other hand, for vegetation models data are available, though they are still sparse. This problem can sometimes be solved by using chronosequences: vegetation data are measured in the present for different stages of vegetation development. With the model, the present-day situation can be predicted with the initialization in the past, for instance when succession began or forests were planted. In this way the model can be validated for different vegetation stages.

Even if a model can be validated with an independent data set, the problem of the right result on the wrong grounds, as described above, remains. We will argue in the following sections that, despite all of these problems, models are valuable tools.

Sensitivity analysis and uncertainty analysis

Sensitivity analysis and uncertainty analysis are powerful tools for gaining insight into the properties and quality of the model and the system modeled

Sensitivity and uncertainty analysis have a lot in common, as both evaluate the effect of input parameters upon the model outcome.

A sensitivity analysis is a relatively simple "what if" study of the effect of changing a parameter, say, by 10% (point sensitivity: δ output/δ input) or in a range between a minimum and a maximum value (range sensitivity). An uncertainty analysis takes into account the uncertainties of the individual input parameters and uses regression to relate input-parameter values to model-outcome values. For example, an input parameter can be drawn from a lognormal distribution with a certain mean (the most likely value) and standard deviation (a measure of the confidence interval).

Sensitivity analysis is a simple but important tool for assessing the relative importance of model parameters: a small change in some parameters may yield a great effect on the output, while this output may be relatively insensitive to changes in other parameters. For example, the viability of a metapopulation may be much more sensitive to the adult survival rate than to the clutch size. The first application we want to mention is that the results of sensitivity analysis can suggest what management measures should be taken. In the example above, measures should be taken that affect the parameter "adult survival rate." Second, results of sensitivity analysis can lead empirical research to focus on the parameter that most affects the output (in the above example, adult survival rate). Both the precision of the model and a general ecological understanding of the system under study will benefit most if knowledge on the most crucial parameter is gathered. As opposed to these general rules, it may be more cost-effective in specific cases to measure or manipulate a less effective input parameter that can be measured or manipulated more easily (and more cheaply). Only an extended sensitivity analysis can point out the most cost-effective option. Third, sensitivity analysis may reveal errors in the model concept or in the computer program. In the example, metapopulation viability should increase monotonically with increasing adult survival rate.

In an uncertainty analysis, the combined effect of the uncertainty in all the input parameters on the model outcome is evaluated, and the contribution of all the individual parameters to this uncertainty. As a result, we can not only give the confidence interval of the model outcome, but also hints to decreasing the model's uncertainty. Insight into the contribution of individual parameters and their confidence intervals to the overall uncertainty reveals which input parameter should be given highest priority to be measured more precisely, resulting in a narrower confidence interval. Again, in specific cases, it may be more cost-effective to measure some parameter other than the one that contributes most to the uncertainty, as some parameters are more easily measured than others. Finally, both uncertainty analysis and sensitivity analysis can point out parameters that are unimportant and can be left out of the model or set to a fixed value.

An uncertainty analysis, unfortunately, requires much more effort than a sensitivity analysis. The simplest way of avoiding non-affecting parameters is by leaving them out beforehand. This is possible when sufficient data are available for the foundation of the model. Before building the model, an analysis of variance or a principal component analysis (PCA) could indicate which parameters are important to be incorporated in the model and which are not.

Scenario studies and comparative use of spatial models

Spatial models are particularly useful for comparative use, such as in scenario studies

Spatial models may be the only objective tools for scenario studies. Translating scenarios into model parameters, for example, metapopulation studies for animals ("turning the knobs"), can simulate effects of, for example, land-use changes. Even when no data quantifying the impact of measures on the input parameters are available, expert guesses and a safety range can be used. Although the exact quantitative model outcome is not necessarily correct or has a high level of uncertainty (large confidence interval), the qualitative results may be robust (insensitive to details in model specification). An example of this is shown by Schouwenberg *et al.* (2000) for the model NTM. They showed that this statistical model had a large uncertainty for a single prediction, but when scenarios were compared the uncertainty was much smaller. Consequently, the best alternative as predicted by the model is likely to be the best one in real life, provided that the model captured the essential qualitative behavior of species and landscape under study. An interesting approach is bringing the science of decision making into conservation ecology (Maguire *et al.* 1987; Possingham, 1997), showing under which conditions a certain decision is the best. Spatial models are probably the most powerful and objective tools we possess to evaluate scenarios.

Predicting (or projecting into) the future

Although we cannot predict the future, we can make projections into the future based upon our knowledge of the present and the past and the processes that cause the change

Considering all the problems and opportunities that have to be taken into account when using spatial models in landscape ecology, we conclude three things. First, we can learn a lot about the systems studied by building and analyzing the models. Second, when dealing with complex spatial phenomena, models are the best tools available for making projections into the future based upon our knowledge of the present, the past, and the processes that caused the changes. Compare to the weather forecast for tomorrow: not being able to predict exactly the weather at a certain time and place is no reason to stop producing weather forecasts. Third, as long as we use models in a comparative way, as in ranking consequences of different future land-use or management scenarios, we do not have to worry too much about the exact quantitative outcome being correct, especially for dynamic population models.

Future research priorities

Bringing together disciplines, bridging the gaps between theory and application, and between models and data

What we postulate above has been said many times before but is still worth repeating. We think not only that the gaps should be bridged, but also that in doing so we should build a sound and comprehensive framework of all available knowledge. Metz (1990; see also Metz and de Roos, 1992) modified May's classification of strategic and tactical models to obtain a better framework for providing a coherent and general picture of robust relations between mechanisms and phenomena, as opposed to the consideration of particular cases only. Within a general and encompassing class of strategic models, Metz distinguishes tactical models with a strategic goal (mathematically as simple as possible and constructed to uncover potential generalities) from tactical models with a practical goal (constructed for prediction or testing and usually incorporating lots of technically awkward detail) (Fig. 9.2). For application of models it is essential to keep this framework in mind. There are always limit cases and simple reference cases that set the frame: point models without space, models with implicit space, spatially explicit models with homogeneous space, models on a torus, models with infinite space. No model result should ever be interpreted as standing alone. It is good scientific practice to compare one's results to others and this is especially important for complex spatial models. On the other hand, all results should be communicated to other scientists for maintaining and supplementing the framework. The building blocks of the framework are not only model results, but also concepts, data, and (other) expert knowledge.

The second research priority is optimization and decision support. Optimization means looking for the best option instead of just evaluating given options. For example, given a certain budget for land acquisition or management, what action will result in the greatest increase in terms of population viability or biodiversity? Or, given the budget for a single ecoduct,

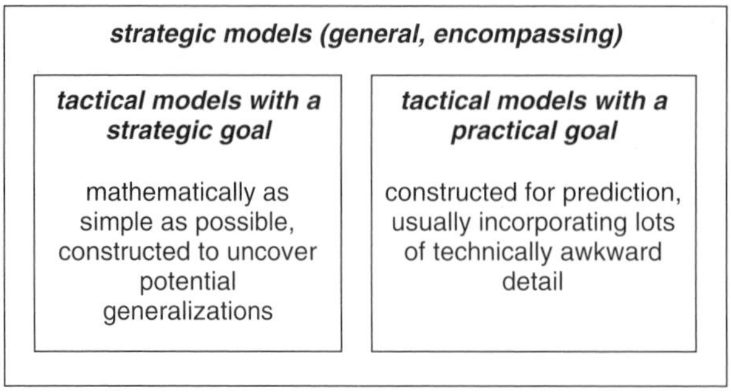

strategic models (general, encompassing)	
tactical models with a strategic goal	**tactical models with a practical goal**
mathematically as simple as possible, constructed to uncover potential generalizations	constructed for prediction, usually incorporating lots of technically awkward detail

FIGURE 9.2
Model classification, after Metz (1990).

where should we plan it; that is, which two populations should it connect? A related approach is multiple criteria evaluation, integrating knowledge of several disciplines. Which policy measure will be most successful under a wide variety of assumptions? Taking into account various aspects such as ground price, costs of management, and public appreciation, what is the best option? Introduce large grazers, change the landscape mechanically, use volunteers, or acquire new area (where?)?

The next generation of models is going to be even more complex than those of today because of more powerful computers, the availability of detailed small-grained GIS data sets, new techniques such as remote sensing, and coupling of existing models into model chains. This development will make all the points raised here, including error propagation, even more relevant.

To end with what we started with, we should aim for a good balance between data gathering and modeling, imbedding new results into the framework provided by existing ones, and performing model uncertainty analyses to provide model outcomes with confidence intervals. Politicians are probably not going to like it when we spend lots of time and effort on uncertainty analysis only to produce less pronounced results. However, that's the way it should be in a world where models are indispensable tools for evaluation and projection and where data and knowledge are sparse.

Epilogue

The two authors, although both involved in spatial modeling in landscape ecology, have very different backgrounds, which made writing this essay together particularly challenging. Whereas JV was been working with dynamic, stochastic, single-species, individual-based, metapopulation models for animals for more than 15 years, WW has mainly worked with statistical, static and dynamic (but multi-species, not individual-based) vegetation models. We discovered many differences between our modeling approaches, associated with the differences in model types, system characteristics, and data availability, to name just a few. These differences in approach and experiences led to many discussions during the process of putting together this essay. Surprisingly, however, we were able to find a solid common ground and there turned out to be more similarities than differences of opinion. We certainly learned a lot from this cooperation and hope our insights have the generality to help others.

References

DeAngelis, D. L., Gross, L. J., Huston, M. A., et al. (1998). Landscape modeling for Everglades ecosystem restoration. *Ecosystems*, 1, 64–75.

Durrett, R. and Levin, S. A. (1994a). Stochastic spatial models: a user's guide to ecological applications. *Philosophical Transactions of the Royal Society of London B*, 343, 329–350.

Durrett, R. and Levin, S.A.(1994b). The importance of being discrete (and spatial). *Theoretical Population Biology*, 46, 363–394.

Levins, R. (1970). Extinction. In *Some Mathematical Questions in Biology: Lectures on Mathematics in Life Sciences*, Vol. II, ed. M. Gerstenhaber. Providence, NY: American Mathematical Society, pp. 77–107.

Maguire, L. A., Seal, S. S., and Brussard, P. F. (1987). Managing critically endangered species: the Sumatran rhino as a case study. In *Viable Populations for Conservation*, ed. M. E. Soulé. Cambridge: Cambridge University Press, pp. 141–158.

May, R. M. (1973). *Stability and Complexity in Model Ecosystems*. Princeton, NJ: Princeton University Press.

Metz, J. A. J. (1990). Chaos en populatiebiologie. In *Dynamische Systemen en Chaos: een Revolutie Vanuit de Wiskunde*, ed. H.W. Broer and F. Verhulst. Utrecht: Epsilon, pp. 320–344.

Metz, J. A. J. and de Roos, A. M. (1992). The role of physiologically structured population models within a general individual-based perspective. In *Individual-Based Models and Approaches in Ecology*, ed. D. L. DeAngelis and L. J. Gross. New York, NY: Chapman and Hall, pp. 88–91.

Mollison, D. (1986). Modelling biological invasions: chance, explanation, prediction.

Philosophical Transactions of the Royal Society of London B, 314, 675–693.

Oene, H. van, van Deursen, E. J. M., and Berendse, F. (1999). Plant–herbivore interaction and its consequences for succession in wetland ecosystems: a modeling approach. *Ecosystems*, 2, 122–138.

Possingham, H. P. (1997). State-dependent decision analysis for conservation biology. In *The Ecological Basis of Conservation*, ed. S. T. A. Pickett, R. S. Ostfeld, M. Shachak, and G. E. Likens. New York, NY: Chapman and Hall, pp. 298–304.

Schouwenberg, E. P. A. G., Houweling, H., Jansen, M. J. W., Kros, J., and Mol-Dijkstra, J. P. (2000). *Uncertainty Propagation in Model Chains: a Case Study in Nature Conservancy*. Alterra Report 001. Wageningen: Alterra.

ter Braak, C. J. F., Hanski, I., and Verboom, J. (1998). The incidence function approach to modelling of metapopulation dynamics. In *Modeling Spatiotemporal Dynamics in Ecology*, ed. J. Bascompte and R. V. Solé. Georgetown, TX: Springer and Landes Bioscience, pp. 167–188.

Verboom, J. (1996). *Modeling Fragmented Populations: Between Theory and Application in Landscape Planning*. Scientific Contribution 3. Wageningen: IBN-DLO.

Wamelink, G. W. W., ter Braak, C. J. F., and van Dobben, H. F. (2003). Changes in large-scale patterns of plant biodiversity predicted from environmental scenarios. *Landscape Ecology*, 18, 513–527.

Wiens, J. A. (1997). Metapopulation dynamics and landscape ecology. In *Metapopulation Biology*, ed. I. A. Hanski and M. E. Gilpin. San Diego, CA: Academic Press, pp. 43–62.

10

The promise of landscape modeling: successes, failures, and evolution

In 1990, Fred Sklar and Robert Costanza began their review of spatial models in landscape ecology with this statement:

> We are at the dawn of a new era in the mathematical modeling of ecological systems. The advent of supercomputers and parallel processing, together with the ready accessibility of time series of remote sensing images, have combined with the maturing of ecology to allow us to finally begin to realize some of the early promise of the mathematical modeling of ecosystems. The key is the incorporation of space as well as time into the models at levels of resolution that are meaningful to the myriad ecosystem management problems we now face. This explicitly spatial aspect is what motivates landscape ecology.

They went on to describe a host of environmental and global issues that, because of their complexity, require spatial analysis and modeling to solve. While their introduction suggests the beginning of *Star Trek*, a popular television and movie series on another type of space exploration, there was a great deal of truth in what they said. The timing of their statement was also prescient. It is now over a decade since Sklar and Costanza and several other papers reviewed the status of landscape change models. Baker (1989) also laid out a useful framework for classifying and thinking about different landscape modeling approaches. While Baker emphasized different spatial and non-spatial methods for modeling changes in land cover classes, Sklar and Costanza (1990) took a somewhat broader view by framing landscape models within prior approaches coming from population models to ecosystem process models. A similar comprehensive review of landscape models at this time would be very useful, as well as a much greater task than it was in the early 1990s. Such a review is not my purpose here.

Nevertheless, the decade in landscape modeling marked out by those reviews spans an incredibly fertile period in the field, as well as a decade of

 Issues and Perspectives in Landscape Ecology, ed. John A. Wiens and Michael R. Moss. Published by Cambridge University Press.

emergence for landscape ecology in general. It is useful, at least, to step back and take a critical view of progress made and unfulfilled. I suggest that landscape modeling has made significant progress since 1990, but has not fulfilled all that Sklar and Costanza envisioned. I also believe that we must assess this progress with a view embedded in the general context of landscape ecology and its evolution. This context includes remaining cognizant of the roots of landscape ecology, as well as the field's placement within the evolving role of science and its relation to management and policy. For a number of reasons, all of this has changed; the successes, shortcomings, and future of landscape modeling must be assessed within this overall change. At the same time, the scale and complexity of many questions and management needs mean that landscape ecology is dependent on simulation models in a unique way. It is generally impossible to carry out landscape experiments (replicated!) at the broad scales relevant to many issues at hand. Especially in landscape ecology, models can be used to test ideas and hypotheses, as well as generate new questions for further research. Even "imperfect" (perhaps "simple" is a better word) models should be used, if their biases and results are clearly stated. These models may be simple conceptual creations, little more than decision diagrams, or complex simulators – all models are, after all, nothing more than systematically composed structures that represent our current knowledge of a system. Many ad hoc management decisions are being made every day with much less information, and less systematically.

The context of landscape models

For this essay, some context is needed to set out the area within landscape ecology and modeling I wish to address, as well as to lay out my personal assumptions. Many others have tried, with many more words than I have here, to describe what landscape ecology is. Indeed, many of the essays in this book take on parts of this task, as well as recent and past journal articles (e.g., Hobbs, 1997; Bastian, 2001; Wiens, this volume, Chapter 35). This continuing discussion is healthy in a relatively young field. We often speak of "North American" and "European" schools of landscape ecology, with the North American school, and particularly that of the United States, having its deepest roots in ecology, more narrowly defined as a branch of biological science. The European school is often described as more strongly derived from the landscape-planning tradition. While useful to some degree, this dichotomy is simplistic. Scanning the literature of landscape ecology over the past two decades certainly reveals influences from both roots in North America and Europe, as well as elsewhere on the globe.

Nevertheless, despite this growing identity it remains true and important to my topic that whatever landscape ecology is, it certainly is a field still described as transdisciplinary, multidisciplinary, or a hybrid discipline. Certainly there are opinions that disagree with this characterization, and indeed some declare that landscape ecology is not a field at all, but merely subsumed more closely within ecology (as "spatial ecology") or a practice that has for decades been carried out as landscape analysis and planning. Gratefully, I am not required to resolve this, and in fact can state my own premise that landscape ecology is all of these things, for better or for worse. Indeed, this seems necessary, because landscape ecology explicitly spans the spectrum from fundamental research to application. I believe that by definition a science that deals often with human-scaled landscapes and effects must integrate research and management.

An interesting and very significant aspect of the evolution of the field during the 1990s was the appearance of the first cohort of students trained first and foremost as "landscape ecologists." Evidence for this can be seen in many places, such as the evolving background of those attending scientific society meetings, as well as hiring within universities, agencies, non-governmental organizations (NGOs), and the private sector for positions explicitly labeled "landscape ecologist." Curricula in landscape ecology, or at least a course or two, have proliferated rapidly at many colleges and universities over the last decade. The first textbooks have also been published. This means that there are now practitioners, researchers, and teachers who have not come to the field after first being trained in another area of ecology, geography, planning, GIS, remote sensing, etc. This is important because of the knowledge and premises this new cohort has taken with them into a variety of professional positions. I believe this reflects a major change in the relation of science to, and its integration with, management. By necessity, models are a part of this change.

What are landscape models?

Staking out this larger framework matters for a discussion of landscape modeling because all of these branches or influences on landscape ecology carry out landscape modeling, often in very different ways. At the broadest level, landscape dynamics can be seen as a continuous loop in which landscape changes drive changes in processes – which can be biological, physical, or social – that in turn feed back and cause further, modified change. While this is indeed a connected loop, for the discussion here it can be useful to examine how different modeling approaches focus on various parts of this loop. This comes back to the varied and highly diverse roots of landscape ecology and its practitioners.

As described by Baker (1989), the simplest, conceptual model of a landscape is one that merely describes the components of a landscape (i.e., land-use or land-cover classes) in quantitative terms; that is, how much of each class is present on a landscape. This can be assessed by simple point sampling, and need not be explicitly spatial. It can be repeated at subsequent points in time to assess change. A progressively more spatial approach addresses not only which classes or how much of them make up a landscape, but where these classes are located. For this purpose a map is needed, and we are implicitly concerned now with the spatial distribution of classes, the size and shapes of patches or polygons, and their juxtaposition. Such a map may be in the form of cells (pixels) or polygons. We accept here the conceptual argument that classes of a landscape can be observed or measured in some way that allows patches or cells of a landscape to be placed into classes.

But needs, questions, and interests vary. For example, population and community ecologists operating within landscape ecology are often most interested in the effects of landscape structure and change on animal and plant species abundance, movement, and fecundity. Those with more of an ecosystem/process focus may emphasize the need to model influences and changes in water, carbon, and nutrient fluxes across time and changes in spatial structure. Some models may be simulators, but built simply to assess theoretical questions. Also, landscape modeling is growing into areas that require linking ecological processes with social drivers to address management questions.

In this essay, I am focusing on models of landscape change *per se*, rather than, for example, models of individual species change. Most landscape models of the type I treat here have some similar basis operationally. Most often, the landscape is represented as a grid of cells. Most landscape models project the state of the cells of a landscape at time $t + 1$ from their state at time t. At a minimum, projecting the state of a landscape at time $t + 1$ and later states requires information on the land-cover class or habitat type present in a cell at time t. Additional attributes about the cells in a landscape at time t or earlier states may be relevant. Such models can be defined as *spatially explicit*, because they operate on a map or a spatial representation of a landscape.

All change or dynamic models also operate based on rules. These rules can be simple, qualitative rules (e.g., "If state x at time t, then state y at time $t + 1$"), statistical relationships (such as those derived from empirical data and applied, for example, in a regression equation), or more complex mathematical relations. More complex landscape models also include information about adjacent patches or pixels in deciding how a given pixel will change. These interactions can vary a great deal in complexity, reflecting many interacting equations with multiple parameters and probability functions. Beside being *spatially explicit*, these latter models and can be defined as *spatially dynamic*, because spatial interactions between cells are considered in changing a given cell over a model time step. Both types can be *temporally dynamic*, and as a result are generally simulation models. Simulation models require execution by a computer, carry out multiple iterations, and do not have a single solution. These models have been seen to hold the promise of being able to integrate complex, interacting phenomena, including complex feedbacks. Such models, in any field, also include the potential for misuse, high uncertainty, and significant error. This latter topic is treated by various modeling texts. It is an important topic that needs more research, especially for complex landscape models.

Evolution of landscape modeling

Given this context, I return to my thesis that landscape modeling has advanced significantly since 1990 but has not been able to meet the hopes that were laid out by Sklar and Costanza (1990) and others. What changing factors account for the advances, and which have made progress slower than we wish?

North American landscape ecology in particular has strong roots in ecosystem science, which in turn largely derives from the International Biological Program (IBP) of the 1960s and 1970s. These roots have helped to drive modeling in landscape ecology that is oriented to problem-solving. The IBP program had a large component of simulation modeling of complex ecological systems. Of mixed success, it probably came of vision ahead of both ecosystem understanding and computational tools available at the time. For many researchers, landscape ecology was the obvious forum for the next stage of this type of work, adding a more explicit spatial component. Also, many ecosystem ecologists were involved in the development of US landscape ecology in the latter half of the 1980s. But to move beyond the point where things left off in the early 1980s required technical advances as well as conceptual growth in the science, and more data on ecosystems.

The growth in landscape ecology of the 1990s probably could not have occurred without the concurrent growth in computer power and accessibility. This does not mean that landscape ecology is primarily based on geographic information systems (GIS), remote sensing imagery, and simulation models, although much work in the field makes use of these tools. Yet, with the explicit consideration of space that landscape ecology has pushed to the forefront, few researchers or practitioners could carry out their work without this growth in technical capability, which we have quickly taken for granted. The computer power that we now have easily available on desktops and even in laptop computers has fostered a dramatic increase in creative applications and methods that underlie landscape ecology. Certainly one of these is modeling. The seductive potential of being able to simulate and represent spatially and visually future states of a landscape has great intuitive appeal and potential value, as well as pitfalls.

A useful example is the development of the LANDIS model (Mladenoff *et al.*, 1996; He and Mladenoff, 1999; Mladenoff and He, 1998, 1999) in my own lab. Around 1991–2, with several colleagues, we determined that many of the questions we wished to address in our research required a spatial model of forest change that included disturbance, management, and succession interactions operating at scales broader than a single stand.[1] We needed to address several issues that all model developers and users must consider to have even a chance of success. These included (1) what information and scale of mechanisms needed to be included in the model; (2) what was computationally possible on generally available desktop computers (Unix or Windows); (3) did adequate knowledge exist for parameterization; (4) did adequate input data of a starting forest landscape and its environment exist, or could it be reasonably created; and (5) could we develop parameter and input data requirements that would allow the model to be used in a variety of ecosystems and locations? We also decided that (6) the model would be built using a modular code structure in C++ that would facilitate iterative improvements and additions to the model.

There is a danger in relating this effort after the fact, in that it may appear more straightforward and organized than it really was. This was not the case. It was a slow, error-prone evolution and learning process. Several approaches were tried, including using a simple but innovative polygon or patch model (LANDSIM) developed at that time by Dave Roberts (Roberts, 1996). In the end, we built on much of his conceptual work, but opted for greater spatial and mechanistic complexity than a patch model could computationally or conceptually provide, and developed the LANDIS model, which is grid-cell based.

As a prototype began to evolve that addressed our needs and the necessary compromises, however, it became clear that the evolving design still far exceeded the current computational capacity of the computers we wanted to use. The final decision took advantage of one of the albatrosses associated with model development – it takes much longer than you hope. We planned out in more detail an attainable, operational model, taking advantage of Moore's Law of computer speed, namely that the speed of available computer processing chips doubles approximately every 18 months. In effect, we designed a model that we knew would need three of these computer speed increases (and associated increases in memory and storage capacity), a model

---

[1] Since that time, I have sometimes been accused of being a "modeler." I wish to state that I am not now nor have I ever been a "modeler." I was (and am) an ecologist who needed a model.

that would approach usable functionality, beyond a prototype, in three to four years. A prototype was presented in 1993 (Mladenoff *et al.*, 1993, 1996), and a full application of the model made in 1998 (He and Mladenoff, 1999). Since that time, the model has continued to evolve, and is being applied in new locations and with added modules (e.g., wind, fire, harvesting, disease, biomass) and other changes. Many of these are described in a special issue of *Ecological Modelling* (for example, Scheller and Mladenoff, 2004).

Taking advantage of new tools in creative ways to answer new questions and solve problems is a manifestation of human nature and how the enterprise of science works, despite its difficulties. Evolving computer capability and accessibility have certainly been factors advancing model use in research. As individuals we also bring a particularly broad array of scientific training, approaches, and opinions to the landscape modeling table. As mentioned earlier, many of us active in landscape modeling in the last decade were trained in other areas of ecology. In some ways, this has meant that great amounts of resources and time have been used, often to develop differing, complex modeling approaches to similar problems.

However, as I think my own example above shows, it is individual investigator-initiated research that drives innovation, although on the surface this may seem inefficient. By this I mean research that comes about in a "bottom up" fashion, with scientists developing and proposing ideas for research, rather than programmatic, "top down" research agendas, often bureaucratically imposed. This is not in opposition to collaboration, but suggests how fruitful small-group collaboration occurs, and why it is not more common. Even though ecosystem ecology has the tradition of working in collaborative groups, so far this has not resulted in a great deal of broad collaboration *across* groups that could produce a more commonly applied (and understood) "modeling toolbox." However, it should be noted that some examples exist and some groups are grappling with this.

Science, models, and management

While this evolution has been occurring within the science, the optimistic promises laid out by Sklar and Costanza (1990) and others have not gone unnoticed by management agencies and policy makers. Models that were only quirky research tools 10 or 15 years ago are now often being used in applied research either by or in collaboration with managers. Attempts to estimate environmental effects of human changes to the biosphere, especially effects of long-term climate change, have put broad-scale spatial models in front of everyone, from scientists, managers, and policy makers to daily news consumers. Not everyone believes or understands how these models work, but their projections are presented as scientific results (Aber, 1997). As model-users in all fields of science know, it is difficult to present simulation results that do not imply, or are often taken to be, truth (Dale and Winkle, 1998).

This struggle to link science with land and resource management is reflected in the various attempts at terminology that have evolved in the last decade and a half. These include terms such as "new forestry," "ecosystem management", and "sustainability." While they are dismissed as buzzwords by some, underlying these terms are efforts and trends to link and reshape how science is applied to management. I believe that landscape modeling is at the center of these efforts.

The needs of society will only increase the demand on landscape modeling from managers, environmentalists, and policy makers to provide answers to ecological questions and problems that can result in tangible recommendations. In different ways, this is the general problem in ecology of the putative dichotomy of "pure" versus "applied" science. This is a simplification, as these terms represent extremes of a continuum rather than a dichotomy. Nevertheless, most ecologists did not engage in applied research over most of the second half of the twentieth century. Engaging in applied research was looked down upon by most ecologists, even though such work can often address important scientific questions as well as provide guidance for environmental management. This situation began to change only slowly during the environmental movement of the late 1960s and 1970s. Greater involvement of scientists in advocacy also grew from this movement, although this is still an area of vigorous debate. Only 15 or 20 years ago, the journals *Ecological Applications* and *Conservation Biology* did not exist. Today, the difference between content of the journals *Ecology* and *Ecological Applications* is still detectable, but blurred. More recently, the newer journal *Ecosystems* is a continuation of this blurring of fundamental and applied research. I believe these changes are necessary and inevitable and will continue, and I suggest that landscape ecology took root in these changes. The explicit treatment of space on human-scaled landscapes it brought to the forefront helped to drive this growing link between ecological science and management.

Where does this leave us?

Model use, capability, and expectations have changed over the last decade. Disagreement between model users and non-users will continue, and this may be helpful. Even within modeling, different approaches, such as empirical or more conceptual process-based approaches, will all continue to find appropriate use. I have tried to show that any scientific field is stuck in its own unique context in time and will be affected by both good and less valuable influences that prevail at the time. They will all change together – the concept of feedbacks fits in such systems as well. This is true particularly of landscape ecology and landscape modeling. Modeling in ecology of all kinds has had its proponents and critics. Landscape modeling is perceived to have perhaps the greatest promise, often because of the addition of spatial ("real") interactions and visual representation. As modelers know, models must be based on some empirical data, even if only to reinforce the logic of simple rules incorporated into the model. It is often not clear where on this model-complexity continuum a given approach lies. Furthermore, I think there is some tendency to dismiss too quickly descriptive studies in landscape ecology. I believe landscape ecology can in part be compared to community ecology through the 1950s and 1960s: a young field requires a breadth of descriptive work, capitalizing on new, quantitative capabilities, to provide the basis for clear questions and hypotheses that might be addressed by experiments. This is how important processes and mechanisms are identified and empirical information is generated for modeling and decision-making. At the same time, the scale and complexity of many questions and management needs means that landscape ecology often is dependent on simulation models in a unique way. It is worth repeating that it is generally impossible to carry out landscape experiments at the broad scales required. In many systems this is an issue that no increase in funding can assist.

The need to address in research, and convey in results, what it is that models can actually do, and the uncertainty associated with model projections, is a need that others have expressed before. In many ways the current state of landscape modeling can be seen as a simple evolution of ecological modeling over time. The current context of this evolution, though, has contained several significant factors that have emerged rather quickly: (1) the promise of confronting explicitly spatial problems with spatial approaches, (2) the increase in computational capability available to nearly all researchers, (3) the intuitive appeal of visual, 2D and 3D representations, (4) increasing demand on the part of society to solve environmental problems, and (5) resulting demand from managers and policy makers to apply these appealing models and provide solutions.

Just as science in general continues to evolve, so will landscape modeling. Science is always a product of its changing social milieu, reflecting that context. Landscape modeling is also embedded in its own time and within the larger science of landscape ecology and the greater social context. They evolve both incrementally and with sudden shifts. For landscape modeling, the growth and advancement in the 1990s was the culmination of change within ecology and society since the 1960s that spawned the fertile link between North American and European influences. Advances in computer power, availability, and ease of use then furthered this burst in growth as a science and the link with applying science to management.

In a sense, the very strength of the landscape ecology paradigm – the importance of explicit consideration of space and its importance in ecological processes – is also its Achilles heel. Although the paradigm is profound, it inevitably leads to the conclusion that the science and modeling of landscapes is profoundly difficult. This leads to a major need for landscape modeling, one that has been acknowledged by Urban *et al.* (1999) and Baker and Mladenoff (1999). Better methods of testing landscape models, evaluating model uncertainty, and presenting results to both scientists and non-scientists are needed (Urban *et al.*, 1999; Schneider, 2001; Gardner and Urban, 2003). This issue has been better addressed for non-spatial models. Some approaches exist for spatial models, but they have not had widespread use or evaluation. This is clearly a priority, both for the role that models can play in scientific advancement and for their role in providing guidance for management and policy.

It is also inevitable that models will grow in complexity, as empirical knowledge, improved data, and computational capacity allow. But more mechanistic complexity is not necessarily a goal in itself. My current and former students probably now roll their eyes when I repeat that "any fool can make a model better by making it more complex." By that I mean that it seems to be our nature to see where things such as models can be improved by adding our favorite mechanism or details. Yet this quickly yields an unwieldy, useless beast, even if it can be parameterized. The framework of hierarchy theory suggests that we seek mechanistic explanation most commonly at a level below the focal level, or level in a system where our questions lie. In a general sense I believe this is true. Another of my often-repeated mantras is "we don't need to model what all the stomates are doing to predict forest change on a landscape." In part, this statement reflects a philosophical point of view. But it is also meant to raise a reminder that there are real limits to our knowledge and technical capabilities. These must be balanced with the need to find answers.

Related to this is the idea that no single model is best for a wide range of scales. The LANDIS model is one that can be customized to the scale and resolution desired, to a degree. Yet, when I receive inquiries from others concerning potential use of the model, the biggest problem is that users often want to use the model at a scale or for questions for which the model is not appropriate. My third common mantra is "different questions, different scales, different models." This fact is another reason why developing a common model "toolbox" is difficult. Nevertheless, this "toolbox" idea needs to remain as a goal, and is solvable. Landscape models are and will be imperfect.

At the same time, they will continue to be refined and become more common. The need for landscape models and the expectations placed on them continue to grow. The models and their context will continue to evolve. At the same time, landscape models have a great deal to contribute to research and management, as long as they are used appropriately.

References

Aber, J. D. (1997). Why don't we believe the models? *Bulletin of the Ecological Society of America*, 78, 232–233.

Baker, W. L. (1989). A review of models of landscape change. *Landscape Ecology*, 2, 111–133.

Baker, W. L. and Mladenoff, D. J. (1999). Progress and future directions in spatial modeling of forest landscapes. In *Spatial Modeling of Forest Landscape Change: Approaches and Applications*, ed. D. J. Mladenoff and W. L. Baker. Cambridge: Cambridge University Press, pp. 333–349.

Bastian, O. (2001). Landscape ecology: towards a unified discipline? *Landscape Ecology*, 16, 757–766.

Dale, V. H. and Winkle, W. V. (1998). Models provide understanding, not belief. *Bulletin of the Ecological Society of America*, 79, 169–170.

Gardner, R. H. and Urban, D. L. (2003). Model validation and testing: past lessons, present concerns, future prospects. In *Models in Ecosystem Science*, ed. C. D. Canham, J. C. Cole, and W. K. Lauenroth. Princeton, NJ: Princeton University Press, pp. 184–203.

Hobbs, R. (1997). Future landscapes and the future of landscape ecology. *Landscape and Urban Planning*, 37, 1–9.

He, H. S. and Mladenoff, D. J. (1999). Dynamics of fire disturbance and succession on a heterogeneous forest landscape: a spatially explicit and stochastic simulation approach. *Ecology*, 80, 81–99.

Mladenoff, D. J. and He, H. S. (1998). Dynamics of fire disturbance and succession on a heterogeneous forest landscape. US–International Association of Landscape Ecology, Annual Meeting, March 1998, E. Lansing, MI. Abstracts: 121.

Mladenoff, D. J., and He, H. S. (1999). Design and behavior of LANDIS, an object oriented model of forest landscape disturbance and succession.

In *Spatial Modeling of Forest Landscape Change: Approaches and Applications*, ed. D. J. Mladenoff and W. L. Baker. Cambridge: Cambridge University Press, pp. 125–162.

Mladenoff, D. J., Host, G. E., Boeder, J., and Crow, T. R. (1993). LANDIS: a model of forest landscape succession and management at multiple scales. Proceedings of the Annual US Landscape Ecology Symposium, Oak Ridge, TN, March 1993. Abstracts: 77.

Mladenoff, D. J., Host, G. E., Boeder, J., and Crow, T. R. (1996). LANDIS: a spatial model of forest landscape disturbance, succession, and management. In *GIS and Environmental Modeling: Progress and Research Issues*, ed. M. F. Goodchild., L. T. Steyaert, and B. O. Parks. Fort Collins, CO: GIS World Books, pp. 75–180.

Roberts, D. W. (1996). Modeling forest dynamics with vital attributes and fuzzy systems theory. *Ecological Modeling*, 90, 161–173.

Scheller, R. M. and Mladenoff, D. J. (2004). A forest growth and biomass module for a landscape simulation model, LANDIS: design, validation, and application. *Ecological Modelling*, 180, 211–229.

Sklar, F. and Costanza, R. (1990). The development of dynamic spatial models for landscape ecology: a review and synthesis. In *Quantitative Methods in Landscape Ecology*, ed. M. G. Turner and R. H. Gardner. New York, NY: Springer, pp. 239–288.

Schneider, S. H. (2001). What is "dangerous" climate change? *Nature*, 411, 17–19.

Urban, D. L., Acevedo, M. F., and Garman, S. L. (1999). Scaling up fine-scale processes to large-scale patterns using models derived from models: meta-models. In *Spatial Modeling of Forest Landscape Change: Approaches and Applications*, eds. D. J. Mladenoff and W. L. Baker. Cambridge: Cambridge University Press, pp. 70–98.

PART III

Landscape patterns

11

Landscape pattern: context and process

The analysis of pattern is a fundamental part of landscape ecology. Typically, we view landscape as a mosaic of elements and believe that their spatial arrangement controls or affects the ecological processes that operate within it. Similarly, we claim that landscape pattern itself is generated by other processes operating across such mosaics. As a scientific community, we face the problem that, while we agree about the importance of pattern, we have few theoretical generalizations to help those interested in the conservation or management of landscape resources (Wu and Hobbs, 2000). Much contemporary work on pattern has focused on the analysis or description of spatial geometry and has failed to provide any understanding of the significance or meaning of those patterns. This tendency has been exacerbated by the availability of digital landscape data and GIS algorithms that allow us to rapidly calculate a whole range of landscape metrics.

Some would dispute the claim that landscape ecology has provided few empirical generalizations about pattern. I feel able to make this claim because I too have been tempted down the road of analyzing landscape pattern using the computer-based technologies now widely available (e.g., Haines-Young and Chopping, 1996). My present unease comes from the observation that, while we have had some success in persuading the policy community that landscape ecology should be taken seriously, we have been unable to give much advice about the sensitivity of ecological systems to changes in the structure and composition of landscape mosaics (Opdam, *et al.*, 2001). Nor have we been able to suggest what kinds of landscape mosaic we should try to produce if we are to maintain and promote, say, biodiversity. At least this is the situation in Britain. I think it is the same elsewhere.

So what is the way forward? In this essay, I will take stock of where progress is being made, and then highlight ways in which we can broaden our thinking to address some of the wider practical challenges that face us.

Issues and Perspectives in Landscape Ecology, ed. John A. Wiens and Michael R. Moss. Published by Cambridge University Press.

Pattern and context

The reason why it is so difficult to make generalizations about landscape pattern is that, although it looks pretty interesting, it has little intrinsic meaning or significance. The significance or meaning of pattern only emerges when we consider it in the context of other problems or processes. As a result, the conclusions that we draw about pattern are often specific to particular ecological systems or geographical locations.

What can we say, for example, about habitat fragmentation? Certainly we can measure it, but pattern indices have little value unless we consider them in relation to the species that occur in the landscape. Some species may be affected adversely by fragmentation but others might be encouraged. Some might be neutral in their response. The message for landscape ecology is that pattern is an "explanatory variable" and we have to know what it is that we want to explain before we measure it. No measurement is "theoretically neutral." We cannot simply take a pattern index "off the shelf" and hope it will show something fundamental about landscapes. The analysis of pattern must start with consideration of ecological process. As Wu and Hobbs (2000) have suggested, "to make landscape metrics truly metrics of landscape, we must 'get inside' the numerical appearance of metrics to find their ecological essence."

Many excellent case studies show the value of pattern analysis when used as an explanatory rather than a descriptive tool. Jonsen and Fahrig (1997), for example, have shown how pattern can have quite different consequences for specialist and generalist insect herbivores in agricultural landscapes. Following their study of epigeic invertebrates in South Africa, Ingham and Samways (1996) have also emphasized both how different individual species' responses can be, and how they can differ from human perceptions of landscape pattern. More recently, Lawler and Edwards (2002) have shown how landscape pattern may be used to predict the occurrence of cavity-nesting birds in the Uinta Mountains of Utah.

Such studies illustrate that once we approach pattern in the context of process, landscape ecology can begin to make significant progress. Moreover, the development of models that link pattern and process could clearly enable the discipline to make a more valuable practical contribution. So is this where the future of pattern studies lies, in the more detailed analysis of structural pattern and process?

The use of pattern as an explanatory tool is a productive area of research and it will continue to develop. However, as we look to the future we need to broaden our thinking because, despite progress, recent work is limited in at least three respects. First, much of it is confined to landscapes that have a distinct spatial structure. What happens in landscapes where gradients rather than patches predominate? Second, while we are beginning to understand the consequence of pattern, we also need to understand what factors control the development of landscape pattern itself. This is important in a management context, when we seek to influence the development of landscapes. Finally, while biophysical models can be helpful for planning, landscape pattern also has meaning or significance in a cultural context. How do we deal with pattern in landscapes where people rather than nature are the dominant force?

Landscapes with fuzzy geometries

Although many indices of landscape pattern are available, most are of little value when we are faced with fuzzy landscapes, that is, landscapes that depart from Forman's (1995) patch–corridor–matrix model. We could deal with them by creating patches, using thresholds of various kinds, but this approach probably obscures many important processes.

Several studies are beginning to emphasize the importance of understanding the pattern of gradients in a landscape. Pickup and his co-workers used remotely sensed data to characterize grazing gradients on rangeland ecosystems in Australia. They showed that both the existence and steepness of environmental gradients can be essential to understanding ecological process in these areas (Pickup *et al.*, 1998).

Another example of what might be observed is shown in Fig. 11.1. These data come from a study that sought to model density of a wading bird, Dunlin (*Calidris alpina*), on the peat-covered landscapes of the Flow Country of Scotland (Lavers *et al.*, 1996). The density of small pools in the peat surface was found to be an important factor explaining spatial variations in bird numbers during the breeding season. Pools occur in clusters, and as the density of pools declines outwards from the cluster center, the density of Dunlin also falls. However, the character of the vegetation surface in which the pools are set also controls bird numbers. Thus, the rate of decline in density with distance depends on the position of the pools on a gradient related to vegetation composition and structure. Such data have been used to estimate the width of buffer zone that should be left around pool systems to minimize the impact of forestry on bird numbers in different parts of the study area.

It has been suggested that changes in gradient structure in fuzzy landscapes can be explored using texture measures (Musick and Grover, 1991). Such approaches lend themselves to the analysis of patterns using remotely sensed imagery. In forest or rangeland landscapes, for example, changes in management regime may affect the gradient structure and thus the distribution of species that map onto these surfaces. But such techniques of gradient analysis are still in their infancy. For the future we need a wider range of

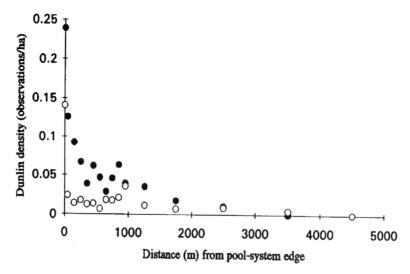

FIGURE 11.1
Dunlin density with distance from the edge of pool systems in the Flow Country, Scotland. Two sets of sites are shown, each drawn from different parts of a major vegetation gradient: solid circles = pool systems that are set in a low-biomass vegetation matrix dominated by *Calluna vulgaris* and *Tricophorum cespitosum*; and open circles = pool systems set in a higher-biomass vegetation matrix, dominated by *Calluna vulgaris* and *Molinea caerulea*. After Lavers *et al*. (1996).

techniques that can be used both to identify the existence of gradients and to classify and map them according to their ecological characteristics.

The dynamics of pattern

Until now we have considered the importance of analyzing landscape pattern as a step in explaining other ecological process. Of equal importance is an understanding of how landscape pattern itself is generated. Indeed, it could be argued that the study of the reciprocal relationship between process and patterns is now one of the key themes emerging in contemporary ecology (Perry, 2000).

Although landscape ecologists often stress the dynamic nature of landscapes, dynamics have rarely been used for landscape classification. Instead, we have tended to concentrate on the structure at a point in time in the hope that it gives an insight into the processes that generated it. Alternatively, we have stacked up a series of historical maps and hoped that the sequence will give us the necessary insight into pattern. The closest we have come to a dynamic analysis is, perhaps, through studies of "patch dynamics." But rarely has such work gone on to make a classification of landscape in terms of the spatial domains in which different disturbance regimes operate.

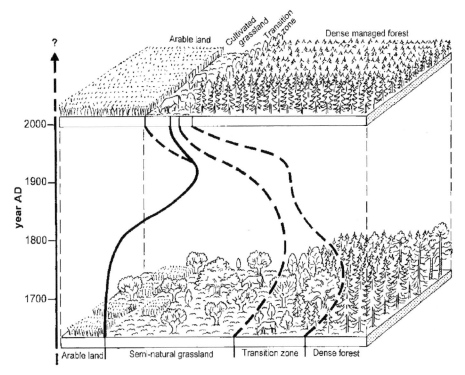

FIGURE 11.2
Landscape changes in Virestad, south Sweden. Modern grasslands pick out an important transition zone between arable land and forest. Biodiversity can be higher in this transition zone because of the land-use history profiles of cover types in these landscapes. From Skånes (1996), reproduced with permission.

The need for a classification based on the dynamics of pattern is particularly important where people are a dominant force in the landscape. Increasingly, we have come to recognize that landscapes have "memory," in the sense that the characteristics we see today are often carried over from previous management regimes. Moreover, it is also clear that the sequence transitions by which the modern mosaic is produced may also be important in constraining what managers can do.

The landscapes of Virestad, south Sweden, are good examples of why we need to understand the dynamics of pattern (Fig. 11.2). In today's landscapes, cultivated grasslands are an important reservoir of biodiversity. Such grasslands are often confined between arable field and commercial forest. However, historical analysis shows they are often a relic of a much wider semi-natural grassland transition zone that existed between the farmed and forested elements of previous landscapes. The biodiversity of the modern forest margins can be higher where they have replaced the older semi-natural grasslands, particularly where spontaneous succession has occurred.

Studies such as those in south Sweden show how "land-use history profiles" can be used to characterize the dynamics of landscape pattern. Such information is important as we seek to recreate or restore habitats that have, for example, been damaged by intensive farming or forestry. We need an understanding of the types of cover change that have occured or will occur, and the extent to which such transformations are reversible.

If we are to achieve more sustainable forms of landscape management, we must explore ways of characterizing the landscape as a set of "process-response" units, rather than a simplistic collection of structural elements. It is a useful exercise to consider how the structural boundaries shown in Fig. 11.2 might be modified if we think about the dynamics of pattern in this way. Such exercises, I suggest, could usefully become the focus of future work in landscape ecology.

As the recent review by Perry (2000) has emphasized, an understanding of the dynamics of pattern is particularly important in the context of emerging models of non-equilibrium landscapes. For, while it is widely accepted that spatial heterogeneity can be explained by reference to the magnitude and frequency of disturbances that operate upon landscapes, there is little evidence to suggest that many landscapes ever achieve a "steady state shifting mosaic," in the sense that the proportions of the different patch types generated by the disturbances are constant. Given the existence of medium- to long-term climate change, the character of natural disturbance regimes is unlikely to be constant over time. Furthermore, in landscapes where people are a significant influence, cultural and economic development will mean that rarely will anything like an equilibrium condition be established. In such situations, the study of pattern is fundamental to our understanding of how landscape change occurs, and what that change means for the structure and dynamics of ecological systems.

Cultural landscapes and qualitative pattern

In broadening our thinking about pattern, a final area that we should consider is the way to deal with cultural patterns and the associated qualitative characteristics of landscapes. I have argued that one future direction for pattern analysis is to represent a landscape as a set of process-response units. The suggestion is not entirely academic because, for some of us, such classifications are already here – in the form of various geographical policy frameworks devised by various national agencies concerned with countryside or rural issues. The problem is they have been imposed from outside the discipline, and we have to learn how to deal with them.

For example, *The Character of England* is a map published jointly by two of our government agencies, as a strategic planning framework for those interested in the English landscape, its wildlife, and natural features (Countryside Agency & English Nature, 1996; Countryside Agency, 2002). The map divides England into a set of "coherent landscapes types" or "character areas," whose borders do not follow administrative boundaries but pick out "associated patterns of wildlife, natural features, land use, human history and other cultural values."

The interesting thing about such a map for landscape ecology is that, while it has very little scientific basis, it is not without "authority" or "meaning." The boundaries were drawn by consultation, negotiation, and compromise between various stakeholder groups. The aim was to capture people's sense of place, rather than to produce a formal scientific classification. It is argued that the framework of the character areas enables people to understand their local context and be better able to judge the significance of landscape change.

The map of the English landscape is a visionary statement rather than a scientific one. However, as scientists we have to take such visions seriously, for they constitute part of the "world view" of our policy customers. Such ideas shape their questions and affect their judgments of our scientific work. Thus, while these character areas are not formal process-response units, we would be foolish to dismiss them. Since we cannot presently build a classification that takes account of all aspects of pattern and process, from the ecological through to the cultural, I suggest that we should adopt a pragmatic approach. We should use these socially constructed visions of landscape as frameworks in which to develop and apply ideas about pattern.

In the short term, such frameworks as the Character Areas of England allow us to take the analysis of pattern beyond geometric issues, to a consideration of the patterns of association between the qualitative aspects of landscape that give an area its local identity or significance for people. In the long term, by testing whether in fact such frameworks describe real landscape units, with some kind of functional integrity, we may be able to provide better ways of representing landscapes. Most significantly, we need to provide an understanding of how the ecological patterns and processes associated with such areas relate to the goods and services that people value or depend upon, and the boundary conditions over which these ecosystem services can be sustained. As I have argued in more detail elsewhere (Haines-Young, 2000) it seems unlikely that, in the context of sustainability, optimal landscape patterns can ever be defined (Forman, 1995; Wu and Hobbs, 2000) because of the "trade-offs" or compromises that we have to make in terms of the different ecological outputs that are required from a contemporary, multi-functional landscape. A key challenge for the future is to use our understanding of pattern and process to show the range of landscape configurations that would sustain the mixes of goods and services that the different stakeholder groups present in an area identified as important. As a result, we will be able to better define the ecological "choice space" within which environmental management decisions are made.

Conclusion

The object of landscape ecology is not to describe landscapes, but to explain and understand the processes that occur within them. Thus, the description of landscape pattern as an end in itself is limited. It is certainly misguided, given the need to find more sustainable forms of landscape management. Recent work has shown the value of using pattern to explain ecological process in landscapes with clearly defined spatial structures. For the future, we must extend our thinking to other types of landscape and begin to understand more about the dynamics of pattern itself. Most of all, we have to extend our thinking to the analysis of pattern in a cultural context. Only then can we meet the challenge of helping people understand the significance of pattern for the landscapes in which they live and work.

References

Countryside Agency (2002). *Countryside Character Initiative.* www.countryside.gov.uk/ LivingLandscapes/countryside_character.

Countryside Commission and English Nature (1996). *The Character of England: Landscape, Wildlife and Natural Features.* Cheltenham: Countryside Commission.

Forman, R. T. T. (1995). *Land Mosaics: The Ecology of Landscapes and Regions.* Cambridge: Cambridge University Press.

Haines-Young, R. (2000). Sustainable development and sustainable landscapes: defining a new paradigm for landscape ecology. *Fennia*, 178, 7–14.

Haines-Young, R. H. and Chopping, M. (1996). Quantifying landscape structure: a review of landscape indices and their application to forested landscapes. *Progress in Physical Geography*, 20, 418–445.

Ingham, D. S. and Samways, M. J. (1996). Application of fragmentation and variegation models to epigaeic invertebrates in South Africa. *Conservation Biology*, 10, 1353–1358.

Jonsen, I. D. and Fahrig, L. (1997). Response of generalist and specialist insect herbivores to landscape spatial structure. *Landscape Ecology*, 12, 185–197.

Lavers C. P., Haines-Young, R. H., and Avery, M. I. (1996). The habitat associations of dunlin (*Calidris alpina*) in the Flow Country of northern Scotland and an improved model for predicting habitat quality. *Journal of Applied Ecology*, 33, 279–290.

Lawler, J. J., and Edwards, T. C. (2002). Landscape patterns as habitat predictors: building and testing models for cavity-nesting birds in the Uinta Mountains of Utah, USA. *Landscape Ecology*, 17, 233–245.

Musick, H. B. and Grover, H. D. (1991). Image texture measures as indices of landscape pattern. In *Quantitative Methods in Landscape Ecology*, ed. M. G. Turner and R. H. Gardner. New York, NY: Springer, pp. 77–103.

Opdam, P., Foppen, R., and Vos, C. (2001). Bridging the gap between ecology and spatial planning in landscape ecology. *Landscape Ecology*, 16, 767–779.

Perry, G. L. W. (2000). Landscapes, space and equilibrium: shifting viewpoints. *Progress in Physical Geography,* 26, 339–359.

Pickup, G., Bastin, G. N., and Chewings, V. H. (1998). Identifying trends in land degradation in non-equilibrium rangelands. *Journal of Applied Ecology*, 35, 365–377.

Skånes, H. (1996). Landscape change and grassland dynamics: retrospective studies based on aerial photographs and old cadastral maps during 200 years in south Sweden. Doctoral dissertation, Stockholm University Department of Physical Geography. University Dissertation Series, 8, III.1–III.51.

Wu, J. and Hobbs, R. (2000). Key issues and research priorities in landscape ecology: an idiosyncratic synthesis. *Landscape Ecology*, 17, 355–365.

KEVIN MCGARIGAL

SAMUEL A. CUSHMAN

12

The gradient concept of landscape structure

The goal of landscape ecology is to determine where and when spatial and temporal heterogeneity matter, and how they influence processes (Turner, 1989). A fundamental issue in this effort revolves around the choices a researcher makes regarding how to depict and measure heterogeneity, specifically, how these choices influence the "patterns" that will be observed and what mechanisms may be implicated as potential causal factors. Indeed, it is well known that observed patterns and their apparent relationships with response variables often depend upon the scale that is chosen for observation and the rules that are adopted for defining and mapping variables (Wiens, 1989). Thus, success in understanding pattern–process relationships hinges on accurately characterizing heterogeneity in a manner that is relevant to the organism or process under consideration.

In this regard, landscape ecologists have generally adopted a single paradigm – the patch mosaic model of landscape structure (Forman, 1995). Under the patch-mosaic model, a landscape is represented as a collection of discrete patches. Major discontinuities in underlying environmental variation are depicted as discrete boundaries between patches. All other variation is subsumed by the patches and either ignored or assumed to be irrelevant. This model has proven to be quite effective. Specifically, it provides a simplifying organizational framework that facilitates experimental design, analysis, and management consistent with well-established tools (e.g., FRAGSTATS; McGarigal and Marks, 1995) and methodologies (e.g., ANOVA). Indeed, the major axioms of contemporary landscape ecology are built on this perspective (e.g., patch structure matters, patch context matters, pattern varies with scale). However, even the most ardent supporters of the patch-mosaic paradigm recognize that categorical representation of environmental variables often poorly represents the true heterogeneity of the system, which may

112 *Issues and Perspectives in Landscape Ecology*, ed. John A. Wiens and Michael R. Moss. Published by Cambridge University Press.
© Cambridge University Press 2005.

consist of continuous multidimensional gradients. Yet alternative models of landscape structure based on continuous environmental variation are poorly developed.

We believe that further advances in landscape ecology are constrained by the lack of methodology and analytical tools for effectively depicting and analyzing continuously varying ecological phenomena at the landscape level. Our premise is that the truncation of landscape-level environmental variability into categorical maps collapses the measurement resolution of continuously varying attributes, resulting in a substantial loss of information and troublesome issues of subjectivity and error propagation. We suggest that the traditional focus on categorical map analysis, to the exclusion of other perspectives, limits the flexibility and efficiency of quantitative analysis of spatially structured phenomena, and contributes to the persistent disjunction between the methods and ideas of community and landscape ecology, as well as slowing the integration of powerful geostatistical and multivariate methods into the landscape ecologist's toolbox.

Accordingly, we believe that the recent attention to scale in ecology (Wiens 1989; Peterson and Parker 1998) has focused too much on "grain" and "extent" issues, and has ignored the nonspatial aspect of observation scale associated with the *map legend*, representing the rules that are followed in defining what is measured and the resolution at which it is measured. The measurement resolution represents the degree of environmental variation discriminated by a given variable. A single variable may be recorded at any number of resolutions. For example, soil temperature may be coarsely measured as either high or low, or by 1 degree, or 0.01 degree increments. An important distinction is whether the measurement scale is *categorical* or *continuous*. The choice of measurement scales and resolution has dramatic influences on the types of associations that can be made and on the nature of the patterns that can be mapped from that variable. We suggest that adopting a perspective that explicitly considers measurement scale and resolution as a third attribute of scale and conducting investigations over appropriate ranges of this attribute (e.g., from simple categorical representations to more complex continuous surfaces) will facilitate the resolution of some of the difficulties described above, and lead to a more robust and flexible analytical science of scale.

The gradient concept of landscape structure

We believe that choosing an appropriate resolution measure for each variable is just as important as choosing a pertinent grain and extent. A priori, we see no reason to assume that environmental variability is usually categorical or that organisms or ecological processes respond categorically to it. Indeed, it seems less tenuous to assume that most environmental factors are inherently continuous and that many of them are perceived and responded to as such by organisms and ecological processes. Accordingly, we propose a conceptual shift in landscape ecology akin to that which occurred in community ecology in the decades following Gleason's (1926) seminal statements on the individualistic response of species in a community and their refinement by Whittaker (1967). Thus, to supplement the current patch-mosaic paradigm, we believe it will be useful for landscape ecologists to adopt a gradient perspective, along with a new suite of tools for analyzing landscape structure and the linkages of patterns and processes under a gradient framework. This framework will include, where appropriate, categorically mapped variables as a special case, and can readily incorporate hierarchical and multi-scaled conceptual models of system organization and control. In the sections that follow we outline how a gradient perspective can be of use in several areas of landscape ecological research.

Gradient attributes of categorical patterns

Even when categorical mapping is appropriate, conventional analytical methods often fail to produce unbiased assessments of organism responses. We propose that organisms experience landscape structure, even in categorical landscapes, as pattern gradients that vary through space according to the perception and influence distance of the particular organism. Thus, instead of analyzing global landscape patterns, for example as measured by conventional landscape metrics for the entire landscape, we would be better served by quantifying the local landscape pattern across space as it may be experienced by the organisms of interest, given their perceptual abilities. Until recently, no tools were readily available to accomplish this. However, FRAGSTATS (McGarigal et al., 2002) now contains a moving-window option that allows the user to set a circular or square window size for analyzing selected class- or landscape-level metrics. The window size should be selected such that it reflects the scale at which the organism or process perceives or responds to pattern. If this is unknown, the user can vary the size of the window over several runs and empirically determine the scales to which the organism is most responsive. The window moves over the landscape one cell at a time, calculating the selected metric within the window and returning that value to the center cell. The result is a continuous surface that reflects how an organism of that perceptual ability would perceive the structure of the landscape as measured by that metric (Plate 1). The surface then would be available for combination with other such surfaces in multivariate models to predict, for example, the distribution and abundance of an organism continuously across the landscape.

Gradient analysis of continuous field variables

When patch mosaics are not clearly appropriate as models of the variability of particular environmental factors, there are a number of advantages to modeling environmental variation as individually varying continuous gradients. First, it preserves the underlying heterogeneity in the values of variables through space and across scales. The subjectivity of deciding on what basis to define boundaries is eliminated. This enables the researcher to preserve many independently varying variables in the analysis, rather than reducing the set to a categorical description of boundaries defined on the basis of one or a few attributes. In addition, the subjectivity of defining cut points for categorization of the variability is eliminated. Imprecision in scale and boundary sensitivity is not an issue, as the quantitative representation of environmental variables preserves the entire scale range and the complete gradients to test against the response variables. The only real subjectivity is the increment or resolution at which to measure variability. By tailoring the grain, extent, and resolution of the measurements to the hypotheses and system under investigation, researchers can capture a less equivocal picture of how the system is organized and what mechanisms may be at work. An important benefit is that one can directly associate continuously scaled patterns in the environment, space, and time with continuous response variables such as organism abundance. A specific advantage is that by not truncating the patterns of variation in the landscape variables to a particular scale and set of categories, a scientist can use a single set of predictor variables to simultaneously analyze a number of response variables, be they species responding individualistically along complex landscape gradients or ecological processes acting at different scales.

When modeling environmental variation as continuous gradients, the landscape is represented as a continuous surface or several surfaces corresponding to different environmental attributes (Plate 1). The challenge lies in summarizing the structure of this surface in a metric. The two fundamental attributes of a surface are its height and slope. The patterns in a landscape surface that are of interest to landscape ecologists are emergent properties of particular combinations of surface heights and slopes across the study area. The challenge is to develop metrics that describe meaningful attributes of surface height and slope that can be used to characterize surface patterns and to derive variables that are effective predictors of organismic and ecological processes.

Geostatistical techniques have been developed that allow us to summarize the spatial autocorrelation of such a surface (Webster and Oliver, 2001). While such measures (e.g., correlograms and semi-variograms) can provide information on the distance at which the measured variable becomes statistically independent and reveal the scales of repeated patterns in the variable (if they exist), they do little to describe other interesting aspects of the surface. Fortunately, a number of gradient-based techniques that summarize these and other interesting properties of continuous surfaces have been developed in the physical sciences for analyzing three-dimensional surface structures. We will briefly describe three promising techniques. Detailed descriptions of these techniques and their potential applications can be found in the sources cited below.

Surface metrology

In the past 10 years, researchers involved in microscopy and molecular physics have developed the field of surface metrology (Stout *et al.*, 1994; Barbato *et al.*, 1995; Villarrubia, 1997). In surface metrology, several families of surface-pattern metrics have become widely utilized. These have been implemented in the software package SPIP (SPIP, 2001). One so-called family of metrics quantifies intuitive measures of surface amplitude in terms of its overall roughness, skewness and kurtosis, and total and relative amplitude. Another family records attributes of surfaces that combine amplitude and spatial characteristics such as the curvature of local peaks. Together, these metrics quantify important aspects of the texture and complexity of a surface. A third family measures certain spatial attributes of the surface associated with the orientation of the dominant texture. The final family of metrics is based on the surface-bearing area-ratio curve, also called the Abbott curve (SPIP, 2001). The curve describes the distribution of mass in the surface across the height profile. Several indices have been developed from the proportions of this cumulative height–volume curve that describe structural attributes of the surface (SPIP, 2001).

Many of the classic landscape metrics for analyzing categorical landscape structure have ready analogs in surface metrology (Plate 1). For example, the major compositional metrics such as patch density, percent of landscape, and largest patch index are matched with peak density, surface volume, and maximum peak height. Major configuration metrics such as edge density, nearest-neighbor index, and fractal-dimension index are matched with mean slope, mean nearest-maximum index, and surface fractal dimension. Many of the surface-metrology metrics, however, measure attributes that are conceptually quite foreign to conventional landscape pattern analysis. Landscape ecologists have not yet explored the behavior and meaning of these new metrics; it remains for them to demonstrate the utility of these metrics, or to develop new surface metrics better suited for landscape ecological questions.

Fractal analysis

Fractal analysis has been well developed for the analysis of two-dimensional surface patterns, but is just as suited for analyzing continuous variables as three- or higher-dimensional surfaces. Fractal analysis provides a vast set of tools to quantify the shape complexity of surfaces. There are many algorithms in existence that can measure the fractal dimension of any surface profile, surface or volume (Mandelbrot, 1982; Pentland, 1984; Barnsley, 2000). In addition, there are surface equivalents to lacunarity analysis of categorical fractal patterns. Lacunarity measures the gapiness of a fractal pattern (Plotnick *et al.*, 1993). Several structures with a given fractal dimension can look very different because of differences in their lacunarities. The calculation of measures of surface lacunarity is a topic that deserves considerable attention. It seems to us that surface lacunarity will be a useful index of surface structure, one which measures the "gapiness" in the distribution of peaks and valleys in a surface, rather than holes in the distribution of a categorical patch type.

Spectral and wavelet analysis

Spectral analysis and wavelet analysis are ideally suited for analyzing surface patterns. The spectral analysis technique of Fourier decomposition of surfaces could find a number of interesting applications in landscape-surface analysis. Fourier spectral decomposition breaks up the overall surface patterns into sets of high, medium, and low frequency patterns (Kahane and Lemarie, 1995). The strength of patterns at different frequencies and the overall success of such spectral decompositions can tell us a great deal about the nature of the surface patterns and what kinds of processes may be acting and interacting to create those patterns. Similarly, wavelet analysis is a family of techniques that has vast potential applications in landscape surface analysis (Bradshaw and Spies, 1992; Chui, 1992; Kaiser, 1994; Cohen, 1995). Traditional wavelet analysis is conducted on transect data, but the principle is easily extended to two-dimensional surface data. There have been great advances in wavelet applications in the past few years, with many software packages now available for one- and two-dimensional wavelet analysis. For example, comprehensive wavelet toolboxes are available for S-Plus, MATLAB,

and MathCad. Wavelet analysis has the advantage that it preserves hierarchical information about the structure of a surface pattern while allowing for
pattern decomposition (Bradshaw and Spies, 1992). It is ideally suited to
decomposing and modeling signals and images, and is useful in capturing,
identifying, and analyzing local, multi-scale, and non-stationary processes
(Bradshaw and Spies, 1992).

Conclusions

Landscape ecology has emerged over the past several decades as the
study of spatial and temporal heterogeneity, and under what circumstances
pattern matters to organisms, communities, and ecological processes (Turner
et al., 2001). The patch-mosaic model of landscape structure has become the
operating paradigm of the discipline. While this paradigm has provided an
essential operating framework for landscape ecologists and has facilitated
rapid advances in quantitative landscape ecology, we believe that further
advances in landscape ecology are somewhat constrained by its limitations.
We advocate the expansion of the paradigm to include a gradient-based
concept of landscape structure that subsumes the patch-mosaic model as a
special case. The gradient approach we advocate allows for a more realistic
representation of landscape heterogeneity by not presupposing discrete structures, facilitates multivariate representations of heterogeneity compatible
with advanced statistical and modeling techniques used in other disciplines,
and provides a flexible framework for accommodating organism-centered
analyses.

Perhaps the greatest obstacles to the adoption of gradient approach are the
lack of familiarity with tools for conducting gradient-based landscape analyses and inexperience in the application of surface metrics to landscape-
ecological questions. While familiar tools now exist for conducting gradient
analyses of categorical map patterns (e.g., moving-window analysis in
FRAGSTATS), landscape ecologists have not yet fully taken advantage of
these. In addition, while numerous surface metrics have been developed for
characterizing continuous landscape surfaces, and the software tools for
computing them are now available, it remains for landscape ecologists to
investigate how these metrics behave and what information they provide in
landscape-surface analysis and to develop additional metrics that quantify
specific surface attributes of importance in landscape ecology. This is an
interesting and important challenge, and until such measures are understood
in the context of landscape analysis, and until additional metrics are tailored to
the specific needs of landscape ecologists, the full potential of gradient-based
methods will not be realized. We believe that landscape ecology, as a discipline,

is poised on the verge of tremendous advances; the gradient concept is an organizational and methodological construct that we believe will facilitate these advances.

References

Barbato, G., Carneiro, K., Cuppini, D., *et al.*, (1995). *Scanning Tunneling Microscopy Methods for the Characterization of Roughness and Micro Hardness Measurements*. Synthesis report for research contract with the European Union under its programme for applied metrology. CD-NA-16145 EN-C. Brussels, Luxembourg: European Commission.

Barnsley, M. F. (2000). *Fractals Everywhere*. San Diego, CA: Elsevier.

Bradshaw, G. A. and Spies, T. A. (1992). Characterizing canopy gap structure in forests using wavelet analysis. *Journal of Ecology*, 80, 205–215.

Chui, C. K. (1992). *An Introduction to Wavelets: Wavelet Analysis and its Applications*. San Diego, CA: Academic Press.

Cohen, A. (1995). *Wavelets and Multiscale Signal Processing*. New York, NY: Chapman and Hall.

Forman, R. T. T. (1995). *Land Mosaics: The Ecology of Landscapes and Regions*. Cambridge: Cambridge University Press.

Gleason, H. A. (1926). The individualistic concept of the plant association. *Bulletin of the Torrey Botanical Club*, 53, 7–26.

Kahane, J. P. and Lemarie, P. G. (1995). *Fourier Series and Wavelets*. Studies in the Development of Modern Mathematics, vol. 3. London: Taylor and Francis.

Kaiser, G. (1994). *A Friendly Guide to Wavelets*. Boston, MA: Birkhauser.

Mandelbrot, B. B. (1982). *The Fractal Geometry of Nature*. New York, NY: Freeman.

McGarigal, K. and Marks, B. J. (1995). *FRAGSTATS: Spatial Analysis Program for Quantifying Landscape Structure*. USDA Forest Service General Technical Report PNW-GTR-351. Portland, OR: USDA Forest Service.

McGarigal, K., Cushman, S. A., Neel, M. C., and Ene, E. (2002). *FRAGSTATS: Spatial Pattern Analysis Program for Categorical Maps*. Amherst, MA: University of Massachusetts.

Pentland, A. P. (1984). Fractal-based description of natural scenes. *IEEE Transactions on Pattern Analysis and Machine Intelligence*, 6, 661–674.

Peterson, D. L., and Parker, V. T. (1998). *Ecological Scale: Theory and Applications*. New York, NY: Columbia University Press.

Plotnick, R. E., Gardner, R. H., and O'Neill, R. V. (1993). Lacunarity indices as measures of landscape texture. *Landscape Ecology*, 8, 201–211.

SPIP (2001). *The Scanning Probe Image Processor*. Lyngby, Denmark: Image Metrology APS.

Stout, K. J., Sullivan, P. J., Dong, W. P., *et al.* (1994). *The Development of Methods for the Characterization of Roughness on Three Dimensions*. EUR 15178 EN. Luxembourg: European Commission.

Turner, M. G. (1989). Landscape ecology: the effect of pattern on process. *Annual Review of Ecology and Systematics*, 20, 171–197.

Turner, M. G., Gardner, R. H., and O'Neill, R. V. (2001). *Landscape Ecology in Theory and Practice*. New York, NY: Springer

Villarrubia, J. S. (1997). Algorithms for scanned probe microscope, image simulation, surface reconstruction and tip estimation. *Journal of the National Institute of Standards and Technology*, 102, 435–454.

Webster, R. and Oliver, M. (2001). *Geostatistics for Environmental Scientists*. Chichester: Wiley.

Whittaker, R. H. (1967). Gradient analysis of vegetation. *Biological Review*, 42, 207–264.

Wiens, J. A. (1989). Spatial scaling in ecology. *Functional Ecology*, 3, 385–397.

THOMAS R. LOVELAND
ALISA L. GALLANT
JAMES E. VOGELMANN

13

Perspectives on the use of land-cover data for ecological investigations

An important ingredient of many research applications in landscape ecology is land-cover data. Land-cover databases reflect the patterns of vegetation, the extent of anthropogenic activity, and the potential for future uses and disturbances of the landscape. These databases are essential for studies of landscape spatial configuration and investigations of ecological status, trends, stresses, and relationships. The evolution of land-cover databases and landscape applications is an iterative process, driven by new developments at both ends. There is a strong demand at all scales for land-cover data, and those developing such data sets must constantly work toward improvements in data content, quality, and documentation to meet the diverse needs of scientific users.

The development of land-cover databases is a major focus of the US Geological Survey (USGS) National Land-cover Characterization Program. Projects span local, to regional, to global venues (e.g., Loveland *et al.*, 1991, 2000; Vogelmann *et al.*, 2001) and the results contribute to a wide range of applications (e.g., Jones *et al.*, 1997, 2001; DeFries and Los, 1999; Hurtt *et al.*, 2001; Maselli and Rembold, 2001). While some of the applications are quite innovative, we find others worrisome, considering the limitations of the source materials, mapping technologies, and expertise inherent in data development. These limitations are important to landscape ecologists because the resultant imperfections in the data sets affect the accuracy, consistency, and credibility of the analyses applied to them. In this chapter we highlight major issues in the application of land-cover data for environmental analyses, including the derivation of land-cover data sets, accuracy, scale, minimum mapping unit, thematic content, data structure, and temporal representation. As might be expected, these issues are interrelated and it is difficult to discuss one without referring to others.

120 *Issues and Perspectives in Landscape Ecology*, ed. John A. Wiens and Michael R. Moss. Published by Cambridge University Press.
© Cambridge University Press 2005

Derivation of land-cover data sets

Most land-cover products are interpreted from remotely sensed data, although some local land-cover maps may be based on field mapping. In all cases, land-cover data sets are the result of interpretations of observations of landscape conditions at a particular period (or set of periods) in time. The interpretations are dependent upon the characteristics and quality of the data, the methods used to assess and map land cover from the data, and the abilities of the interpreters doing the analyses. Land-cover data products are models, not gospel, and this should be kept in mind. For a review of the technical characteristics of remotely sensed data from a landscape ecology perspective, readers may consult Quattrochi and Pelletier (1990).

One form of remotely sensed data, aerial photography, is usually interpreted using manual mapping techniques where a suite of variables visible in the photo, including color or tone, pattern, texture, size, shape, location, and association, are considered. With satellite imagery, such as from Landsat and SPOT, computer-assisted techniques are commonly (though not exclusively) used to map land cover. In this case, the relationship between land cover and spectral characteristics is the starting point for determining land-cover types. Different satellites collect data in different portions of the electromagnetic spectrum, with different frequencies of overflights. The suitability of the data for land-cover mapping depends on the specific spectral region and the number of spectral bands collected by the particular sensor, as well as the timing of the sensor overpass. In addition, a number of artefacts, including atmospheric variables and instrument noise, can act to hinder interpretability of the data. With either manual or computer-assisted interpretation, the outcomes are the direct result of interpreter decisions and there can be significant variability among interpreters (McGwire, 1992).

Accuracy

The most obvious measure of land-cover mapping quality is classification accuracy. It is essential that all land-cover data sets produced for scientific application have accuracy statements (Estes and Mooneyhan, 1994). In the past, accuracy assessments of land-cover products were uncommon (see Foody, 2002), often due to physical logistical or budget constraints. This has been particularly true for large-area classifications. Recently, greater emphasis has been placed on this issue. As realistic accuracy statements are produced, database developers and users must collectively define the acceptable accuracy standards that guide decisions regarding the use of a particular data set in an ecological assessment.

Our experience has shown that when mapping general land-cover characteristics for large areas using computer-assisted interpretation of satellite data, overall classification accuracy of approximately 75% should be expected (see Kroh *et al*, 1995; Homer *et al.*, 1997; Vogelmann *et al.*, 1998). While there are many examples in the remote-sensing literature of accuracy at 90% or better, those figures typically represent small-area methodological tests that seldom yield such impressive results when applied over large geographic areas. Perhaps more importantly, accuracy numbers will be directly related to the number of classes. Is a two-class map with 95% accuracy better than an eight-class map with 80% accuracy? Consider, also, that the accuracy of land-cover maps varies significantly from category to category. While high accuracy levels can be attained when mapping water, consistent differentiation of mixed forests from needleleaf or broadleaf forests is very difficult, so confusion among these classes will be common.

People often assume that an accuracy value somehow provides a sort of panacea. In actuality, accuracy values can often give the wrong impression. It is seldom that we are concerned about any single pixel in land-cover classification work; more often, we are interested in patterns of pixels, or groups of similar pixels. Curiously, most accuracy assessments are done at the single pixel level. These estimates will not necessarily provide the information that is appropriate for conveying the utility of the data to users. Single-pixel assessments are needlessly stringent and often produce deceptively low levels of accuracy. Alternative approaches for conveying accuracy include consideration of spatial resolution (e.g., single pixel versus groups of pixels; Yang *et al.*, 2001), thematic resolution (e.g., Anderson Level 1 versus Level 2 classes; Zhu *et al.*, 2000), and magnitude of misclassification error (Foody, 2002).

It is important to think about the cost of misclassification error with respect to the intended application of the land-cover data. A study by DeFries and Los (1999) showed that a global land-cover data set having an overall accuracy level of 78% actually has a climate modeling application accuracy greater than 90% because some types of misclassification are "acceptable" (i.e., they have no negative effect on the parameterization of land—atmosphere interaction models, as they do not affect the derivation of surface roughness or leaf area index parameters). In an example by Wickham *et al.* (1997), the impacts of classification accuracy and spatial consistency on landscape metrics were considered.

Accuracy statements may provide insight into the appropriate scale of use for the data. What is key is that sufficient information on accuracy accompanies the classification products to enable flexible tailoring of data sets for different applications. Landscape ecologists should insist on land-cover accuracy statements that provide information on the sampling procedures used to assess accuracy, the characteristics of the reference ("truth") data, and the statistics used to estimate accuracy (Stehman, 2001; Foody, 2002). Ecologists must then evaluate those statements in the context of the particular research application.

Scale and minimum mapping unit

These two characteristics are often misunderstood and should be considered in the context of each other. Scale is communicated as the representative fraction between earth and map distance (for example, 1 : 24 000 means that one unit of measurement on a map equals 24 000 of the same units on the earth). Scale is a term of confusion between mappers/geographers and landscape ecologists because they use the term in opposite ways. To the former, a large-scale (large representative fraction) map covers a small geographic area and typically provides detailed land-cover information. In general, the larger the scale, the more spatial and thematic detail can be represented in the map. Thus, a 1 : 24 000-scale land-cover map will depict smaller occurrences of land cover and more detailed land-cover categories than a 1 : 250 000-scale map.

Minimum mapping units (MMUs) define the smallest land areas represented in a database. As map scale decreases (meaning the information content becomes more general but covers larger geographic areas), the MMU increases. When calculating landscape metrics corresponding to landscape configurations, scale and MMU become important. Generally, smaller scales and larger MMUs result in simpler measures of complexity. We should note that this concept is typically understood in studies in which our land-cover data are applied. However, the 1970s vintage land-use and land-cover data (commonly known as LUDA or Land Use Data Analysis data) produced by the USGS are often applied without consideration of the MMU. The MMU of this data set varies with land-cover category. Classes representing human activity have a 10-acre (4 ha) MMU, whereas other classes have a 40-acre (16 ha) MMU (Anderson et al., 1976). Thus, measures of landscape fragmentation and complexity will be affected by a mapping decision to represent some classes at a finer spatial detail. Interpretation of statistics generated from these data must consider this issue.

A special note about pixels, or picture elements, is necessary. Pixels are the smallest geographic unit in digital satellite images; however, they do not represent the effective MMU in a land-cover data set interpreted from digital images. Because of a number of technical issues corresponding to land surface–atmosphere–energy interactions, sensor operation, and image processing methods, the actual MMU is typically greater than the pixel dimensions. For example, the USGS Land-cover Characterization Program AVHRR land-cover data set covering the globe has 1-km pixels, but the smallest resolvable geographic feature is more likely about 4 km by 4 km (Loveland et al., 2000). Thus, landscape features that are mapped from these data must have a spatial extent of approximately 16 km². So even though we assign land-cover attributes to pixels, we rarely interpret land cover at that spatial resolution. Rather, we are concerned primarily with documenting the spatial patterns made by similar pixels. Moreover, all pixels represent an internal mix of land-cover elements at some spatial or thematic scale. We point to observations by Quattrochi and Pelletier (1990) that concepts of heterogeneity and homogeneity are scale-dependent because they describe how individual land-cover components or processes are interrelated across a landscape. For any given study there is an appropriate scale for analysis that corresponds with the size of the study area, the landscape patterns being investigated, and the maps that capture patterns of land cover.

Thematic content

Land-cover maps typically comprise categories of land cover, land use, and/or environmental condition. It is not uncommon to find all three types of categories occurring in the same classification scheme, as when "graminoid/ herbaceous" (a cover type), "cropland" (a land use), and "emergent wetland" (a condition related to hydrologic regime) are included as classes. All three represent herbaceous vegetation cover, but distinctions are made because of planned or projected uses of the land-cover data set. Thematic inconsistencies such as these can lead to inconsistencies in the execution of the classification process. For example, emergent wetlands that occur within cropped fields in the midwestern USA may be plowed and planted in crops for a portion of the growing season. These part-time wetlands can be functional for some eco-logical processes, but not others. This leads to a conceptual issue relative to the definition of "wetland" (if the wetland is used as cropland part of the year, is it still a wetland?) and a logistical issue relative to the timing of remote data collection (which cover feature was present at the time of sensor overpass?). Both will affect the classification product.

Because land-cover data sets most often comprise discrete classes, many users infer that land-cover types are spectrally and conceptually discrete. Spectral data, however, are ambiguous because of a multitude of influences, including vegetation phenological processes, relationships between vegeta-tion canopy densities and soil background brightness, shadowing due to clouds, terrain features, sun angle, and sensor height and angle, and local effects (moisture from recent rainstorms or irrigation, haze/smoke, harvesting ...). Given appropriate (or perhaps inappropriate) conditions, very different cover types can appear spectrally indistinguishable. There are conceptual challenges as well. In reality, land cover is a continuum, and gradations of cover types and management practices can be readily observed. This becomes increasingly problematic as mapping projects incorporate larger and larger areas. In the semiarid western United States, for instance, gradients of management exist where land is seeded and irrigated for pasture, irrigated but not seeded for pasture, seeded but not irrigated for pasture, not seeded or irrigated but used as pasture at certain times of the year or in certain years. So, what is an appropriate and discrete definition for "pasture"?

Generally, thematic content is based on hierarchical classification schemes such as the USGS Anderson system (Anderson *et al.*, 1976) or the National Vegetation Classification Standard produced by the Federal Geographic Data Committee (1997). Theoretically, scale is closely tied to classification systems, and small-scale maps usually use very general land-cover classes. In practice, land-cover maps are typically mapped to the most detailed level possible, often varying from class to class so that the resulting map may include categories from all levels of the hierarchy. Thus, maps may have inconsistent thematic detail – which translates to variable spatial complexity. As with variable MMUs, this will introduce bias in measurements of landscape complexity.

Data structure

Land-cover maps derived from remote sensing are developed from either raster images or photos. Manual interpretation from photos produces smooth, clean lines and polygons, with the amount of spatial detail determined by the interpreter. Two interpreters working on adjacent areas may use different decision rules regarding line generalization. Even when a concerted attempt is made to hold the decision rules constant, differences among interpretations can be considerable (Plate 2). Land-cover maps classified using digital remotely sensed imagery typically have mapping units defined by statistical criteria, and therefore have the potential to be applied more consistently. However, because of ambiguities between spectral data and land cover, digital classifications are inherently noisy, with jagged-edge map regions and "salt-and-pepper" pixel patterns. Although the results look complex, the complexity may be an artefact of the mapping techniques (as well as the relatively finer spatial scale, i.e., pixel, at which the classification rules are applied). Comparison of landscape metrics calculated for land-cover maps derived from analog versus digital sources, captured as lines or vectors versus pixels, is problematic (Plate 3), and can yield highly mislead-ing results

Temporal representation

All land-cover data are specific to a particular time that corresponds with the dates the source data were collected. For local-area studies, remotely sensed data typically represent a specific date. However, as the area mapped becomes larger, the time period of the source imagery becomes broader because more time is required for overpasses of aircraft or satellites and cloud-free conditions may be more difficult to achieve. In some cases, several years may be required to compile a relatively cloud-free data set. During this time, changes in land cover can occur. For example, our 1-km Global Land-cover Characterization database was interpreted from satellite data collected over a 12-month period (Loveland *et al.*, 2000), whereas our 30-m US land-cover data set is based on satellite images collected over several years (Vogelmann *et al.*, 2001). The differences in phenological conditions may result in land-cover databases with internal inconsistencies. Currently, this problem is unavoidable, but it should be con-sidered when interpreting landscape metrics.

Summary and future directions

Basically, there are no perfect land-cover data. It is therefore important to understand the strengths and weaknesses of the data that you are consider-ing for your study. Because image interpretation is both an art and a science, there are subjective aspects to the process that can result in inconsistent interpretations. Understanding the nature of the inconsistencies is important to the wise use of the data and ensures that valuable analyses ensue.

We have described a number of issues regarding land-cover data sets that affect outcomes of environmental analyses. Our purpose is to encourage data users to become better informed about what these data sets represent. Data sources and method of classification, thematic suitability, effective accuracy, and informational and spatial resolution of the land-cover data are important considerations for intended applications. Applying caution and careful inter-pretation to analytical results will lead to more sound scientific statements.

We hope for ongoing dialogue between land-cover mappers and landscape ecologists regarding data strengths and weaknesses, and the development of more useful and innovative databases in the future. We see some important trends in land-cover programs that will affect the land-cover databases avail-able for future scientific applications. Anticipate increases in:

Available land-cover data.   The USGS Land-cover Characterization Program will continue producing national and global land-cover databases on both an operational and an experimental basis. The USGS Gap Analysis Program will also provide detailed vegetation data sets for the nation on a cyclic basis (Scott *et al.*, 1993). International programs, such as the Global Observation of Forest Cover of the Committee on Earth Observation Satellites, will work toward improvements in land-cover data needed for environmental treaty compliance (Ahern *et al.*, 1998).

Quantitative and/or continuous attributes of land-cover, including tree canopy density, leaf area index, other physiognomic variables, and percent impervious surface.

Dimensionality of land-cover products,   including multi-resolution, multi-attribute (i.e., different land-cover legends, physiognomic variables, floristic descriptions), and multi-temporal (i.e., phenology) elements. The added dimensions should improve the suitability of land-cover products for a wider range of applications.

Emphasis on the use of appropriate metadata standards   that provide the necessary evidence of data quality and heritage. Included in this are accuracy statements.

A variety of factors, including improvements in satellite and airborne sensors, computing capabilities, acceptance of geographic information systems as analytical tools, and advancements in integrated environmental modeling and assessments, are combining to provide the impetus for innovation and expansion in operational land-cover characterization programs. For these programs to be successful, ongoing dialogue and collaboration between land-cover data producers and users are crucial.

Acknowledgments

The authors thank Limin Yang and Jesselyn Brown for their helpful reviews of this manuscript.

References

Ahern, F., Belward, A., Churchill, P., *et al.* (1998). *A Strategy for Global Observation of Forest Cover*. Ottawa: Committee on Earth Observation Satellites.

Anderson, J. R., Hardy, E. E., Roach, J. T., and Witmer, R. E. (1976). *A Land Use and Land-cover Classification System for Use with Remote Sensor Data*. US Geological Survey Professional Paper 964. Reston, VA: US Geological Survey.

DeFries, R. S. and Los, S. O. (1999). Implications of land-cover misclassification for parameter estimates in global land surface models: an example from the Simple Biosphere Model (SiB2). *Photogrammetric Engineering and Remote Sensing*, 65, 1083–1088.

Estes, J. E. and Mooneyhan, D. W. (1994). Of maps and myths. *Photogrammetric Engineering and Remote Sensing*, 60, 517–524.

Federal Geographic Data Committee (1997). *Vegetation Classification Standard*. FGDC-STD-005. Reston, VA: US Geological Survey.

Foody, G. M. (2002). Status of land-cover classification accuracy assessment. *Remote Sensing of Environment*, 80, 185–201.

Homer, C. G., Ramsey, R. D., Edwards, T. C. Jr., and Falconer, A. (1997). Landscape cover-type modeling using a multi-scene Thematic Mapper mosaic. *Photogrammetric Engineering and Remote Sensing*, 63, 59–67.

Hurtt, G. C., Rosentrater, L., Frolking, S., and Moore, B. (2001). Linking remote-sensing estimates of land-cover and census statistics on land use to produce maps of land use of the conterminous United States. *Global Biogeochemical Cycles*, 15, 673–685.

Jones, K. B., Riitters, K. H., Wickham, J. D., et al. (1997). *An Ecological Assessment of the United States Mid-Atlantic Region: a Landscape Atlas*. EPA/600/R-97/130. Washington, DC: US Environmental Protection Agency, Office of Research and Development.

Jones, K. B., Neale, A. C., Nash, M. S., et al. (2001). Predicting nutrient and sediment loadings to streams from landscape metrics: a multiple watershed study from the United States mid-Atlantic region. *Landscape Ecology*, 16, 301–312.

Kroh, G. C., Pinder, J. E. III, and White, J. D. (1995). Forest mapping in Lassen Volcanic National Park, California using Landsat TM data and a geographic information system. *Photogrammetric Engineering and Remote Sensing*, 61, 299–305.

Loveland, T. R., Merchant, J. W., Ohlen, D. O., and Brown, J. F. (1991). Development of a land-cover characteristics database for the conterminous U.S. *Photogrammetric Engineering and Remote Sensing*, 57, 1453–1463.

Loveland, T. R., Reed, B. C., Brown, J. F., et al. (2000). Development of a global land-cover characteristics database and IGBP DISCover from 1-km AVHRR data. *International Journal of Remote Sensing*, 21, 1303–1330.

Maselli, F. and Rembold, F. (2001). Analysis of GAC NDVI data for cropland identification and yield forecasting in Mediterranean African countries. *Photogrammetric Engineering and Remote Sensing*, 67, 593–602.

McGwire, K. C. (1992). Analyst variability in labeling of unsupervised classifications. *Photogrammetric Engineering and Remote Sensing*, 58, 1673–1677.

Quattrochi, D. A., and Pelletier, R. E. (1990). Remote sensing for analysis of landscapes: an introduction. In *Quantitative Methods in Landscape Ecology*, ed. M. G. Turner and R. H. Gardner. New York, NY: Springer, pp. 51–76.

Scott, J. M., Davis, F., Csuti, B., et al. (1993). Gap analysis: a geographic approach to protection of biological diversity. *Wildlife Monographs*, 123.

Stehman, S. V. (2001). Statistical rigor and practical utility in thematic map accuracy assessment. *Photogrammetric Engineering and Remote Sensing*, 67, 727–734.

Vogelmann, J. E., Sohl, T., and Howard, S. M. (1998). Regional characterization of land-cover using multiple sources of data. *Photogrammetric Engineering and Remote Sensing*, 64, 45–57.

Vogelmann, J. E., Howard, S. M., Yang, L., Larson, C. R., Wylie, B. K., and Van Driel, N. (2001). Completion of the 1990s National Land-cover Data Set for the conterminous United States from Landsat Thematic Mapper data and ancillary data sources. *Photogrammetric Engineering and Remote Sensing*, 67, 650–662.

Wickham, J. D., O' Neill, R. V., Riitters, K. H., Wade, T. G., and Jones, K. B. (1997). Sensitivity of selected landscape metrics to land-cover misclassification and differences in land-cover composition. *Photogrammetric Engineering and Remote Sensing*, 63, 397–414.

Yang, L., Stehman, S. V., Smith, J. H., and Wickham, J. D. (2001). Thematic accuracy of MRLC land-cover for the eastern United States. *Remote Sensing of Environment*, 76, 418–422.

Zhu, Z., Yang, L., Stehman, S. V., and Czaplewski, R. L. (2000). Accuracy assessment for the U.S. Geological Survey regional land-cover mapping program: New York and New Jersey region. *Photogrammetric Engineering and Remote Sensing*, 66, 1425–1435.

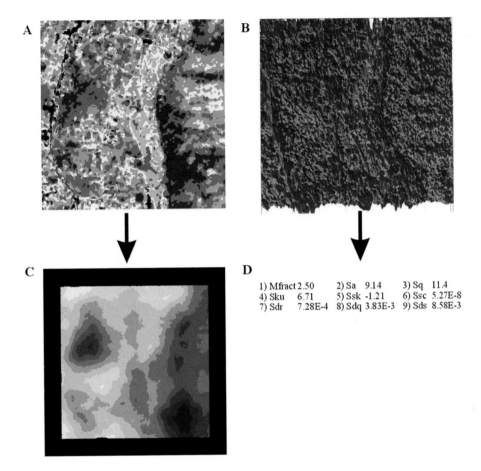

A

B

C

D

1) Mfract 2.50	2) Sa 9.14	3) Sq 11.4
4) Sku 6.71	5) Ssk -1.21	6) Ssc 5.27E-8
7) Sdr 7.28E-4	8) Sdq 3.83E-3	9) Sds 8.58E-3

PLATE 1

Comparison of categorical and gradient mapping of the normalized difference vegetation index (NDVI) for a 25-km^2 landscape in western Massachusetts. (A) The landscape classified into nine discrete classes using a natural-breaks classification criterion. (B) The same landscape depicted as a three-dimensional surface whose height is proportional to the NDVI value at each pixel (15-m cell size). (C) A moving-window calculation of the Aggregation Index (AI) for the categorical map in (A) based on a 500-m radius circular window. AI measures the aggregation of like-valued cells and is computed as a percentage based on the ratio of the observed number of like adjacencies to the maximum possible number of like adjacencies, given maximum clumping of classes. There is a border classified as "no data" around the edge of the landscape to a depth of the selected neighborhood radius. Higher AI values are dark, lower values are light. Note that the global AI value for the entire landscape is 84.87. (D) Calculation of nine surface-pattern metrics for the continuous surface shown in (B). The nine surface-pattern metrics include: Mfract – mean profile fractal dimension, which is the mean fractal dimension of 180 profiles taken at 1-degree increments across the surface; Sa – average deviation of the surface height from the global mean; Sq – variance in the height of the surface; Sku – peaked-ness (kurtosis) of the surface topography; Ssk – asymmetry (skewness) of the surface height distribution histogram; Ssc – average of the principal curvature of the local maximums on the surface; Sdr – ratio of the surface area to the area of the flat plane with the same x–y dimensions; Sdq – variance in the local slope across the surface; and Sds – number of local maximums per area.

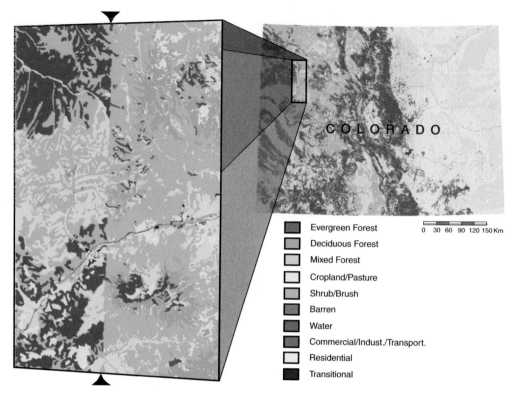

Evergreen Forest
Deciduous Forest
Mixed Forest
Cropland/Pasture
Shrub/Brush
Barren
Water
Commercial/Indust./Transport.
Residential
Transitional

0   30 60 90 120 150 Km

PLATE 2

Land cover mapped for Colorado as part of the LUDA data set. The pointers in the inset map show a "seam" where the products of different image interpreters working on adjacent geographic areas were merged. These interpreters had comparable source material and were following the same land-classification criteria.

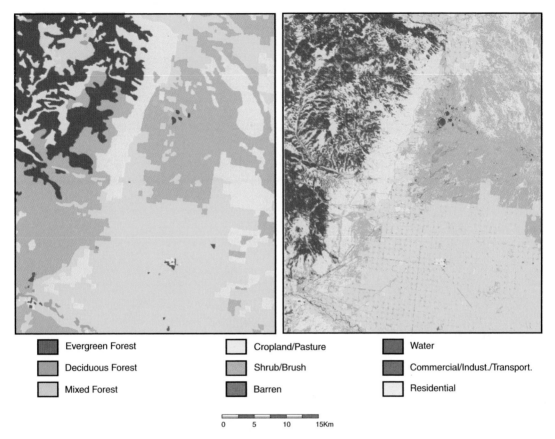

▦	Evergreen Forest	▢	Cropland/Pasture	▦	Water
▦	Deciduous Forest	▦	Shrub/Brush	▦	Commercial/Indust./Transport.
▦	Mixed Forest	▦	Barren	▢	Residential

0   5   10   15Km

PLATE 3
Land-cover maps derived from the late 1970s analog data and processing techniques (left)
versus. early 1990s digital imagery and processing techniques (right). A comparison of change
in relative abundance of cover types or pattern characteristics for the two time periods would
lead to faulty interpretations. Differences in land-cover characteristics between the images
might be due to differences in image grain, processing methods, interpreter bias, land-cover
class definitions, classification accuracy, and/or actual changes in land cover.

PART IV

Landscape dynamics on multiple scales

14

Landscape sensitivity and timescales of landscape change

Ideas concerning what is now usually termed "landscape sensitivity" have been a part of geomorphological thinking for half a century, illustrated by the concepts of *biostasie* and *rhexistasie* formulated by Erhart (1955) to describe the switch from biogeochemical equilibrium and chemical sedimentation to conditions of erosion and clastic sedimentation. However, the term was first used explicitly by Brunsden and Thornes (1979) to assist understanding of episodes of accelerated erosion and sedimentation as they affect the natural landscape. Although widely employed, the concept has received less attention than might have been expected, and was not widely reviewed until D. Thomas and Allison (1993) brought together a series of papers to show the impacts of environment and land-use changes on landscapes. More recently, another symposium has reviewed the concept and its applications (M. Thomas and Simpson, 2001).

The notion of sensitivity is related to the concept of erosion thresholds and to other aspects of systems analysis, widely discussed since the publication of papers by Knox (1972), Schumm and Parker (1973), and Schumm (1977, 1979) in the 1970s. But "landscape" is a complex entity that has proved difficult to subject to systems analysis. Most geomorphologists have felt more at home with research into fluvial and hillslope systems, and issues concerning landscape per se have received less attention. Often this has implied a lack of emphasis on the role of the vegetation cover and much greater concern with stream channels than with interfluves and hillslopes.

As methods of monitoring natural systems have advanced, systems thinking and the concepts of threshold and sensitivity have been absorbed into scientific writing (Phillips, 1999, 2003; Thomas, 2001; Thomas and Simpson, 2001). But there is increasing recognition that landscape sensitivity cannot be discussed solely in terms of threshold-crossing events lasting nanoseconds, and that periods of record (usually decades) are also too short. Two important

Issues and Perspectives in Landscape Ecology, ed. John A. Wiens and Michael R. Moss. Published by Cambridge University Press.

reasons for this situation are, first, that landscape instability is unlikely to be triggered by a single threshold-crossing event, and second, that the sensitivity of a landscape to change is influenced by past changes and prior development over varying time periods, often embracing 10^4 years. The idea that the timescale of enquiry influences our understanding of the factors that control change was emphasized by Schumm and Lichty (1965), and geomorphologists have frequently returned to this theme (Brunsden and Thornes, 1979; Cullingford *et al.*, 1980; Thomas, 2004). The timescales of climate and environmental change have also become widespread concerns across many disciplines (Driver and Chapman, 1996). This has come about on the one hand because it is apparent that our period of record is too short to encompass all significant events in the formation of landscape, and on the other hand because proxy evidence of Quaternary environmental change has revealed the importance of millennium- and century-scale climate fluctuations to our understanding of human history and landscape change.

Landscapes as non-linear dynamic systems

Landscapes are maintained by complex, non-linear, dynamic natural systems, and Phillips (1999, 2003) has pointed out that when they experience threshold-crossing events leading to rapid change they behave in a non-linear fashion. Natural systems are largely controlled by energy inputs that are subject to complex temporal and spatial variations due to secular trends, cyclical fluctuations, and stochastic variations in climate. Erosion thresholds are crossed when force (stress) exceeds resistance, but the sensitivity of natural systems to stress can change significantly over time and at widely varying rates. Across a complex landscape not all elements will have equal sensitivity to change, and this spatial heterogeneity is central to rates of landscape change. In the face of this complexity, Ruxton (1968) referred to "order and disorder" in landforms, the disorder being due to the multicomplexity of process and to inheritance. Strategies for understanding this complexity need (*inter alia*) to focus on the time and spatial scales of change (Thomas, 2004).

Landscape sensitivity and timescales of change

Landscape instability is expressed in geomorphic terms by episodes of erosion and sedimentation, and the sediments stored in the landscape reveal much about its history and evolution. This evolution is not steady but is punctuated by the impacts of extreme events and major climate changes, those of the post-glacial period possibly being the most relevant (Fig. 14.1).

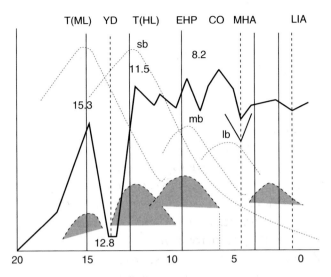

FIGURE 14.1
Some Quaternary climate change indicators of relevance to landscape sensitivity studies. The firm line follows a schematic temperature curve for the last 20 000 years (20 ka). Open dotted curves show sediment yields in formerly glaciated landscapes, indicating the paraglacial decline following glacial Termination (T).T(ML) applies to mountain glaciation in middle latitudes; T(HL) applies to ice-sheet glaciation in high latitudes. Separate curves for T(HL) are shown for small basins (sb), medium-sized basins (mb), and large basins (lb), to indicate the delays in arrival of sediment pulses down major river catchments. The shaded curves show the timing of major sediment pulses through small and medium-sized catchments in tropical west Africa. YD – Younger Dryas; EHP – early Holocene pluvial; CO – Climatic Optimum; MHA – mid-Holocene arid phase in the tropics and subtropics; LIA – Little Ice Age. All numbers refer to cal ka. The vertical scales are arbitrary. Incorporates information from Church and Ryder (1972), Church and Slaymaker (1989), Ballantyne (2002a, 2002b), Thomas and Thorp (1995).

Sources of non-linearity in natural systems were formalized by Phillips (2003), and include threshold-crossing events, effects of sediment storage and sediment exhaustion, other depletion effects in weathering and soil development. Self-limiting processes involve negative feedback that leads, for example, to saturation in groundwaters or infilling of depocentres with sediment (especially lakes). Positive feedback in natural systems causes acceleration and/or spatial extension. This occurs when gully incision reinforces subsurface water flow, which leads to gully extension. However, a longer time frame may reveal gully extension to be a self-limiting process, as the upslope catchment is reduced in area and/or sediment supply becomes exhausted. Gully advance is also usually episodic due to changing storm size and frequency, over decades or centuries. On a millennial timescale the healing or extension of gullying may depend on changes in annual and seasonal rainfall totals and their impacts on vegetation cover.

Since the end of the last glaciation, river systems in both temperate high latitudes and the tropics have experienced switches between braided, bedload-dominated behavior and more stable, meandering activity, involving accumulation of overbank suspended sediment. This switching probably involved many threshold-crossing events, not all of them associated with the river channel, and not all taking place synchronously, but cumulatively they lead to a fundamental change in fluvial behavior, often over a millennium time period. According to Werritty and Leys (2001), fluvial systems may be described as "robust" or "responsive." The former undergo internal readjustment within a persistent landform assemblage, crossing only internal (or intrinsic) thresholds, while the latter respond to environmental perturbations by making fundamental changes to their morphology, crossing external (or extrinsic) thresholds to create new landform assemblages. What determines whether a fluvial system will be "robust" or "responsive" to short-term environmental perturbations may involve long-term (millennium-scale) preparation for episodes of rapid change.

Issues that complicate this topic relate to the possibility that internal readjustments within the fluvial system following disturbance will lead to stratigraphies that have no direct correlation to the original environmental perturbation, so-called "complex response" (Schumm, 1977, 1979). But many studies have shown that consistent, basin-wide responses to environmental changes can be distinguished from local complexities of self-organisation (Knox, 1993, 1995; Blum et al., 1994). It has also proved possible to distinguish climatic influences from human impacts on river systems (Macklin and Lewin, 1993; Brown, 1996, 1998).

In some studies, the impacts of recent land use can be seen in the context of late Quaternary climate change. For example, slope deposits and alluvium in the Bananal area of southeastern Brazil show that widespread colluviation took place around 12–13 cal k yr BP (Coelho-Netto, 1997) and that after 9 cal k yr BP the landscape was stable until the era of European coffee plantation 200 years ago. In the Piracema Valley, sedimentation rates reached 1485 m^3 km^{-2} yr^{-1} during the Pleistocene–Holocene aggradation cycle, equivalent to local lowering of 1.5 mm yr^{-1}. In the last 200 years that rate has been 0.75 mm yr^{-1}, and has produced only a thin veneer of new sediment.

Episodes of rapid change or destabilisation in the landscape taking place over years to decades may result from complex changes to natural systems that have taken centuries or millennia to become effective. Such issues raise the question of what we mean by "abrupt" or "rapid" change in natural landscape systems. In the late Quaternary (10^4 yr), fluvial systems appear to have switched behavior from braiding to meandering channel patterns, on a millennial timescale (10^3 yr) (Starkel, 1995; Lewis et al., 2001; Vandenberghe and Maddy, 2001; Veldkamp and Tebbens, 2001). At first, it is tempting to see this observation merely as an artefact of our sampling and dating resolution. However, both empirical studies of river sediments and oceanographic research have revealed a clear millennium-scale cyclicity of environmental change, comprising cold Heinrich events (recurring every 5000–7000 yr) and Dansgaard–Oeschger warming episodes within Bond cycles of 1400–1500 years duration (Heinrich, 1988; Dansgaard *et al.*, 1993; Bond *et al.*, 1997; Bard *et al.*, 1997, 2000; Ganapolski and Rahmsdorf, 2001). The GRIP and GISP2 ice cores have also revealed similar periodicities in climate and document rapid warming episodes over 10^1–10^2 yr, followed by gradual cooling over 10^3 yr (Stuiver, *et al.*, 1995). The global importance of Heinrich events has been demonstrated from ocean drilling off the northeastern Brazilian coast, where the chemical signature of pulses of terrigenous sediment has been related to landward impacts of climate change (Arz *et al.*, 1998), and off the Iberian peninsula (Sanches-Goni *et al.*, 2000, 2002; Hinnov *et al.*, 2002). Chappell (2002) has demonstrated the importance of Heinrich events to sea-level changes recorded by coral terraces in Papua New Guinea .

Extreme events in the context of Quaternary climate change

The study of extreme events usually lasting hours or days demonstrates the reality of energy and sediment pulses passing through the landscape. But the integration or "coupling" of the different parts of the landscape is a far more complex issue (Church and Slaymaker, 1989; Harvey, 2002). Sediment shed from headwater reaches of river systems may be stored downstream in channel bars and in floodplains for long time periods, and immediate coupling of hillslope processes to stream channels is mainly restricted to mountainous areas. This ensures that the landscape is a mosaic of different forms and deposits of varying ages, and sediment stores can be dated to episodes of landscape instability throughout the Quaternary (2 Ma), and by inference beyond. Paleoflood analyses, often using evidence from slackwater deposits (Baker, 1987), have also revealed the distribution of extreme events on Quaternary timescales (Brakenridge, 1980). Synchronous sedimentary units in many floodplains can be considered in this context as evidence of periods of strong sediment transport interspersed with periods of reduced flows during the late Quaternary. The nature of the sediments and the character of the river channels also supply information regarding the status of catchment protection by the vegetation cover and the seasonality or flood regime of the rivers.

Studies of cyclones and similar storms establish direct connections between the rainfall inputs and the system response such as slope erosion, slope failure, flooding, and sedimentation. But not all, and perhaps not many, such events lead to major system changes that transform entire landscapes. This is because most extreme events occur within a spectrum of similar occurrences (over 10^1–10^2 yr) and the landscape is already configured to accommodate these. On alluvial fans, for example, this does not mean that destruction and loss of life will not be a consequence of channel changes; rather, it implies that shifting channels are part of the environmental system, which is adjusted to receive large quantities of water and coarse sediment. In the course of a major flood event many thresholds will be crossed, enabling huge boulders to be tossed around, buildings undermined and the position of channel bars to be altered. However, if a single event is big enough and areally extensive, then major landscape change can result.

The environmental context of landscape change becomes complex, however, when the occurrence of extreme events is placed within cycles or periods of sustained climate change, because extreme events of a given magnitude are likely to have different impacts on landscapes according to their sensitivity to perturbation and change. It is also probable that the magnitude and frequency of extreme events will vary within the time spectrum of decades to millennia.

Climate deterioration over centuries or millennia will cause the progressive depletion of plant cover, and the sensitivity of the landscape to extreme events may be gradually increased. Sediment yield from slopes will increase if rainfall intensities remain high although annual totals are reduced. At the same time stream power is reduced and this could mean that eroding and meandering rivers will become choked with debris, and braided plains and fans start to form. There is empirical evidence for this type of lagged or delayed response to climate change from tropical rainforest areas. In Africa and South America, Maley (1992) has documented rainforest decline from c. 28 ka, while similar pollen work in northeast Queensland (Kershaw, 1992; Moss and Kershaw, 2000) has shown decline in the vine forests after 38 ka, with further rainfall decline after 27 ka.

The landscape response in terms of erosion and sedimentation appears to have lagged the vegetation changes by several millennia. In Queensland, streams draining the east-facing escarpment into the Coral Sea around Cairns began fan-building around 30 ka, which continued until c. 14 ka (Nott et al., 2001; Thomas et al., 2001).

In West Africa, very few river sediments and no embedded wood are recorded after 24/22 ka (Thomas and Thorp, 1980, 1995), around the Last Glacial Maximum (LGM) for 5000–6000 years. In both cases, reduced discharges caused loss of stream power, and increased seasonality is thought to have led to long periods of very low flows. In West Africa, low gradients and an absence of highland catchments led to an almost complete cessation of deposition for several millennia, while in Queensland (and in many other areas) torrential streams formed large alluvial fans. Climate warming in the postglacial period began around 17 ka, and continued for around 4000 years until the interruption of the Younger Dryas (YD), which was cool and dry in the tropics and subtropics. Only after this interval did the Holocene climate reach a peak of humidity, followed by final recovery of the forest after 10.6 ka. The alluvial record indicates that the response of rivers at the West African sites was to leave coarse gravel bars containing large tree trunks from *c.* 15 to 13.5 ka. Only with the recovery of the rainforest after the YD did rivers convert to meandering, single-thread channels and deposit thick overbank silts. Pulses of energy and sediment showing the impact of Holocene climate fluctuations are recorded in floodplain sedimentation (Fig. 14.1). Published evidence (see Thorp and Thomas, 1992; Thomas and Thorp, 1995, 2003; Thomas, 2001) indicates that similar responses have occurred widely in tropical rivers.

In glaciated areas, there is a limited preglacial legacy relevant to the issue of landscape sensitivity, and the process of (the last) deglaciation itself was a unique episode in the formation of present-day landscapes. In some limited areas this was a multiple event as ice-sheets re-advanced during the YD. The withdrawal of the ice in mountain areas led to almost catastrophic instability, as slopes failed due to loss of support, glacial oversteepening and subsequent unloading, melting of ground ice, and the operation of sub-aerial processes on largely unvegetated slopes. Rockfalls and other slope failures at this time are well documented from Europe and the United States (González Díez *et al.*, 1996; Soldati, 1996; Berrisford and Matthews, 1997; Soldati: *et al.*, 2004). Large tracts of land were also subject to glacio-fluvial outwash and deposition. Subsequent evolution of these terrains has arguably been strongly influenced by the continuing readjustment of the landforms, the so-called "paraglacial" effect (Church and Ryder, 1972; Church and Slaymaker, 1989; Ballantyne, 2002a, 2002b). This paraglacial relaxation continues after more than 10 k yr have elapsed in most areas, but it followed a curve of rapid non-linear decay, most of the readjustment taking place within 1–2 k yr (Fig. 14.1). Early Holocene vegetation was sparse and we know that tree pollen were not abundant before *c.* 9.5 [14]C k yr BP (10.8 cal k yr BP) at Hockham Mere, Norfolk, and that Scots Pine did not appear in northern Scotland until *c.* 9 cal k yr BP (see Wilson *et al.*, 2000). The frequency of slope events, fluvial development, and lacustrine sequences were all modulated by later Holocene climate and vegetation changes (see Ballantyne, 2002a, 2002b). Studies have shown that, following the early major slope failures, subsequent evolution has either continued the same pattern of development or has been in the form of small-scale slope instability. There is also some evidence for sediment exhaustion occurring on hillslopes denuded in the early Holocene of most loose sediment left by the ice age. This implies that parts of the landscape may develop a *reduced* sensitivity to erosion with time. This can occur if sediment sources are depleted, but in northern Britain the spread of blanket peat has also protected the ground surface from erosion.

The course of Holocene erosion in Britain and Ireland has been reviewed by Edwards and Whittington (2001), based on the analysis of lake sediments and the variable relationship between landscape change and rates of sedimentation. In many cases there were delays in system response, but overall, lakes were found to be valuable indicators of landscape sensitivity. Clusters of dates recording rises in sedimentation at 26 sites at *c.* 5.3–5.0 k yr BP, 4.5–4.2 k yr BP and 3.0–2.8 k yr BP were thought to be related to phases of woodland clearance from the Neolithic to the Bronze Age and no climatic inferences were made. The dates were thought to indicate when "catchment soils ... around a particular site were pushed beyond an erosional threshold" (Edwards and Whittington, 2001). According to the robustness or sensitivity of the catchment, the "age" of the sediments would range from before vegetation change was found in the pollen record until some time afterwards. It is clear that lake data at the century scale for the Holocene incorporate the combined effects of climate change, human impact, and delayed response. The resultant "noise" makes interpretation very difficult.

The transformation of river channels is another aspect of late Quaternary landscape change that has been noted. Many lowland rivers in Europe switched from braided to meandering habits as catchments became forested in the early Holocene (Starkel, 1995; Lewis *et al.*, 2001). In areas not covered by ice during the YD, there is evidence that the duration of this period (*c.* 800 yr) was not long enough to transform river systems from established patterns. For similar reasons, the erratic and poorly defined Little Ice Age (LIA) is associated with some increases in certain types of event, but not with widespread fluvial reorganisation.

Different kinds of system behavior are implied by these examples. The pre-LGM preparation of landscapes for major instability and change in many extra-glacial areas shows a trend toward more open vegetation, accelerating toward the LGM. System behavior was progressively altered by the changes in climate and vegetation, and landscape sensitivity to extreme events probably increased with time elapsed along the curve of change. When climate and vegetation recovered after the LGM, it took several millennia before these same landscapes were stabilized (Thomas, 2004). Increased rainfall was effective from at least 15.3 ka in Africa and other parts of the tropics, for example, but full recovery of the rainforest was delayed until after the YD interval of cold dry climates, post 11 ka.

Two important principles can be drawn from late Quaternary landscape histories. First, some major landscape changes appear to lag behind climate changes by significant periods of time, often on a millennial scale. Second, the impact of extreme events will depend not only on the inherent sensitivity of the landscape system to change, but also on their occurrence within the longer time spectra of change. It is also important to return to the earlier assertion, that our perception of "rapid" change and the nature of that change are scale dependent. In the present context, this implies that, while small changes will be observed in natural systems (landscapes) over short time periods, major landscape transformations are likely to be observed after extended periods of 10^3 yr. Some exceptions to this generalization have been noted.

Spatial aspects of landscape systems

How the spatial dimension of landscape change can be understood within this temporal framework clearly requires further elaboration. One way in which we can attempt this is to look again at patterns of erosion and sedimentation. Events of a certain magnitude will trigger changes in landscape elements or components of a given sensitivity, but as event magnitude increases so more and more landscape elements will become affected, providing that event duration and rate of application of stress remain similar. Also, we can expect that as more and more elements of the landscape become incorporated into a process of catastrophic change, the greater will be the likelihood that the impact of these changes will endure. An example of such an event was a storm that hit the Serra des Araras in eastern Brazil in 1967. According to Jones (1973) a 3.5-hour storm delivered 275 mm rainfall and "laid waste … a greater landmass than ever recorded in geological history," involving more than 10 000 landslides, mostly debris flows, in an area of 180 km². There were 1700 deaths and there was total disruption of road and rail transport and the power infrastructure. The scars of this event remain clear after more than 30 years, partly because the landsliding involved a mantle of weathered rock (saprolite) that was largely removed from the multi-convex hills, converting convex slopes to linear debris flow scars and concave valley heads. Very little forest recovery is evident in the area. Most individual landslide scars are persistent over decades, and many will experience renewed activity over centuries.

Landslide-prone areas, however, show distinctive patterns of landslide occurrence, and even well-forested slopes may conceal many landslide scars and deposits. Results from Hong Kong (Lumb, 1975; Au, 1993) and from Puerto Rico (Larsen and Simon, 1993) show that slope failure as a response to rainfall events can be predicted. But the actual location and volume of future landslides is much more difficult to determine. Reasons for this spatial problem illustrate some issues in studies of landscape sensitivity. Rainfall intensity during a storm probably exhibits stochastic variations across complex terrain. Moreover, the inherent sensitivity of slopes to failure does not only depend on easily mapped criteria such as inclination and length, although these remain important. Other factors include regolith thickness, which may partly reflect variations in time elapsed since the previous landslide at different locations, the existence of hidden structures and fracture patterns, and the location of unmapped older landslides. The existence of large paleo-landslide scars is widespread, and smaller modern slides may be nested within the older features and represent a process of slope relaxation over 10^2–10^3 yr following an earlier catastrophic event. The recurrence interval of slope failures will also vary greatly between different slope elements and may decrease where regolith properties and thickness promote instability or where slope relaxation within older landslides continues.

All these factors combine to promote "divergence" between landscape elements over time, but this trend does not always continue indefinitely, because stabilization can occur. This is exemplified by the formation of stony soils in semiarid regions such as southeastern Spain (Alexander *et al.*, 1994; Cammeraat and Imeson, 1999). Exposure to infrequent intense rainfalls may result from overgrazing or other pressures on plant cover, leading to loss of fines and emergence of stones (bedrock pieces, calcrete fragments). The stones then form a lag that has many functions: shading the soil and conserving moisture, protecting soil from raindrop impact, and impeding surface sediment transport but possibly promoting formation of rills and gullies. In these landscapes, deep-rooting bushes grow at intervals of a few meters, allowing organic accumulation and surface moisture conservation. Such slopes adopt a quasi-stable pattern over a time period of decades. Only when the period is extended to millennia is the destabilization and differentiation of the landscape focused. Gullies have formed and extended into still earlier valleys during the period of settlement (wall building) and this has triggered groundwater flow beneath interfluves. The high sodium content of the marls has led to widespread dispersion of fines and opening of subterranean pipe/tunnel systems, many of which have collapsed. This implies that surface landscape patterns, which may be stable over decades, are linked to instability on longer timescales, during which the system gradually approaches collapse and rapid change (see Poesen and Valentin, 2003). Many such examples can be cited. This also illustrates the point that in many cases where pollen spectra appear unchanged for long periods, the system that maintains the vegetation pattern may be converging over centuries or millennia with thresholds for rapid, even catastrophic, change.

Other instances of such system behavior include the lags between climate change, vegetation change, and sediment yield already noted, where rises in the amount and caliber of sediment shed from slopes depend on changes to precipitation patterns and to the structure of the plant cover. Under natural conditions, vegetation is likely to change slowly. Kadomura (1995) has suggested that many former forested areas of the tropics gradually became forest–savanna mosaics approaching the LGM, the savanna areas being found on plateau tops and interfluves, where moisture stress and possibly fire would be limiting factors. Most pollen records are unable to infer landscape patterns at this spatial scale (Sugita et al., 1999). The use of fire by immigrant human groups probably accelerated such changes. This has been inferred from the pollen record at Lynch's Crater, northern Queensland (Turney et al., 2001), where the rise in charcoal corresponds with a long-term decline in the Auracarian vine forests (Kershaw, 1992). This site is close to the area of fan accumulation previously described. We do not know whether human impact could have been the trigger for major landscape instability in this area .

The coupling and divergence of landscape elements

Two important spatial concepts emerge in this context: coupling and divergence. Hillslope–channel coupling has been frequently discussed since it was introduced within the landscape sensitivity concept (Brunsden and Thornes, 1979). In a recent review Harvey (2002) considers the effective timescales in terms of: "(i) the frequency of (threshold exceeding) events, (ii) the recovery time, (iii) the propagation time (of changes that are not damped out)." Landscape changes propagated from one spatial element to another are dependent on the coupling or transfers of energy and matter (usually sediment) between them. At the local scale, these processes operate on short timescales from hours to decades, but as the spatial scale enlarges so the applicable temporal scales for understanding change are extended (Harvey, 2001, 2002; Thomas, 2001, 2004). Harvey also stresses that propagation from above is likely to be driven by climate changes and event frequencies on Quaternary timescales, whereas propagation of change from below will result from more gradual base-level influences, usually over much longer time periods. The propagation of change throughout a landform–landscape system is fundamental to understanding landscape sensitivity (Thomas, 2001) and should guide our perception of problems such as erosion or landslide hazards. It is possible to enter a local landscape subject to severe gullying and degradation and yet misunderstand the danger of uncontrolled extension of these conditions. In some badland areas, gullies exhibit a reticulated pattern.

But in others, they are confined to sensitive elements of the landscape. The well-known gullies at St. Michael's Mission, Zimbabwe, illustrate this point. Visitors concerned with erosion issues are likely to be shown this site, where valleys drain between two topographic levels. Presumed Quaternary climate changes have led to the accumulation of unconsolidated, stratified sediments up to 10 m thick, and the gullies are carved into them (Stocking, 1984; Thomas, 1994). Toward the valley margins, the sedimentary fill thins and the gullies die out, but in many areas of the tropics, sensitive colluvium is more extensive. It is also clear that the gullying at this site is only the current phase of recurrent instability in a sensitive landscape location .

These ideas also govern how we understand diversity in landscapes, which arises from three sets of linked factors: (1) spatial heterogeneity in landscape foundations of rocks and major landforms, (2) divergence between landscape elements arising from differences in process rates, and (3) long-term developmental trends in erosion and accumulation. The order in which we consider these is significant, because, by setting out the framework (1) for landscape diversity we set aside the notion of change in favor of stability over long time periods. This is not realistic where "new" land is formed by vulcanism or coastal progradation, nor where unconsolidated materials underlie extensive tracts of land, as in loess areas and some deserts. But if hills and plains are considered in this way, then the geological basis of landscape variety is acknowledged. On this model, surface process systems operate differentially to ensure divergence and increasing complexity so long as local and regional base levels present no limits to erosion and sedimentation. Successive generations of erosion scars, fans, and terrace surfaces are formed over 10^5 yr periods and are often complicated (or replaced) by forms and deposits resulting from glacial or eolian interruptions. Repeated sea-level change during the last 2 million years, together with the rising continental "freeboard" during the last 100 million years, has ensured that the long-term trend towards the ultimate destruction of major relief forms has been frequently interrupted. But on the land surfaces of the oldest cratons, found in South America, Africa, India, and Australia, relief is often subdued and dominated by widely spaced residual hills. These Gondwanaland plains have been isolated from continental base-level controls in the center of a super-continent for 10^8 yr. Yet, on and below their unexciting surfaces the deposits and weathering profiles are extremely complex. The complexity, however, is limited to a microtopography comprised of resistant materials that have survived removal, over significant periods of earth history (10^6–10^8 yr), and to the intricacies of the weathered mantle. The properties of these ancient regoliths remain fundamental to the understanding of the soil and vegetation patterns developed on them, and their long-term stability is responsible for many land resource issues, such as groundwater salinity and the concentration of economic mineral species. Such areas have had no connectivity (coupling) to sites of rapid landscape change over very long time periods .

The question of inheritance

Divergence and fragmentation of the landscape lead to spatial differentiation and to survival of landscape elements inherited from past climates (Thomas, 2001). This inheritance is an inevitable product of differential rates of change, as some elements of the landscape change more rapidly, while others remain little altered. Some inherited features can be extremely stable elements in the landscape; duricrusted hills and benches, and some forms of till, might be examples. On the other hand, overprinting and replacement of landscape properties can occur, so that a new set of features blankets and conceals the older ones. Sedimentation into a subsiding delta or other depocenter is an obvious example in geology, the growth of peat a process from pedology (Thomas, 2001).

Concluding remarks

The relevance and application of different timescales of enquiry to landscape sensitivity is dependent on the context of study. Increasing awareness of the inability of process monitoring alone to provide an adequate time frame for the understanding of climate-change impacts in the future has focused attention on the detailed proxy records available for the understanding of the Quaternary. These records also permit the reappraisal of events in the history of human civilization and settlement and provide added impetus to new historical enquiry. The timescales of relevance to different problems in landscape sensitivity may span seven orders of magnitude and an attempt is made here to outline their connections to landscape processes and change (Table 14.1).

Much of the terminology used to describe landscape sensitivity has emerged from geomorphology and related earth sciences, but the subject of landscape change is the province of many other research groups from the natural and historical sciences. The study of erosion and sedimentation over different time periods focuses attention on energy flows and rates of change. The spatial dimensions of landform study also raise fundamental issues concerning connectivity and coupling between different landscape elements, and these in turn lead to related questions concerning differential rates of change and divergence to produce landscape patterns. Some of these patterns have their origins in remote geological time periods, but in this study concepts are developed that can be applied within the 10^5 year time frame of the last glacial cycle, for which we now have abundant data (Table 14.2).

Table 14.1. Climate change and landscape sensitivity over a Quaternary glacial cycle, indicating the most appropriate timescales of enquiry

	Timescale of enquiry (years)						
	10^5	10^4	10^3	10^2	10^2–10^1	10^1–10^{-1}	10^{-1}–10^{-2}
Climate change	Glacial cycles; orbital changes (Milankovitch)	Major stadials; orbital changes	Glacial stades; cooling; Heinrich events (HE); Bond cycles; marine isotope stages (MIS)	Rapid warming episodes (GISP2) (D–O events)	Solar variability (complex)	Southern Oscillation (ENSO events)	Extreme events
Typical frequency	Eccentricity (glacial/interglacial cycles) 140 kyr	Obliquity 41 kyr; precession 23/19 kyr	Climate cycles/ D–O interstades 1.5–3 kyr; HE every 5–7 kyr	Occur within sub-Milankovitch cycles of 10^3 kyr duration	11, 22, ~88; 140, 220 yr; solar period 420 yr	SO index varies over years to decades	10, 50, 100 yr probabilities typically used
Duration	100–120 kyr	10^3 kyr	HE 1–3 kyr	D–O measured in decades	Decades to centuries	Typically 9–12 months	Days, hours
Climate and hydrology	Major temperature and precipitation changes	Temperature – 5–7°C; precipitation loss; ice sheets	Cooling, glacier advance; rainfall changes; reduced stream flow	Increased rainfall, storminess (?); erosion, floods	Rainfall fluctuations; floods; droughts	Regional impacts on rainfall and floods	Landslides, floods, cyclones

Vegetation cover	Major biome changes and replacement	Major biome changes	Changes in species composition and vegetation structure	Local changes; possible expansion of forests	Obscured by complex time series	Local patterns; gap dynamics	Local destruction of land cover
Landscape sensitivity issues	No direct connection	Influence on regional vegetation patterns	Millennium-scale triggers for landscape change	Possible association with energy pulses	Influence on magnitude and frequency of extreme events	Immediate influence on regional storm intensities	Erosion–sedimentation events
Landscape stability concepts	No direct connection	Lagged response	Paraglacial instability; switching of river behavior	Energy pulses; decadal flood variation	Periods of slope and channel instability	Episodes of slope and channel instability	Threshold-crossing events; disturbance of equilibria

Table 14.2. Geomorphic concepts and phenomena associated with landscape instability within the Quaternary timescale. Process–time relationships (allocation to cells in table) indicate the most relevant timescales of enquiry; arrows indicate where processes operate over a range of timescales. Note the importance of the millennial timescale.

Timescale of enquiry (years)						Geomorphic and sedimentary examples
⇐ 10^5 ⇐	10^4 ⇑	10^3 ⇑	10^2 ⇑	10^1 ⇑	10^{-1}–10^{-2} ⇒	
Quasi-cyclical landform evolution						Multiple glaciation
						Major depositional forms: fans, terraces
						Regional loess sequences
						Weathering phenomena
⇐	Progressive landform change		⇑			Slope forms and curvature
						Sediment accumulation: fans, coastal barriers
	Non-linear decay→ depletion			⇑		Weathering/soil systems
						Sediment exhaustion (mainly paraglacial)
						Post-glacial sea-level rise: Holocene deltas

⇐?

⇓

⇐?

Relaxation time→
new equilibria

⇒

⇒?

Channel patterns
Slope erosion: sediment
accumulation
Fining-upward sediment
sequences

Lags
Coupling
Propagation

⇒

⇒ ?

Rainfall→ vegetation→
sediment yield
Rill→ gully network
Slope→channel coupling

Enhanced or reduced
flow regimes

⇒

Sedimentary units
Incision→terraces

⇐

Energy pulses
Punctuated
equilibria

⇒

Floods→channel bars
Slope failure→colluvium
Fining-upward sediment
sequences

→

Equilibria
Thresholds
Self-
organization

Slope, channel patterns

References

Alexander, R. W., Harvey, A. M., Calvo, A. C., James, P. A., and Carda, A. (1994). Natural stabilisation mechanisms on badland slopes. In *Environmental Change in Drylands: Biogeographical and Geomorphological Perspectives*, ed. A. C. Millington, and K. Pye. Chichester: Wiley, pp. 85–111.

Arz, H. W., Pätzold, J., and Wefer, G. (1998). Correlated millennial-scale changes in surface hydrography and terrigenous sediment yield inferred from last-glacial marine deposits off northeastern Brazil. *Quaternary Research*, 50, 157–166.

Au, S. W. C. (1993). Rainfall and slope failure in Hong Kong. *Engineering Geology*, 36, 141–147.

Baker, V. R. (1987). Palaeoflood hydrology and extraordinary flood events. *Journal of Hydrology*, 96, 77–99.

Ballantyne, C. K. (2002a). Paraglacial geomorphology. *Quaternary Science Reviews*, 21, 1935–2017.

Ballantyne, C. K. (2002b). A general model of paraglacial landscape response. *The Holocene*, 12, 371–376.

Bard, E., Rostek, F., and Sonzogoni, C. (1997). Interhemispheric synchrony of the last deglaciation inferred from alkenone palaeothermomotry. *Nature*, 385, 707–710.

Bard, E., Rostek, F., Turon, J. -L., and Gendreau, S. (2000). Hydrological impact of Heinrich events in the subtropical northeast Atlantic. *Science*, 289, 1321–1324.

Berrisford, M. S. and Matthews, J. A. (1997). Phases of enhanced rapid mass movement and climatic variation during the Holocene: a synthesis. *Paläoklimaforschung Palaeoclimate Research*, 19, 409–440.

Blum, W. B., Toomey, R. S. III, and Valastro, S. Jr. (1994). Fluvial response to late Quaternary climatic and environmental change, Edwards Plateau, Texas. *Palaeogeography, Palaeoclimatology and Palaeoecology*, 108, 1–21.

Bond, G., Showers, W., Cheseby, M., *et al.* (1997). A pervasive millennial-scale cycle in North Atlantic Holocene and Glacial climates. *Science*, 278, 1257–1266.

Brakenridge, G. R. (1980). Widespread episodes of stream erosion during the Holocene and their climatic cause. *Nature*, 283, 655–656.

Brown, A. G. (1996). Human dimensions of palaeohydrological change. In *Global Continental Changes: the Context of Palaeohydrology*, ed. J. Branson, A. G. Brown, and K. J. Gregory. Geological Society Special Publication 115 London: Geological Society, pp. 57–72.

Brown, A. G. (1998). Fluvial evidence of the medieval warm period and the late medieval climatic deterioration in Europe. In *Palaeohydrology and Environmental Change*, ed. G. Benito, V. R. Baker, and K. J. Gregory. Chichester: Wiley, pp. 43–52.

Brunsden, D. and Thornes, J. B. (1979). Landscape sensitivity and change. *Institute of British Geographers, Transactions*, 4, 463–484.

Cammeraat, L. H. and Imeson, A. C. (1999). The evolution and significance of soil–vegetation patterns following land abandonment and fire in Spain. *Catena*, 37, 107–127.

Chappell, J. (2002). Sea level changes forced ice breakouts in the last glacial cycle: new results from coral terraces. *Quaternary Science Reviews*, 21, 1229–1240.

Church, M. and Ryder, J. M. (1972). Paraglacial sedimentation: a consideration of fluvial processes conditioned by glaciation. *Geological Society of America Bulletin*, 83, 3059–3071.

Church, M. and Slaymaker, O. (1989). Disequilibrium of Holocene sediment yield in glaciated British Columbia. *Nature*, 337, 452–454.

Coelho-Netto, A. L. (1997). Catastrophic landscape evolution in a humid region (SE Brazil): inheritances from tectonic, climatic and land use induced changes. *Geografia Física e Dinâmica Quaternária*, Suppl. **III**, 21–48.

Cullingford, R. A., Davidson, D. A., and Lewin, J. (1980). *Timescales in Geomorphology*. Chichester: Wiley.

Dansgaard, W., Johnsen, S. J., Clausen, H. B., *et al.* (1993). Evidence for general instability of climate from a 250 kyr ice-core record. *Nature*, 364, 218–220.

Driver, T. S. and Chapman, G. P. (1996). *Timescales and Environmental Change*. London: Routledge.

Edwards, K. J. and Whittington, G. (2001). Lake sediments, erosion and landscape change during the Holocene in Britain and Ireland. *Catena*, 42, 143–173.

Erhart, H. (1955). Biostasie et rhexistasie: esquise d'une théorie sur le rôle de la pédogenèse en tant que phénomène géologique. *Comptes Rendues Academie des Sciences Française*, 241, 1218–1220.

Ganapolski, A. and Rahmsdorf, S. (2001). Rapid changes of glacial climate simulated in a coupled climate model. *Nature*, 409, 153–158.

González Díez, A., Salas, L., Díaz de Terán, J. R., and Cendrero, A. (1996). Late Quaternary climate changes and mass movement frequency and magnitude in the Cantabrian region, Spain. *Geomorphology*, 15, 291–309.

Harvey, A. (2001). Coupling between hillslopes and channels in upland fluvial systems: implications for landscape sensitivity, illustrated from the Howgill Fells, northwest England. *Catena*, 42, 225–250.

Harvey, A. (2002). Effective timescales of coupling within fluvial systems. *Geomorphology*, 44, 175–201.

Heinrich, H. (1988). Origin and consequences of cyclic ice rafting in the northeast Atlantic Ocean during the past 130 000 years. *Quaternary Research*, 29, 142–152.

Hinnov, L. A., Schulz, M., and Yiou, P. (2002). Interhemispheric space-time attributes of the Dansgaard–Oeschger oscillations between 100 and 0 ka. *Quaternary Science Reviews*, 21, 1213–1228.

Jones, F. O. (1973). Landslides of Rio de Janeiro and the Serra das Araras Escarpment, Brazil. *US Geological Survey Professional Paper*, 697.

Kadomura, H. (1995). Palaeoecological and palaeohydrological changes in the humid tropics during the last 20 000 years, with reference to equatorial Africa. In *Global Continental Palaeohydrology,* ed. K. J. Gregory, L. Starkel, and V. R. Baker. Chichester: Wiley, pp. 177–202.

Kershaw, A. P. (1992). The development of rainforest–savanna boundaries in tropical Australia. In *Nature and Dynamics of Forest-Savanna Boundaries*, ed. P. A. Furley, J. Proctor, and J. A. Ratter. London: Chapman and Hall, pp. 255–271.

Knox, J. C. (1972). Valley alluviation in southwestern Wisconsin. *Annals of the Association of American Geographers*, 62, 401–410.

Knox, J. C. (1993). Large increases in flood magnitude in response to modest changes in climate. *Nature*, 361, 430–432.

Knox, J. C. (1995). Fluvial systems since 20 000 years BP. In *Global Continental Palaeohydrology*, ed. K. J. Gregory, L. Starkel, and V. R. Baker. Chichester: Wiley, pp. 87–108.

Larsen, M. C. and Simon, A. (1993). A rainfall intensity–duration threshold for landslides in a humid tropical environment, Puerto Rico. *Geografiska Annaler*, 75**A**, 13–23.

Lewis, S. G., Maddy, D., and Scaife, R. G. (2001). The fluvial system response to abrupt climate change during the last cold stage: the Upper Pleistocene River Thames fluvial succession at Ashton Keynes, UK. *Global and Planetary Change*, 28, 341–359.

Lumb, P. (1975). Slope failures in Hong Kong. *Quarterly Journal of Engineering Geology*, 8, 31–65.

Macklin, M. G. and Lewin, J. (1993). Holocene river alluviation in Britain. *Zeitschrift für Geomorphologie*, 88, 109–122.

Maley, J. (1992). The African rainforest vegetation and palaeoenvironments during the Quaternary. In *Tropical Forests and Climate*, ed. J. Myers. Dordrecht: Kluwer.

Moss, P. T. and Kershaw, A. P. (2000). The last glacial cycle from the humid tropics of northeastern Australia: comparison of a terrestrial and a marine record. *Palaeogeography, Palaeoclimatology and Palaeoecology*, 155, 155–176.

Nott, J., Thomas, M. F., and Price, D. M. (2001). Alluvial fans, landslides and Late Quaternary climatic change in the wet tropics of northeast Queensland. *Australian Journal of Earth Sciences*, 48, 875–882.

Phillips, J. D. (1999). *Earth Surface Systems*. Oxford: Blackwell.

Phillips, J. D. (2003). Sources of non-linearity and complexity in geomorphic systems. *Progress in Physical Geography,* 27, 1–23.

Poesen, J. and Valentin, C. (2003). Gully erosion and global change. *Catena*, 50, 87–564.

Ruxton, B. P. (1968). Order and disorder in landform. In *Land Evaluation*, ed. G. A. Stewart. Melbourne: Macmillan, pp. 29–39.

Sanches-Goni, M. F., Turon, J. L., Eynaud, F., and Gendreau, S. (2000). European climatic response to millennia-scale changes in the atmosphere–ocean system during the last glacial period. *Quaternary Research,* 54, 394–403.

Sanches-Goni, M. F., Cacho, I., Turon, J. L., *et al*. (2002). Synchronicity between marine and terrestrial responses to millennial-scale climatic variability during the last glacial period in the Mediterranean region. *Climate Dynamics*, 19, 95–105.

Schumm S. A. (1977). *The Fluvial System.* Chichester: Wiley.

Schumm, S. A. (1979). Geomorphic thresholds: the concept and its applications. *Institute of British Geographers, Transactions,* 4, 485–515.

Schumm, S. A. and Lichty, R. W. (1965). Time, space and causality in geomorphology. *American Journal of Science*, 263, 110–119.

Schumm, S. A. and Parker, R. S. (1973). Implications of complex response of drainage systems for Quaternary alluvial stratigraphy. *Nature*, 243, 99–100.

Soldati, M. (1996). Landslides in the European Union. *Geomorphology*, 15, 364.

Soldati, M., Corsini, A., and Pasuto, A. (2004). Landslides and climate change in the Italian Dolomites since the late glacial. *Catena*, 55, 141–161.

Starkel, L. (1995). Palaeohydrology of the temperate zone. In *Global Continental Palaeohydrology*, ed. K. J. Gregory, L. Starkel, and V. R. Baker. Chichester: Wiley, pp. 223–257.

Stocking, M. A. (1984). Rates of erosion and sediment yield in the African environment. In *Challenges in African Hydrology and Water Resources* (Proceedings of the Harare Symposium, 1984). IASH Publication 144, pp. 285–293.

Stuiver, M., Grootes, P. M., and Brazunas, T. F. (1995). The GISP2 δ^{18} climate record of the past 16 500 years and the role of the sun, ocean and volcanoes. *Quaternary Research*, 44, 341–354.

Sugita, S., Gaillard, M. J., and Broström, A. (1999). Landscape openness and pollen records: a simulation approach. *The Holocene*, 9, 409–421.

Thomas, D. S. G. and Allison, R. J. (1993). *Landscape Sensitivity*. Chichester: Wiley.

Thomas, M. F. (1994). *Geomorphology in the Tropics*. Chichester: Wiley.

Thomas, M. F. (2001). Landscape sensitivity in time and space: an introduction. *Catena*, 42, 83–98.

Thomas, M. F. (2004). Landscape sensitivity to rapid environmental change: a Quaternary perspective with examples from tropical areas. *Catena,* 55: 107–124.

Thomas, M. F. and Simpson, I. (2001). Landscape sensitivity: principles and applications in cool temperate environments. *Catena*, 42, 81–386.

Thomas, M. F. and Thorp, M. B. (1980). Some aspects of the geomorphological interpretation of Quaternary alluvial sediments in Sierra Leone. *Zeitschrift für Geomorphologie*, N.F., Supplementband, 36, 140–161.

Thomas, M. F. and Thorp, M. B. (1995). Geomorphic response to rapid climatic and hydrologic change during the Late Pleistocene and Early Holocene in the humid and sub-humid tropics. *Quaternary Science Reviews*, 14, 193–207.

Thomas, M. F. and Thorp, M. B. (2003). Paleohydrological reconstructions for tropical Africa since the Last Glacial Maximum: evidence and problems. In *Paleohydrology: Understanding Global Change*, ed. K. J. Gregory and G. Benito. Chichester: Wiley, pp. 167–192.

Thomas, M. F., Nott, J. M., and Price, D. M. (2001). Late Quaternary stream sedimentation in the humid tropics: a review with new data from NE Queensland, Australia. *Geomorphology*, 39, 53–68.

Thorp, M. B. and Thomas, M. F. (1992). The timing of alluvial sedimentation and floodplain formation in the lowland humid tropics of Ghana, Sierra Leone, and western Kalimantan (Indonesian Borneo). *Geomorphology*, 4, 409–422.

Turney, C. S. M., Kershaw, A. P., Moss , P ., *et al.* (2001). Redating the onset of burning of Lynch's Crater (North Queensland): implications for human settlement in Australia. *Journal of Quaternary Science*, 16, 767–771.

Vandenberghe, J. and Maddy, D. (2001). Editorial: the response of rivers to climate change. *Quaternary International*, 79, 1–3.

Veldkamp, A. and Tebbens, L. A. (2001). Registration of abrupt climatic changes within fluvial systems: insights from numerical modelling experiments. *Global and Planetary Change*, 28, 129–144.

Werritty, A. and Leys, K. F. (2001). The sensitivity of Scottish rivers and upland

valley floors to recent environmental change. *Catena*, 42, 251–273.

Wilson, R. C. L., Drury, S. A., and Chapman, J. L (2000). *The Great Ice Age*. London: Routledge.

DONALD A. DAVIDSON

IAN A. SIMPSON

15

The time dimension in landscape ecology: cultural soils and spatial pattern in early landscapes

Contributors to this volume have been invited to write personal statements and perspectives on their particular area of landscape ecology, and we accept this challenge even though we appreciate that our views may well be controversial. Our overall perspective is that landscape ecology is a science that primarily depends upon spatial analysis in order to elucidate landscape processes. The roots of the subject lie in landscape classification systems, an emphasis evident in many of the other essays in this volume. More flexible approaches are now evident, given that the notion of landscapes is largely a cultural concept. Such flexibility has been fostered by the application of GIS and image analysis techniques, and by incorporating economic methods of analysis. Nevertheless, landscape ecology is focused primarily on spatial rather than temporal differentiation as the analytical core. This is not to deny that temporal dimensions are explicitly included in the many definitions of landscape ecology, or that much research has been done on landscape change through sequential sampling, the analysis of aerial photographs, or other remote-sensed imagery.

The essential thrust of this essay is to argue that landscape ecology as a spatial science needs to find ways of interfacing with such subjects as environmental archaeology and history in order to combine spatial and temporal analysis. It is only with such a linkage to longer timescales that landscape ecologists can begin to understand long-term landscape processes and build robust models for predicting future landscapes.

Though much landscape ecology lacks temporal analysis of any significant duration, environmental archaeology, history, or environmental science often faile to produce the necessary spatial resolution. There are, for example, considerable difficulties in reconstructing regional or local patterns of vegetation at various times in the past based on the analysis of pollen as retrieved from peat stratigraphies at a limited number of sites. An environmental record of change

Issues and Perspectives in Landscape Ecology, ed. John A. Wiens and Michael R. Moss. Published by Cambridge University Press. © Cambridge University Press 2005.

through time is inevitably site-specific and poses spatial interpolation problems. Documentary sources for reconstructing environmental history may well be excellent at providing aggregated data based on administrative or management units, but often cannot be applied to determine precisely what was going on at particular points in the landscape in the past. The most satisfactory form of record is often maps, but this record frequently lacks appropriate detail and spatial resolution. Given the limitations of these conventional approaches to long-term landscape change, an alternative approach to the question of providing detailed spatial resolution of earlier landscapes is required, particularly over the last *c.* 250 years that are critical to tracing the development of present-day landscapes. Such an alternative is to be found in the identification and analysis of soil properties, an approach recognizing that soils reflect the landscape in which they have been formed and that landscape history, particularly human activity, is imprinted in soil properties. The challenge for the pedologist working in this context is to recognize those properties in soils that reflect past landscape patterns and processes, a theme that we now elaborate with reference to our own particular research interests, cultural soils.

Cultural soils and landscape ecology

Soils vary in four dimensions: spatially (three dimensions) and temporally (one dimension). As a result, soils offer a unique opportunity in landscape ecology to investigate spatial and temporal patterns. The traditional approach to investigating soil spatial patterns is through a soil survey. The vast majority of published soil maps are based on the landscape or free-survey approach, whereby landscape units are delimited using aerial photo and field evidence. The essential assumption is that variability in soil types and properties will be less within such landscape units than between them. Much research has demonstrated the broad validity of such an approach, at least at scales less detailed than 1 : 25 000. Increasing research is being done using geo-statistical techniques for spatial interpolation of individual soil properties. Central to such an approach is the quantification of spatial dependence using variograms, which are central to the process of kriging. For the traditional landscape approach to soil survey, the central concept is that soils co-vary with landscape units. Thus, the emphasis in many soil surveys has been to interpret the "naturally occurring" soil types within landscape units rather than basing mapping on soil properties as they actually exist. In fairness, there has been a growing use of classification systems such as the US *Soil Taxonomy* (Soil Survey Staff, 1996), which requires field and laboratory-derived data to remove or at least minimize soil type interpretation by surveyors. Soil property approaches have also been used to classify and define the quality of agricultural land in England and Wales.

The analysis of soil spatial patterns is comparatively simple because, ignoring practical problems, soils can be sampled at any place and depth. Difficulties arise when consideration is given to the time dimension. Soils are not like neat accumulating sediments with a resultant stratigraphy, but instead possess a range of properties, many resulting from processes that operated at differing times in the past. Soils are continually stirred by faunal or physical mechanisms including tillage. Soil is essentially a living entity with scars, attributes, and characteristics that reflect the history of the soil. Furthermore, such properties will react in different ways and timescales to changes in the soil-forming environment. We argue that, despite the considerable challenges to research on soil change through time, soils very much need to be addressed through a realization that many current properties will be relict from earlier conditions, and that these properties can be used to reconstruct and interpret landscapes of the past.

Human activity in the past is often of particular importance in terms of inducing soil change. Imagine a group of students and their instructor round a soil profile at any location within the settled part of the world, with the aim to consider soil development. After an overview of the general environmental setting, there would be discussion on the impact on the profile of past and present human-related activities. Such activities include vegetation change, compaction, drainage, tillage, manuring, disposal of waste, construction, cropping, soil import, and stone removal. These are examples of direct impacts and there can also be indirect ones such as changes in flood or drought regimes, or acid input. These are all human-related activities and thus all soils, to varying extents, can be considered as cultural or anthropogenic soils. Cultural is a better word since it implies the influence of a range of human-related activities, whilst anthropogenic suggests a more limited range of processes with soil improvement as the key objective. Anthrosols are soils which have been modified by human activities, primarily from agricultural practices and settlement. They can be subdivided into anthropogenic soils, which have been intentionally modified, and anthropic soils, which were modified unintentionally. In practice, such a distinction is often difficult to apply. All soils in the settled part of the world have cultural attributes reflecting human history and use. They can thus provide an excellent focus in landscape ecology when the aim is to integrate spatial and temporal analysis. Plaggen soils are examples of cultural or anthropogenic soils and are discussed in outline below, demonstrating how they may be applied to questions of long-term landscape change.

Plaggen soils are named after the German term *Plaggenboden*, also known in Germany first as *Esch* soils and now as *Plaggenesche*, in the Netherlands as *Enk* soils, and in Belgium as *Plaggen-gronden*. They correspond to *Fimic Anthrosols* in the FAO–UNESCO system (FAO, 1988) or *Plaggepts* in the US *Soil Taxonomy* (Soil Survey Staff, 1996). Plaggen are turves which were cut heath or grass sods, and which after drying were used as bedding in byres and stables (Spek, 1992; Blume, 1998). This material was accumulated in a dung or midden heap and then other materials may have been added, for example domestic and hearth waste or calcareous sand. The result was then spread onto fields as manure, again with other potential materials such as seaweed, as a means of maintaining arable soil fertility. In the Netherlands, plaggen turves were cut every 5–15 years with 5–10 ha heathland being needed to supply 1 ha of arable land. Turves cut from heathland resulted in the formation of black topsoil, whilst a brown color was the consequence from grassland turf. The turves when cut also included mineral material, both within the organic layers and at the base where there was the interface between the organic and more mineral horizons. The result of this process is the gradual accumulation at a rate of *c.* 1 mm per year to produce a diagnostic topsoil up to *c.* 1 m in depth in northwest Europe. In Europe the process was most widespread in areas of inherently poor-quality soils, for example, in areas underlain by fluvioglacial sands and gravels. Plaggen soils are extensive in northern Germany, the Netherlands, northern Belgium, and southwestern Denmark, with distinctive occurrences also in France, southwest England, southern and southwestern coastal areas of Ireland, the remoter islands of Scotland (Orkney and Shetland), and in the far north of Norway (Lofoten Islands). Extensive deepened soils known as *Terra Preta* are present in Amazonia (Woods and McCann, 1999) and raised *Camellón* field systems have been identified in Inter-Andean Valleys in Ecuador (e.g., Wilson *et al.*, 2002). In the Netherlands and Scotland, plaggen formation took place predominantly from the thirteenth century and continued up to the early twentieth century in the remoter parts of Shetland (Davidson and Simpson, 1994; Davidson and Smout, 1996). Archaeological evidence suggests that plaggen soil formation was present in the Netherlands by 500 BC to AD 100. A buried plaggen soil on Sylt in the north Friesen islands (Germany) occurs under a Late Bronze Age mound (Blume, 1998). Small areas (*c.* 1 ha) of fossil plaggen soils associated with settlement sites from the Bronze Age and buried under calcareous wind-blown sands have also been identified in Orkney and in Shetland. Here grassy turves, peat ash, and human manures were used to stabilize highly erodible soils and enhance soil fertility, allowing cultivation in a highly marginal environment (Simpson *et al.*, 1998). Thus, plaggen soil formation has been occurring, not necessarily on a continuous basis, for more than 3000 years in northwest Europe. Areas of plaggen soils in the Netherlands are distinctive because they are raised by the order of 1 m, giving them local relief. The diagnostic plaggen epipedon, known as the *Eschhorizont* in Germany, is usually 50–100 cm in thickness, homogeneous in field morphology and color (dark brown or black), with organic content in the range 1–8%, usually high in sand content, and phosphate-rich if animal excrement was added to the turves. Highly fragmented artefacts of tiles or pottery are often present in this topsoil, again indicating inputs during the period when the material accumulated in midden heaps.

Detailed analysis of plaggen soils in the West Mainland of Orkney through the synthesis of relict soil properties, including thin-section micromorphology, organic biomarkers, phosphorus chemistry, and particle size distributions, has begun to demonstrate marked temporal and local spatial variability in the development of these soils (Simpson, 1997). Such shifts can be demonstrated to reflect variation in cultural landscape processes. These soils cover an area of some 7 km² and are relict features of infield management between the late Norse period and the agricultural improvements of the late nineteenth century. Soil properties reflect a simple and successful, though labor-intensive, process of maintaining and enhancing soil fertility in these arable areas. Turves were stripped from the unenclosed podzolic hill-land, causing significant damage to summer grazing areas, and composted with varying proportions of domestic ruminant and pig manures prior to their application on the arable area. Minor amounts of seaweed were also applied, but there is no evidence to support exploitation of other landscape resources for use in these arable infield areas. Relict soil properties indicate that the intensity of manure application was greater with proximity to the farmstead and became greater as the cultural soil developed, perhaps reflecting greater demand for produce from an increasing population. It is clear from the soil properties that the management of these infield areas was not uniform and varied both temporally and spatially, becoming more organized as the cultural soil developed, although earlier detailed patterns may have been lost through post-depositional pedogenesis. The level of cultivation intensity of these soils was moderate, plowed rather than spaded, as it did not result in substantial down-slope and down-profile movement of fine material.

These cultural soils represent areas in the cultural landscape where nutrients were concentrated for the purposes of arable activity, suggesting a collective organization of landscape resources, integrating arable and livestock husbandry practices. In Orkney, turf for the infield came only from the hill-land, on which livestock would have been grazed during the summer, and not from the grassland areas of the enclosed township. Although this caused substantial damage to the hill-land and gave major problems for reclamation during the subsequent early modern improvements, it meant that the enclosed grassland and meadow areas could be maintained for the provision of winter grazing and fodder. This in turn made available the animal manures that were applied to the infield and which would have been collected by housing the animals, at least overnight if not throughout the winter period. Under such a scenario, the ratio of arable to enclosed grazing land becomes important to the maintenance and enhancement of infield fertility levels. In West Mainland Orkney, this ratio is approximately 1 : 4.6 and, on the basis of relict soil-property indicators, would appear to be at a level which could more than adequately maintain arable-land soil fertility where manures were used in conjunction with turf.

Similar detailed patterns of relict soil properties in cultural contexts are evident in other areas of northwest Europe. In Lofoten, northern Norway, relict soils dating from *c.* AD 700 to the late 1900s provide opportunities to identify land-management practices in landscapes climatically marginal for agriculture (Simpson and Bryant, 1998). Here it is evident from field survey and soil micromorphology that there was deliberate management of erodible sandy soils in sloping locations to create small areas of cultivation terrace, and that cultivation and manuring practice also took place in more gently sloping locations. A range of materials including wet turf, dry turf, fish wastes, and domestic animal manures was used to stabilize the accumulated soil, enhance fertility, and secure subsistence-level barley production in an early cultural landscape dominated by livestock production and fishing activity. Such detailed studies serve to emphasize the spatial and temporal variability of relict soil properties evident in cultural soils, overturning the notion that such areas of land were static and uniformly managed features in early cultural landscapes. It also serves to demonstrate that relict soil properties clearly have a role to play in establishing and explaining the complexities of both manuring and cultivation in cultural landscapes, together with the associated patterns of landscape organization.

The example of plaggen soil formation and distribution emphasizes the importance of a longer timescale perspective than is conventionally the case in landscape ecology. It also permits the conclusion to be drawn that relict soil properties in general, and cultural soil properties in particular, can provide a means by which a spatially explicit analysis of early landscape pattern and process becomes possible. Soils permit integration of spatial, temporal, and anthropogenic considerations in landscape ecology. They give an appreciation of the interplay between natural processes of soil formation, systems of land management and cropping in the past, changing patterns of human populations, and the need to sustain increasing numbers at particular times and in areas of low inherent fertility. Landscape ecology badly needs a greater time depth to confirm and enhance its disciplinary status and to give it credibility in wider policy and academic communities. A soils-based approach to the historical dimensions of landscape ecology offers a realistic yet challenging way forward.

References

Blume, H. P. (1998). *History and Landscape Impact of Plaggen Soils in Europe*. Montpellier: World Congress of Soil Science.

Davidson, D. A. and Simpson, I. A. (1994). Soils and landscape history: case studies from the Northern Isles of Scotland. In *History of Soils and Field Systems*, ed. T. C. Smout and S. Foster. Aberdeen: Scottish Cultural Press, pp. 66–74.

Davidson, D. A. and Smout, C. (1996). Soil change in Scotland: the legacy of past land improvement processes. In *Soils, Sustainability and the Natural Heritage*, ed. A. G. Taylor, J. E. Gordon, and M. B. Usher. Edinburgh: HMSO, pp. 44–54.

FAO (1988). *Soil Map of the World*. Reprinted with corrections. World Soil Resources Report 60. Rome: FAO.

Simpson, I. A. (1997). Relict properties of anthropogenic deep top soils as indicators of infield management in Marwick, West Mainland, Orkney. *Journal of Archaeological Science*, 24, 365–380.

Simpson, I. A. and Bryant, R. G. (1998). Relict soils and early arable land management in Lofoton, Norway. *Journal of Archaeological Science*, 25, 1185–1198.

Simpson, I. A., Dockrill, S. J., Bull, I. D., and Evershed, R. P. (1998). Early anthropogenic soil formation at Tofts ness, Sanday, Orkney. *Journal of Archaeological Science*, 25, 729–746.

Soil Survey Staff (1996). *Keys to Soil Taxonomy*, 7th edn. Washington, DC: US Department of Agriculture.

Spek, T. (1992). The age of plaggen soils. In *The Transformation of the European Rural Landscape: Methodological Issues and Agrarian Change 1770–1914*, ed. A. Verhoeve and J. A. J. Vervloet. Belgium: National Fund for Scientific Research. pp. 35–54.

Wilson, C., Simpson, I. A., and Currie, E. J. (2002). Soil management in pre-hispanic raised field systems: micromorphological evidence from Hacienda Zuleta, Ecuador. *Geoarchaeology*, 17, 261–283.

Woods, W. I., and McCann, J. M. (1999). The anthropogenic origin and persistence of Amazonian Dark Earths. *Yearbook, Conference of Latin American Geographers*, 25, 7–14.

HAZEL R. DELCOURT

PAUL A. DELCOURT

16

The legacy of landscape history: the role of paleoecological analysis

Present-day landscape patterns are the outcome of a number of ecological, geological, climatological, and cultural processes occurring over prehistoric and historic time frames (Delcourt and Delcourt, 1991, 2004; Delcourt, 2002). The interactions of these processes change through time and are mediated by changing natural and anthropogenic disturbance regimes (Wiens *et al.*, 1985; Delcourt and Delcourt, 1988; Turner, 1989; Russell, 1997; Foster *et al.*, 1998a). The legacy of long-term landscape history is a lasting overprint upon both natural and cultural landscapes, as the effects of past processes leave a mark on present landscapes that may endure long into the future. This legacy has been understood for a long time in Great Britain (Rackham, 1986) and Europe (Delcourt, 1987; Birks *et al.*, 1988) and it is now increasingly recognized in North America (Abrams, 1992; Russell, 1997; Delcourt *et al.*, 1998; Delcourt and Delcourt, 1998, Foster *et al.*, 1998a, 1998b).

How we view the relevant processes involved in the development of landscape patterning is conditioned by the temporal and spatial window through which we view landscape change as well as by the techniques we use to measure landscape response to physical and biological interactions (Fig. 16.1). Physical constraints on landscape development may be depicted as a nested hierarchy of controlling factors (Urban *et al.*, 1987; Delcourt and Delcourt, 1988). For example, on a timescale of thousands of years, large and predictable changes occur in global and regional climate. As little as 9000 calendar years ago, Northern Hemisphere perihelion occurred in summer rather than in winter as it does today, resulting in higher seasonal contrast (warmer summers, colder winters) that influenced the survival, adaptability, and rates of spread of plant and animal species as they adjusted to postglacial conditions (Bennett, 1996). On this millennial timescale, the landscape matrix may change several times. For example, in response to global warming at the end of the Pleistocene Epoch,

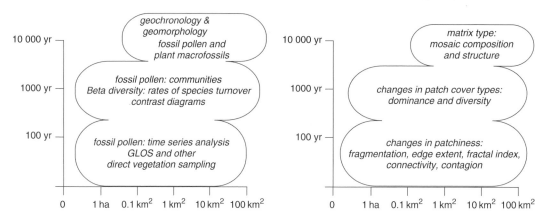

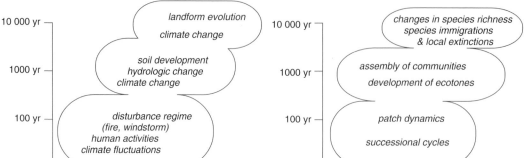

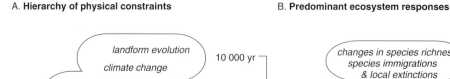

FIGURE. 16.1

Space-time hierarchical diagram for integrated analysis of paleoecological and landscape ecological data on a series of nested scales: (A) hierarchy of physical constraints; (B) predominant ecosystem responses; (C) techniques to measure land-scape response; and (D) predicted changes in landscape heterogeneity. Modified from Delcourt and Delcourt (1988).

in northern temperate regions the landscape changed from glacial ice or bare ground to tundra, then to boreal forest, and finally to temperate forest or grassland (Watts, 1988).

In formerly glaciated regions and along coastal zones, landforms have changed dynamically on a timescale of hundreds to thousands of years, and they continue to change today in response to changes in sea level (Clark, 1986) and lags in uplift of the land with postglacial rebound (Davis and Jacobson, 1985). On this timescale, changes in species richness, immigrations, and extinctions occur as ecosystems undergo dynamic transformations that affect both the composition and structure of the entire landscape mosaic (Prentice, 1986).

Over a timescale of hundreds to thousands of years, soil development, hydrologic changes, and climate changes are all relevant physical factors that affect the assembly of biological communities (Davis *et al.*, 1998) and the development of ecotones (Delcourt and Delcourt, 1992). Ecological implications are changes in composition, dominance, and diversity of cover types ranging in scale from local stands to regional landscapes.

On the timescale of tens to hundreds of years, changes in disturbance regimes, for example in recurrence intervals of fire or of catastrophic windstorms (Foster *et al.*, 1998b), affect the equilibrium state of the landscape (Turner *et al.*, 1993) through feedbacks involving patch dynamics and successional cycles (Delcourt and Delcourt, 1988). On this timescale, changes in patchiness, fragmentation of patches, extent of edge between adjacent cover types, and connectivity within the landscape mosaic may be expected, all occurring within a nested mosaic of landscape development where the top level has cascading effects upon all other levels (Urban *et al.*, 1987).

Paleoecological studies are essential to comprehensive long-term landscape-ecological studies. Measuring the legacy of past processes requires: (1) a conceptual framework of hierarchical relationships and scaling (Delcourt and Delcourt, 1988; Fig. 16.1); (2) integration of appropriate research techniques across temporal scales; (3) making paleoecological inferences spatially explicit; (4) adequate temporal resolution of samples during critical times of landscape change; and (5) quantitative methods of mapping and analyzing landscape mosaics simultaneously through time and space (an extension of "multi-temporal spatial analysis," *sensu* White and Mladenoff, 1994).

The role of paleoecology in reconstructing pattern and process at the landscape scale is illustrated by a case study from our research in the eastern Upper Peninsula of Michigan, USA (Delcourt and Delcourt, 1996; Delcourt *et al.*, 1996, 2002; Petty *et al.*, 1996; Delcourt, 2001). Along the northern shore of Lake Michigan, the Laurentide Ice Sheet receded by 10 600 radiocarbon years ago, leaving behind a freshly deglaciated landscape with a bare-ground mosaic of glacial ice-contact deposits, glacial stream and lake sediments including outwash sands, delta deposits, and lake clays, and highland outcrops of Silurian-age dolomite bedrock forming the Niagara Escarpment (Petty *et al.*, 1996). With the weight of glacial ice removed, postglacial rebound of more than 100 m occurred as the land surface rose upward, rapidly at first, then more slowly after 8000 radiocarbon years ago. Levels of the Great Lakes fluctuated as new drainage outlets were cut and others were dammed. During times of high stands in the position of lake level, such as occurred 6900 radiocarbon years ago, embayments of Lake Michigan extended 10 to 15 km inland from the present-day shoreline.

Beginning 5400 radiocarbon years ago, a climate cycle with a 70-year periodicity began to drive oscillations in the level of Lake Michigan, resulting in coastal accretion of 75 sets of beach ridges and inter-dune swales (Delcourt et al., 1996). The combination of continuous uplift of the land and cyclic fluctuations in Lake Michigan has created a broad swath of gently undulating lake plain that extends as much as 4.5 km inland from the modern shoreline.

In the mid-postglacial interval, between 8000 and 4000 years ago, regional climate warmer and drier than present led to fluctuating soil moisture conditions that resulted in soil leaching and precipitation of iron sesquioxides as a hard pan or ortstein layer in sandy outwash soils. This pedogenic ortstein layer impedes downward percolation of meteoric water through what otherwise are porous and permeable sandy substrates. Development of ortstein between 6900 and 3200 radiocarbon years ago corresponded with the establishment of communities of mesic hardwood trees (Delcourt et al., 2002). Xeric pine-dominated forest was replaced in part by mesic hardwoods after about 4000 radiocarbon years ago as regional climate became cooler and moister.

With a major increase in lake effect precipitation by 3000 radiocarbon years ago, extensive wetlands developed in two contrasting landscape settings: (1) paludified upland depressions forming bog patches up to 5 km × 20 km in extent; and (2) the broad lake plain formed parallel to the present-day shoreline of Lake Michigan (Petty et al., 1996; Delcourt et al., 2002). Prehistoric Native American occupation sites were located on south-facing slopes with gradients of less than 2%, concentrated both on bedrock knolls (for procurement of chert for making projectile points) and on lowland landscapes near the shoreline of Lake Michigan (for proximity to spring spawning areas of sturgeon and for procurement of beaver, moose, deer, and plant resources) (Silbernagel et al., 1997).

As in the case from the eastern Upper Peninsula of Michigan, if there is a change over time in physical baselines such as topographic contrast, hydrologic setting, or extent of terrestrial habitats available for colonization by plants and animals, including humans, then landscape heterogeneity can be expected to change over time intervals ranging from centuries to millennia. Rather than a static edaphic baseline setting the overall expectable level of landscape heterogeneity, and modified only by changes in intensity of disturbance (as postulated by Wiens et al., 1985), we suggest that a much more complex landscape history emerges in which longer-term edaphic changes may occur in cycles (beach-ridge formation) or as discrete events (ortstein development). The resulting changes in landscape heterogeneity are related to edaphic thresholds (for example, rapid paludification) as well as to climate change (increases or decreases in lake-effect precipitation). Future changes in landscape heterogeneity may be difficult to predict from measurement of the landscape configuration at any one point in time because of the complexity of these interacting variables.

Wallin *et al.* (1994) observed that changes in patterning on managed landscapes may lag by decades to hundreds of years behind changes in land-management plans designed to promote specific landscape patterns ("pattern inertia"). In order to predict and manage the future state of landscape heterogeneity, conservationists must therefore take into account not only the legacy of the long-term natural trajectory of change but also the lasting effects of twenty-first-century management practices (Turner *et al.*, 1993; Wallin *et al.*, 1994, Kline *et al.*, 2001). In addition, near-future changes in regional and global climate may result in unprecedented changes in ecosystems and in species distributions (Iverson and Prasad, 1998) in the time frame of the next 50 to 100 years that represents only one rotation cycle of forest cutting (Botkin and Nisbet, 1992; Wallin *et al.*, 1994). From the paleoecological record, we infer that under such circumstances, state variables such as ecosystems or regional landscape types may be inappropriate targets for conservation efforts; instead, relevant processes underlying landscape pattern are the appropriate focus of conservation efforts (Pickett *et al.*, 1992; Delcourt and Delcourt, 1998). Because of the recognition that environmental change may trigger disassembly and reassembly of biological communities, the hierarchy of indicators proposed by Noss (1990) for monitoring biodiversity in the twenty-first century may now be modified (Fig. 16.2) to include the probability that rapid climate change may destabilize ecosystems, particularly along major ecotones (Delcourt and Delcourt, 1992, 2001). The result may be "bifurcation" to alternate landscape states (Turner *et al.*, 1993) with concomitant changes in landscape heterogeneity.

The legacy of landscape history persists as an imprint upon present-day landscapes, which in turn are only a snapshot of the long-term trajectory of landscape change. The challenge is to integrate ecological knowledge across spatial and temporal scales, to understand the processes that are fundamental in producing landscape pattern, and to develop predictive models of future landscape changes that will help in conservation and management of biodiversity and landscape heterogeneity in the face of near-future environmental changes associated with global warming.

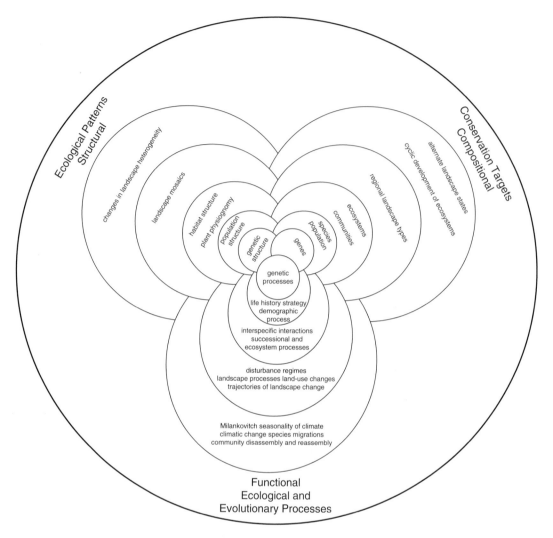

FIGURE 16.2
Ecological and evolutionary processes, ecological patterns, and conservation targets over a hierarchy of levels of biological organization. Modified from Noss (1990).

References

Abrams, M. D. (1992). Fire and the development of oak forests. *BioScience*, 42, 346–353.

Bennett, K. D. (1996). *Evolution and Ecology: the Pace of Life*. Cambridge: Cambridge University Press.

Birks, H. H., Birks, H. J. B., Kaland, P. E., and Moe, D. (1988). *The Cultural Landscape: Past, Present and Future.* Cambridge: Cambridge University Press.

Botkin, D. B., and Nisbet, R. A. (1992). Projecting the effects of climate change on biological diversity in forests. In *Global Warming and Biological Diversity*, ed. R. L. Peters and T. E. Lovejoy. New Haven, CT: Yale University Press, pp. 277–293.

Clark, J. S. (1986). Dynamism in the barrier-beach vegetation of Great South Beach, New York. *Ecological Monographs*, 56, 97–126.

Davis, M. B., Calcote, R. R., Sugita, S., and Takahara, H. (1998). Patchy invasion and the origin of a hemlock–hardwoods forest mosaic. *Ecology*, 79, 2641–2659.

Davis, R. B., and Jacobson, G. L. (1985). Late glacial and early Holocene landscapes in northern New England and adjacent areas of Canada. *Quaternary Research*, 23, 341–368.

Delcourt, H. R. (1987). The impact of prehistoric agriculture and land occupation on natural vegetation. *Trends in Ecology and Evolution*, 2, 39–44.

Delcourt, H. R. (2001). Creating landscape pattern. In *Learning Landscape Ecology*, ed. S. Gergel and M. G. Turner. New York, NY: Springer, pp. 62–82.

Delcourt, H. R., (2002). *Forests in Peril: Tracking Deciduous Trees from Ice-age Refuges into the Greenhouse World*. Blacksburg, VA: McDonald and Woodward.

Delcourt, H. R., and Delcourt, P. A. (1988). Quaternary landscape ecology: relevant scales in space and time. *Landscape Ecology*, 2, 23–44.

Delcourt, H.R., and Delcourt, P.A. (1991). *Quaternary Ecology: a Paleoecological Perspective*. New York, NY: Chapman and Hall.

Delcourt, H. R., and Delcourt, P. A. (1996). Presettlement landscape heterogeneity: evaluating grain of resolution using General Land Office Survey data. *Landscape Ecology*, 11, 363–381.

Delcourt, P. A., and Delcourt, H. R. (1998). Paleoecological insights on conservation of biodiversity: a focus on species, ecosystems, and landscapes. *Ecological Applications*, 8, 921–934.

Delcourt, P. A., and Delcourt, H. R. (1992). Ecotone dynamics in space and time. In *Landscape Boundaries*, ed. A. J. Hansen and F. di Castri. New York, NY: Springer, pp. 19–54.

Delcourt, P. A., and H. R. Delcourt. (2001). *Living Well in the Age of Global Warming*. White River Junction, VT: Chelsea Green.

Delcourt, P. A., and Delcourt, H. R. (2004). *Prehistoric Native Americans and Ecological Change: Human Ecosystems in Eastern North America since the Pleistocene*. Cambridge: Cambridge University Press.

Delcourt, P. A., Petty, W. H., and Delcourt, H. R. (1996). Late-Holocene formation of Lake Michigan beach ridges correlated with a 70-year oscillation in global climate. *Quaternary Research*, 45, 321–326.

Delcourt, P. A., Delcourt, H. R., Ison, C. R., Sharp, W. E., and Gremillion, K. J. (1998). Prehistoric human use of fire, the eastern agricultural complex, and Appalachian oak-chestnut forests: paleoecology of Cliff Palace Pond, Kentucky. *American Antiquity*, 63, 263–278.

Delcourt, P. A., Nester, P. L., Delcourt, H. R., Mora, C. I., and Orvis, K. H. (2002). Holocene lake-effect precipitation in northern Michigan, USA. *Quaternary Research*, 57, 225–233.

Foster, D. R., Motzkin, G., and Slater, B. (1998a). Land-use history as long-term broad-scale disturbance: regional forest dynamics in central New England. *Ecosystems*, 1, 96–119.

Foster, D. R., Knight, D. H., and Franklin, J. F. (1998b). Landscape patterns and legacies resulting from large, infrequent forest disturbances. *Ecosystems*, 1, 497–510.

Iverson, L. R., and Prasad, A. M. (1998). Predicting abundance of 80 tree species following climate change in the eastern United States. *Ecological Monographs*, 68, 465–485.

Kline, J. D., Moses, A., and Alig, R. J. (2001). Integrating urbanization into landscape-level ecological assessments. *Ecosystems*, 4, 3–18.

Noss, R. F. (1990). Indicators for monitoring biodiversity: a hierarchical approach. *Conservation Biology*, 4, 355–364.

Petty, W. H., Delcourt, P. A., and Delcourt, H. R. (1996). Holocene lake-level fluctuations and beach ridge development along the northern shore of Lake Michigan, USA. *Journal of Palaeolimnology*, 15, 147–169.

Pickett, S. T. A., Parker, V. T., and Fiedler, P. L. (1992). The new paradigm in ecology: implications for conservation biology above the species level. In *Conservation Biology: the Theory and Practice of Nature Conservation, Preservation, and Management*, ed. P. L. Fiedler and S. K. Jain. New York, NY: Chapman and Hall, pp. 66–88.

Prentice, I. C. (1986). Vegetation responses to past climatic variation. *Vegetatio*, 67, 131–141.

Rackham, O. (1986). *The History of the Countryside: the Full Fascinating Story of Britain's Landscape*. London: Dent.

Russell, E. W. B. (1997). *People and the Land Through Time: Linking Ecology and History*. New Haven, CT: Yale University Press.

Silbernagel, J., Martin, S. R., Gale, M. R., and Chen, J. (1997). Prehistoric, historic, and present settlement patterns related to ecological hierarchy in the eastern Upper Peninsula of Michigan, USA. *Landscape Ecology*, 12, 223–240.

Turner, M. G. (1989). Landscape ecology: the effect of pattern on process. *Annual Review of Ecology and Systematics*, 20, 171–197.

Turner, M. G., Romme, W. H., Gardner, R. H., O'Neill, R. V., and Kratz, T. K. (1993). A revised concept of landscape equilibrium: disturbance and stability on scaled landscapes. *Landscape Ecology*, 8, 213–227.

Urban, D. L., O'Neill, R. V., and Shugart, H. H. (1987). Landscape ecology: a hierarchical perspective can help scientists understand spatial patterns. *BioScience*, 37, 119–127.

Wallin, D. O., Swanson, F. J., and Marks, B. (1994). Landscape pattern response to changes in pattern generation rules: land-use legacies in forestry. *Ecological Applications*, 4, 569–580.

Watts, W. A. (1988). Europe. In *Vegetation History*, ed. B. Huntley and T. Webb III. Dordrecht: Kluwer, pp. 155–192.

White, M. A., and Mladenoff, D. J. (1994). Old-growth forest landscape transitions from pre-European settlement to present. *Landscape Ecology*, 9, 191–206.

Wiens, J. A., Crawford, C. S., and Gosz, J. R. (1985). Boundary dynamics: a conceptual framework for studying landscape ecosystems. *Oikos*, 45, 421–427.

17

Landscape ecology and global change

We often hear that the world is growing smaller. "Globalization" via rapid air travel, trade agreements, the internet, and a highly migratory global population are rapidly turning the earth into one very large landscape. Land-use change, once thought to be only a local phenomenon, is now of such a scale as to alter the composition of the atmosphere and to affect climate in far distant locations from the original perturbation. Industry across the globe, driven largely by fossil fuel combustion, has altered the composition of the atmosphere and is now clearly warming the earth's climate and producing complex responses and feedbacks between the earth's surface and its atmosphere. The global changes in the atmosphere, oceans, and land surface have forced the development of large-scale models both to understand the responses and feedbacks of change and to "predict" or forecast possible future changes, with the possibility of interventions to forestall or slow the onset of negative consequences. Since the issues of global change are by definition global, the models of atmosphere, oceans, and terrestrial biosphere are constrained to relatively coarse grids, due largely to computational limits. Unfortunately, in all three "spheres" many of the processes that determine the large-scale patterns occur at sub-grid scales. Dynamic Global Vegetation Models (DGVMs), for example, are typically implemented at $0.5°$ latitude–longitude resolution (c. 50-km resolution). Yet most of the patterns and processes fundamental to ecosystem modeling are sub-grid scale (landscape and lower levels), rendering global simulations a challenging enterprise.

The International Geosphere–Biosphere Program (IGBP), now in Phase II, has recognized these problems in the Phase I research plan. Specifically, Activity 2.2 (Landscape Processes) addressed the issues of landscapes and global change. Activity 2.2 was further subdivided into four tasks: (1) landscape-scale responses of vegetation to changing land use and disturbance; (2) fire as a major

Issues and Perspectives in Landscape Ecology, ed. John A. Wiens and Michael R. Moss. Published by Cambridge University Press.

disturbance that will be influenced by global change and will in turn feed back to landscape pattern and processes; (3) the interactions between landscape patterns and species migration in response to climate change; and (4) the effects of landscape pattern on primary ecosystem processes. Two other activities within Focus 2, Patch Dynamics and Global Vegetation Dynamics, also bear directly on landscape patterns and processes and global change. Thus, the entire Focus 2 program was structured around three spatial scales, patch (or stand), landscape, and global, all of which are relevant to landscapes and global change.

As a practitioner within one of these activities, Global Vegetation Dynamics, I understand all too well how easy it is to become too focused on one's particular area (scale; King, this volume, Chapter 4) of immediate research and lose sight of the interconnections among the program elements. Although these large research programs are well designed, integration across the projects (scales) is often difficult. My goal in this essay is to attempt to slice through the issues, across scales, in an integrative way in an attempt to show some of the immediacy and applicability of landscape issues when attempting to build models of global vegetation dynamics. This will not be a discussion of potential impacts on landscapes from global change. Rather, I will present a personal view of some of the landscape issues that must be considered in order to build global-scale models that can be credibly pushed beyond current climate and land-use conditions.

What is a landscape and why do we need a landscape perspective?

According to the IGBP, "landscapes are defined as spatial entities comprising [*sic*] of a set of interacting ecosystems sharing a common broad abiotic environment... and land use system. Usually, the geographic range spans from a few to several hundred km^2." The keywords are "spatial" and "interacting ecosystems." Many important processes operate at scales from leaf to landscape, such as gas exchange, fires, local plant dispersal, and many others. Landscapes up to several hundred km^2 are also common management units, although management of the land surface is itself a hierarchical phenomenon, occurring from local to regional and national scales. Insofar as they are "spatially" considered and contain interacting elements, all of these scales can credibly be considered as landscapes. However, we tend to focus on the traditional landscape scale, in part because it is the most amenable to human experience. Even so, we should not lose sight of the importance of landscape, or spatial, processes at multiple scales. A dung beetle views the landscape quite differently than does a soaring eagle.

Important patterns and scales

Ecosystems span an enormous range of scales in both time and space, from seconds (leaf physiology) to centuries, and from molecules to biogeographic zones (Neilson, 1986). O'Neill *et al.* (1986) nicely describe some of the properties of ecosystem hierarchies:

> The higher level appears as an immovable barrier to the behavior of the lower levels. This constraint is a natural consequence of the asymmetry in rate constants. The rates always become slower as one ascends the hierarchy and, therefore, the lower levels are constrained because they are unable to affect the behavior of the higher level ... Lower-level behaviors are essential to the functioning and persistence of higher-level structure that, in turn, constrains the behavioral flexibility of all lower-level objects.

In a sense, higher-level structure is an emergent property of lower-level processes, but one that also constrains lower-level processes to operate within certain bounds.

This hierarchical premise holds for climate systems as well as ecosystems. For example, climate is traditionally viewed as a slowly changing process (e.g., glacial–interglacial time scales) and can normally be viewed as a constant. Yet the patterns and processes over which global climate is simulated span at least 14 orders of magnitude (Michael Schlesinger, personal communication). Simulation of global climate is not done at the scale of air masses. Rather, modelers simulate the fluid dynamics of the entire global atmosphere at a timestep of about 20 minutes. Large-scale weather and climate patterns are emergent properties that are constrained by the physics of the atmosphere and its interactions with the oceans, cryosphere, topography, and biosphere. Even so, only about three orders of magnitude are currently simulated directly and many sub-grid processes such as cloud dynamics are empirically "parameterized." Sensitivity studies indicate that the nature of the cloud parameterization could produce either positive or negative feedbacks on global warming and that both feedbacks occur, depending on the nature of the clouds.

Similarly, large-scale spatial ecological patterns are emergent properties of interacting processes at multiple scales, as mediated by natural organisms. Ecosystems are organized within slowly changing climate zones that are typically viewed as constant. At the other extreme, fast processes, such as photosynthesis, are normally considered to be stable and can be simulated using simple empirical equations. The importance for global patterns and processes of sub-gridcell (landscape) dynamics is only now beginning to be appreciated.

The simplest and earliest form of biogeographic modeling was to correlate the emergent patterns of climate with the emergent patterns of biogeographic zones or biomes. However, this presumes that the processes that create both climate zones and biomes are stable, neither of which is true under the current conditions of rapid global warming. Climatic zones of today carry certain properties of temperature, humidity, and other characteristics associated with seasonal changes in weather systems. Under climate change, however, these properties will vary, both in quantity and in timing. Hence, there is a need for climate modelers to simulate fundamental processes in order to estimate the "structure" of "new" climatic zones. Similarly, organisms operate differently under higher CO_2 levels, for example, with different rates of photosynthesis and different water-use efficiencies. Thus, attempts to simulate large-scale biotic responses to climate change must begin with fundamental processes at the organismal and lower levels. Fortunately, the organisms performing these functions can be grouped into functional types to simplify simulation of processes.

The unique aspect of global ecosystem modeling in comparison to more traditional ecological modeling is that the emergent, large-scale spatial patterns and their dynamics are the primary points of interest. State-of-the-art biogeographic modeling relies on small-scale processes (leaf to landscape) but is calibrated to large-scale biogeographic and hydrologic patterns (e.g., Neilson, 1995). The challenge is to find the simplest model structure that is sufficient to capture the necessary processes at all the appropriate scales (Verboom and Wamelink, this volume, Chapter 9). In the simplified view of the world that I implemented in the MAPSS biogeography model, I perceive two fundamentally different kinds of upland plants, based on their different rate processes: slowly responding woody plants and rapidly responding grasses and other ephemerals (Mapped Atmosphere–Plant–Soil System; Neilson, 1995). These functional types (grass or woody) have an inferred or explicit allometry and phenological inertia, and the woody overstory competes with the ephemeral understory at a patch level.

The functional types in the MAPSS model interact through competition for common resources – light and water. If the overstory leaf area is sufficiently dense, the understory cannot be supported and the system simplifies to a homogeneous forest or shrubland, at effectively a stand scale. Similarly, if water is sparse and fires abundant, the woody functional type is removed and the system simplifies to homogeneous grassland, also at effectively a stand scale. The structurally and dynamically interesting systems are intermediate (i.e., tree or shrub savannas) and can imply stand to landscape scale, but over a homogeneous substrate.

Positive feedbacks (O'Neill et al., 1986) can operate to enhance differences among adjacent ecosystem types. For example, as one moves from wet to dry along an aridity gradient, the density of the forest will thin to a point where a grassy understory just begins to be supportable with enhanced understory light. Introduction of an understory creates competition for water, which further thins the canopy overstory, thereby allowing even more understory, creating a positive feedback. Additional feedbacks through fire can thin the overstory even more, allowing yet more grass and more fire until an equilibrium is reached. If the woody component is sufficiently dense, the system can be considered as homogeneous woodland (stand scale). However, if the overstory becomes sufficiently thin, then the ecosystem must be considered as biphasic (Whittaker *et al.*, 1979), containing trees with a grassy understory (one phase) and grass with no tree overstory (another phase).

Thus, along this hypothetical aridity gradient, with no topographic complexity, there is an endogenous shift from a homogeneous system (forest) at the wet end to a heterogeneous system (savanna) with increasing aridity and back to a homogeneous system (grassland) with further increases in aridity. With yet further increases in aridity, grasses thin out, fires become infrequent and shrubs can enter the system, introducing a new but different scale of heterogeneity (Ludwig, this volume, Chapter 6). Transitions between these physiognomic shifts in heterogeneity are generally termed ecotones. An example of this gradient would be a transect from the eastern US forests into the Great Plains grasslands (through woodlands) and into the arid southwest semi-desert grasslands and shrublands. These broad-scale emergent biogeographic patterns should be possible to simulate from fundamental processes operating in a global vegetation model. For example, in simulating the distribution of Xeromorphic Subtropical Shrubland (a woody/grass system), the MAPSS model has produced a nearly perfect overlay of the very complex distribution of *Quercus turbinella* (canyon live oak) and its relationship to regional airmass gradients in the arid southwest.

If we interject topographic complexity into the above moisture gradient, the spatial disposition of ecotones can become quite complex along both elevational and horizontal temperature and moisture gradients. For example, a north–south transect along the west slope of the Rocky Mountains from southern Idaho to the Mexican border illustrates the complex shifts in elevational ecotones along latitudinal temperature and moisture gradients (Neilson, 2003). Winter temperature increases from north to south along the transect, as does summer rainfall. The temperature gradient allows upper elevational ecotones to increase in elevation with decreasing latitude, while the summer rainfall gradient allows the lower elevational ecotones to decrease in elevation with decreasing latitude. Thus, these elevationally divergent gradients create a latitudinal "wedge" of ecotones. In the southern part of the transect, the wide elevational separation of ecotones creates the classic ecosystem zonation patterns described by Whittaker and Niering (1965) on the Santa Catalina Mountains of Arizona. At the northern part of the transect, however, the elevational ecotones converge into one elevation. The result is a spatial pattern of complexity that contains both vertical and horizontal gradients of diversity. Peet (1978) described a similar latitudinal gradient along the east slope of the Rocky Mountains.

It is well recognized that diversity tends to increase at ecotones, at least for the dominant organisms (Hansen and di Castri, 1992). Trees and grass, for example, interdigitate at the prairie–forest ecotone, enhancing local diversity. The same type of interdigitation and spatial diversity gradients occur with elevation at the southern end of the transect, for example in the Santa Catalina Mountains. At the northern end of the transect, with the spatial convergence of ecotones, the different vegetation zones sort out on unique topoedaphic facets, compressing the interdigitation of vegetation from the macro scale to the micro scale and creating a wholly new elevational zonation pattern.

Thus, attempts to understand the patterns of local, gradient, and regional diversity at only one end of the transect, for example, would be only partially revealing and would provide little general understanding of the landscape patterns. Descriptive landscape statistics (Haines-Young, this volume, Chapter 11) might accurately describe the patterns at each end of the transect, but would shed little light on the causes of the patterns. The context of the landscape spatial patterns within the regional climatic gradients can, however, help explain the local patterns. Nested scale analyses are very powerful tools for such purposes. The study that led to the description of this "wedge" of ecotones was based on a set of nested-scale experimental seedling transplants along environmental gradients at scales of meters (shrub to intershrub), tens of meters (landscape geomorphic facets), hundreds of meters (elevation), and hundreds of kilometers (regional) (Neilson and Wullstein, 1983).

Simulations at the relatively coarse scale of 10-km resolution (Neilson, 1995) were able to elicit the same regional gradients in ecotones, providing inferences to spatial patterns and processes at landscape-scale resolutions much smaller than the 10-km grid cells (Neilson, 2003). Such regions of convergence of ecotones may tend to concentrate where steep airmass gradients converge. I propose that these "nodes" of air-mass convergence drive a rescaling of ecological gradients, which is most manifest at the landscape scale. Large-scale, homogeneous "grains" of vegetation distal to these nodes become small-scale grains sorting out on topoedaphic microsites in proximity to the nodes (Neilson et al., 1992). The large-scale biogeographic correlations between climate and air masses are reproducible using the new class of models, such as MAPSS. Perhaps more interesting, however, is the possibility of inferring landscape-scale patterns from the coarse-scale, regional patterns simulated by the models.

Important processes and scales

Patterns at all scales change through time and could change very rapidly under global change. Robust predictions of changes in pattern, however, require a solid underpinning of the processes that produce patterns and their changes. Numerous ecological processes occur across a wide range of scales and are critical for global vegetation modeling. Ecosystem physiology controls trace gas and water exchanges across the biosphere–atmosphere interface and must be scaled from leaf to canopy, landscape, and region. Likewise, population processes, including dispersal, establishment, growth, and reproduction and their meta-population equivalents, should be represented. The current suite of DGVMs, however, does not deal well with these population processes, as such models are focused on functional types rather than species. Yet even functional types must reproduce and disperse although they must exhibit the functions and spatial distribution of at least one species.

Ecosystem productivity, carbon balance, nutrient cycling, and water balance are clearly related to the spatial patterns of ecosystem structure at landscape scales. Accurate quantification of these processes becomes difficult with increasing sub-gridcell heterogeneity. Ecosystem disturbances, such as fire and pest infestations, also operate across a range of scales that can span gridcell dimensions. For example, within a gridcell one must somehow keep track of fire intensity and size and the fraction of the cell burned, but fire spread is not directly simulated, nor are fires currently allowed to spread from cell to cell at the coarse gridcell resolution.

Hydrologic processes are strongly coupled to vegetation processes and span scales from local infiltration processes to regional river routing, yet most of the physics occurs at very fine scales. Vegetation and hydrologic modeling grew out of separate disciplines and historically the two sets of processes were rarely coupled, mechanistically. A common assumption in both disciplines was that no model could be calibrated to work well beyond a relatively small domain without re-calibration. Traditionally, a vegetation modeler might construct a very simple water-balance model to meet just the needs of local simulations. When first building the MAPSS model, I attempted just such a simple structure for soil hydrology, but imposed the constraints that a single calibration must work well in every region and landscape of the conterminous United States and that transpiration be driven by leaf and canopy processes. I used four contrasting sub-regions within the country to build and test the model, and quickly discovered that I could calibrate the simple model to any one or two regions, but not to all regions simultaneously. After enhancing the model through several levels of increasing structural complexity, I found the minimal complexity that could be calibrated to all regions. The model was calibrated against observed runoff data from many watersheds with an average area of about 4 km². Thus, the MAPSS model is calibrated as a landscape-scale model, but its structure was imposed by a continental-scale implementation.

Another example of how the constraint of fine-scale processes can affect broad-scale patterns occurred in the structuring and calibration of the transpiration equation in the MAPSS model. There is no consensus on the mathematical formulation of the canopy conductance term in any typical biophysical transpiration equation. Usually, some form for the equation is implemented and the ground surface characteristics are specified. That is, the spatial distribution of leaf area and roughness are imposed. Under such imposed constraints, it is possible to implement any number of forms for the conductance equation, since other components of the conductance (leaf area and roughness) are fixed. In the MAPSS model, however, both leaf area and roughness are emergent properties. In attempting to calibrate the equation, I discovered that the orientation of the prairie–forest border along its entire north–south extent in the conterminous United States was sensitive to the structure of the equation for canopy conductance. If a sub-term in the equation was in one location (as, for example, a linear function), then the location of the ecotone could be properly calibrated in the north but not in the south, and vice versa. That is, over the length of the ecotone it was canted diagonally, rather than being correctly positioned in a primarily north–south orientation. However, with the sub-term in a different location (as, for example, an exponential function), the ecotone was properly oriented. Thus, the use of a broad-scale biogeographic pattern as a constraint forced a specific structure to a leaf-scale physiological process. Had the model been developed over one small landscape or had the biogeographic pattern been imposed rather than an emergent property, these nuances of structure would not have been discovered.

Sub-gridcell heterogeneity: representing the landscape in coarse grids

The landscape scale is inherently a sub-grid problem when one conducts global simulations. Typically, each gridcell is viewed as a homogeneous entity. A topographically induced mosaic of forests and grasslands, for example on opposing aspects, would appear as a savanna in a large gridcell. For some issues the simulated savanna may provide sufficient accuracy, but for others it clearly won't. There are numerous schemes being considered for handling such situations and they range from simple to complex. The most simple is to recognize that there are different entities within the gridcell and that the relative areas of each are known. However, their spatial positions with respect to each other are not known, nor are there explicit interactions among the different "landscape" elements. For example, a gridcell containing a mosaic of forests and grasslands, perhaps scattered among many isolated patches, will be represented as containing only two patch types with aggregate areas summing to the total of the isolated, but similar patches. More complicated schemes would allow interactions among patches and eventually a more spatially explicit rendering of the patches, as discussed below.

The simple biphasic system described earlier (tree–grass versus grass alone) can be handled through explicit simulation of each patch type, while keeping track of the area of each. For convenience, the areas of the forest patch can be estimated from the average landscape-level tree leaf-area index with the area of the grass patch being the balance. Light competition can then be area-weighted within one equation, so that a single patch simulation captures the average behavior of both forested and open-grass patches. One advantage of this aggregated approach is that it allows the root systems of the two types to compete for water and nutrients, while maintaining independent light regimes. In other words, we've explicitly recognized heterogeneity in the above-ground components at the landscape scale, but have preserved a more homogeneous below-ground competitive environment. Different processes within landscapes can operate at very different spatial and temporal scales. Even so, the heterogeneity is implicit in the mathematical structure of a single simulation and does not represent explicit simulation of unique landscape elements. If the tree patches become too sparse, even below-ground competition would be truncated and a wholly new simulation would be required to capture the non-interacting patches. These independent simulations would still be maintained within a single gridcell with a common climate and soil.

The areas of forest and non-forest patches can change over time as a function of disturbance. Fires and other disturbances in the landscape produce significant problems for global simulations. They create a mosaic of uneven-aged patches, with new patches being created as often as each year in some cases. There are numerous structural and process differences between 1-yr-old and 15-yr-old patches. However, the differences between 100-yr-old and 115-yr-old patches may be very marginal when under the same climate and substrate. Thus, one approach is to allow creation of new patches each year and to track them individually, but as they become increasingly similar with age, merge them back together. In an otherwise homogeneous gridcell, these patches initially would be non-interacting and would only be represented uniquely by their areas and ages. In gridcells with complex terrain, these patches could be maintained on unique soils and with unique climates, but again non-spatially. Eventually, there could be some level of spatial interaction among patches, but still without spatially explicit representation within the cell. Even more intensive is to simulate the patches explicitly using nested grid systems or variable grid systems. The grid mesh would be of high resolution in complex terrain and of low resolution in simple terrain. In these situations, new age classes would be accommodated across several cells, rather than within a single cell. These approaches will be very CPU-intensive and will likely require supercomputer technology.

Other schemes are possible, but all carry trade-offs in either spatial detail or temporal dynamics. These approaches will require considerable testing and validation to arrive at the most simple method that accurately captures the necessary level of structural and temporal dynamics over large spatial extents. Of course, the definition of "necessary" is itself variable, depending on the issues under consideration.

Complex dynamics and changing boundary conditions

One of the more exciting features shown by our prototype dynamic vegetation models is the potential for complex dynamics. Complex dynamics may appear chaotic through time, or could show endogenous "rhythms" or increasing oscillatory behavior approaching a "singularity" or critical threshold, rapidly changing the system from one state to another (Verboom and Wamelink, this volume, Chapter 9). It has been shown that simple logistic competitive or predator-prey systems can exhibit complex dynamics (*ibid.*). It should, therefore, be no surprise to see such behavior in simple competitive vegetation systems. The tendency toward this behavior occurs predominantly in transitional systems where positive feedbacks, such as those previously described, tend to push the system away from transitions. That is, those areas that are transitional between woody and grass systems tend to be spatially quite heterogeneous and susceptible to relatively rapid changes among alternative states. Since these areas are climatically determined, they could occur in narrow ecotonal zones or, if regional climate gradients are comparatively flat, they could occur over broad regions. The drier parts of the southern United States are good examples of broad areas that are highly susceptible to rapid change from one state to another, given external perturbations from variable climates, grazing, fire, or other disturbances (Neilson, 1986).

Simulations (unpublished) of woody–grass interactions within the southeastern United States using one of our prototype DGVMs produced endogenous long-wave patterns of oscillating tree–grass dominance over about a 100-year cycle when under a constant climate. Similar simulations in central Texas showed increasing oscillations over the course of decades between grasses and shrubs until the shrubs quite suddenly died out. These preliminary results suggest a sensitivity to initial and boundary conditions, with possible alternative quasi-steady states being initiated or maintained by outside forces, such as grazing, fire, or climate oscillations. In a conceptual sense, landscapes that are biogeographically transitional between homogeneous states, such as forests or grasslands, are clearly near critical thresholds and should exhibit complex dynamics with the possibility for alternative quasi-stable states. Deterministic, process-based models are best suited to simulate such complex situations under changing climate and CO_2 conditions (i.e., altered boundary conditions).

Complex dynamics can also result from interactions among different patch types, in terms of propagules, water, disturbances (fire), and other processes. A clear limitation of current, process-based DGVMs is the lack of interaction among mosaic elements in a landscape context, whether or not they are rendered spatially explicit. These interactions are generally sub-gridcell phenomena, but they could affect the overall gridcell outcome. Such interactions could be included in the present structure, but one would want to test the simplest constructs first. To the extent that complex dynamics resulting from patch interactions cannot be captured (and are viewed as necessary), then the model structure could be enhanced.

Conclusions

Current modeling approaches within IGBP landscape activities are organized around three different scales. Most DGVM modelers are attempting to incorporate the important processes that occur at all three scales: patch (competition, gas exchange), landscape (fire, dynamic heterogeneity), and global (emergent, spatial pattern). It will be very important for practitioners working within one of these three modeling communities to coordinate closely with those working at the other scales. Patch models built around one type of ecosystem or in one region may not be well structured for working in other systems or regions or capable of accurately changing from one ecosystem state to another. Consistency of process should be maintained across scales. If models are to be nested or linked across scales, then their processes should be based upon the same theoretical underpinnings or they may not translate well across scales, as in the examples of different hydrologic and transpiration algorithms and their impacts on large-scale patterns.

An area of research that I believe may have some potential, but that remains largely untapped, is the possibility of downscaling from regional to landscape patterns using coarse-scale information, either from models or from satellite imagery. Insights regarding spatial and temporal patterns of biodiversity, for example, could be inferred and possibly inform managers regarding conservation priorities and strategies (e.g., papers by Crow, Rolstad, Margules, With, this volume, Chapters 20, 21, 23, 24). The example of differing ecotone orientations along the west slope of the Rocky Mountains as determined by large-scale air-mass gradients serves to illustrate some of the possibilities for inferring landscape-scale phenomena (e.g., community and diversity patterns) from coarse-scale information.

The key points of this discussion serve to emphasize the importance of accurate simulation of ecosystem constraints and emergent properties at all relevant scales. Under a rapidly changing climate and with changing physiology under elevated CO_2, constraints normally assumed to be stationary must now be assumed to be dynamic and must be explicitly simulated. Heterogeneous landscapes are among the most complex, yet globally among the most dominant, types of ecosystems. Accurate simulation of landscape patterns and processes under global change requires attention to organism-level and lower processes within the constraints of biome-level dynamic biogeography.

References

Hansen A. J. and di Castri, F. (eds.) (1992). *Landscape Boundaries: Consequences for Biotic Diversity and Ecological Flows*. New York, NY: Springer.

Neilson, R. P. (1986). High-resolution climatic analysis and southwest biogeography. *Science*, 232, 27–34.

Neilson, R. P. (1995). A model for predicting continental-scale vegetation distribution and water balance. *Ecological Applications*, 5, 362–385.

Neilson, R. P. (2003). The importance of precipitation seasonality in controlling vegetation distribution. In *Changing Precipitation Regimes and Terrestrial Ecosystems: a North American Perspective*, ed. J. F. Weltzin and G. R. McPherson. Tucson, AZ: University of Arizona Press, pp. 47–71.

Neilson, R. P. and Wullstein, L. H. (1983). Biogeography of two southwest American oaks in relation to atmospheric dynamics. *Journal of Biogeography*, 10, 275–297.

Neilson, R. P., King, G. A., DeVelice, R. L., and Lenihan, J. M. (1992). Regional and local vegetation patterns: the responses of vegetation diversity to subcontinental air masses. In *Landscape Boundaries*, ed. A. J. Hansen and F. di Castri. New York, NY: Springer, pp. 129–149.

O'Neill, R. V., DeAngelis, D. L., Waide, J. B., and Allen, T. F. H. (1986). *A Hierarchical Concept of Ecosystems*. Princeton, N. J.: Princeton University Press.

Peet, R. K. (1978). Latitudinal variation in southern Rocky Mountain forests. *Journal of Biogeography*, 5, 275–289.

Whittaker, R. H. and Niering, W. A. (1965). Vegetation of the Santa Catalina Mountains, Arizona. (II) A gradient analysis of the south slope. *Ecology*, 46, 429–452.

Whittaker, R. H., Gilbert, L. E., and Connell, J. H. (1979). Analysis of two-phase pattern in a Mesquite Grassland, Texas. *Journal of Ecology*, 67, 935–952.

PART V

Applications of landscape ecology

18

Landscape ecology as the broker between information supply and management application

In this era of very sophisticated and still-developing GIS functionality, and with an as-yet unknown availability of data, some argue that we do not need integrated ecological (land) classification and mapping nor (ecosystem) geographers. In fact, they maintain, we do not need landscape ecology at all, as the knowledge gathered by all the underlying more specialist disciplines makes it a superfluous discipline: the information technicians can easily handle, combine, and provide all the required information, and the policy makers can select the relevant information and draw conclusions by themselves.

Here we have, in my opinion, two mistakes. One is that integrated classification and mapping is old-fashioned and can be done without, and the second is that transdisciplines are superfluous in this era of information technology. I will explain why I consider these to be mistakes. Meanwhile, I will argue that we need landscape ecology as a mind-set or attitude for professionals in spatial planning and in policy analysis even more urgently than as a scientific discipline in its own right. I will refer to recent experiences from my current involvement in river (basin) management. Finally, I will go into some issues that, in my opinion, will require the attention of landscape ecologists in the near future, but without having the necessity of incorporating them into "our discipline".

The stage

Some years ago I wrote that ecological land classification is a quintessential tool to be used in two fields: for land evaluation for land-use planning, and for environmental impact assessment (EIA) in the planning of such activities as infrastructure planning, water resource exploitation, or river management (Klijn, 1997). I recognized these two fields primarily in an

Issues and Perspectives in Landscape Ecology, ed. John A. Wiens and Michael R. Moss. Published by Cambridge Univeristy Press. © Cambridge University Press 2005.

academic environment, but with a view to their application. Meanwhile, I have become primarily a practitioner myself, engaged in water-resources and river-management planning. To the two fields of land evaluation and EIA (both ex-ante evaluation) I would now add monitoring (ex-post evaluation) for policy evaluation. When we extend the applications of ecological land classification to those of landscape ecology, we might also argue that applied landscape ecology involves both the design of the planning measures themselves and their evaluation in a cyclic process of successive optimization (see also Opdam *et al.*, 2001).

The Netherlands, Europe, and the world at large are experiencing rapid changes in three related realms: societal changes, physical changes, and normative changes. Societal changes concern, for example, demography, economy, increased pressure on land due to urban and industrial sprawl, agricultural intensification in some regions and land abandonment in others, but also water (mis)management (Vos and Klijn, 2000). As for the latter topic, we are confronted with vast physical changes related to climate change: an increasing scarcity of water resources of the required quality for drinking water supply, food production, etc., and at the same time increasing flood risks due to increasing flood hazard (magnitude and frequency) and damage potential (number of inhabitants, intensity of land use, and invested capital). Normative changes include changing demands On the quality of the landscape, from a utilitarian viewpoint (including risks), from an esthetic viewpoint (scenery), and from an ethical viewpoint ("intrinsic value" or "partnership with nature"). As for water management, normative changes include a growing dislike of further technical river-management works – high dikes, huge dams, etc., and a revival of "design with nature" principles (McHarg, 1969; WL/Delft Hydraulics, 2000) as exemplified by, for example, the "room for rivers" ideas (Silva *et al.*, 2001; Klijn *et al.*, 2001).

In other words, societal pressure is changing, the environment/landscape itself is changing, and our demands on the landscape change. It is indeed a huge task to guide this development, which seems to be steadily speeding up and which provokes a number of unwanted and sometimes irreversible effects. The complexity of the issue requires, in my opinion, a humble but also firm involvement of landscape ecologists, among others! After all, only those who are professionally engaged with landscapes (and their quality) are sufficiently aware of long-term, delayed, irreversible, and/or off-site effects and can really judge the severity of landscape changes. In addition, landscape ecologists tend to care for landscapes and generally have a tendency toward environmentalism. This implies a certain commitment to "the cause," but not necessarily compromising scientific integrity! I admit that this is a plea for some interference with policy making; I shall come back to it later. First, however, some examples.

Water (resources) management planning

Growing population and growing demands on fresh water exceed its availability in large parts of the world. Climatic change may further influence the availability of freshwater resources. Thus, water increasingly becomes part of the socioeconomic sphere, which is reflected in the term "resource." From a landscape-ecological viewpoint, however, water is not only a resource, but it also provides conditions. Such conditions are (1) for the survival of biotic subsystems, both in their own right (e.g., mangrove forests with e.g., Bengal tigers) or as a resource for local populations through fishing, cutting, or ecotourism; and (2) for direct human use (e.g., for shipping or bathing). This requires a more comprehensive approach to water management than merely seeing it as a resource. It requires due knowledge of vertical ("topological") relationships as well as of horizontal ("chorological") relationships in catchment areas.

Examples of studies tackling questions of groundwater management in a landscape-ecological context are the study for the Netherlands' policy on surface water and groundwater management (Claessen et al., 1994) and the study for the Netherlands' policy on drinking-water supply (Claessen et al., 1996; Van Ek et al., 2000). Both strongly rely on eco-hydrology (Klijn and Witte, 1999), and were based on connected ecological land classifications at the scale of ecotopes (the vegetation response) and ecoseries (response of soil chemistry and physics) (Klijn, 1997). The alternative use of existing, but separately measured, data on soils, groundwater, land use, vegetation, and individual species by simple GIS overlaying proved impossible. It caused the well-known spaghetti problem and the generation of sliver polygons in the case of polygon-GIS, or alternatively the emergence of nonsense combinations in the case of grid-GIS. It once again proved that only specialists in the field of "whole" landscape ecology can evaluate and combine large geographical databases and judge the results of GIS operations.

In the context of surface-water management, the question of environmental flow requirements is gaining attention (e.g., the 2002 Congress of the International Association for Hydraulic Research, held in South Africa). The distribution of water resources amongst users can be modeled relatively easily (e.g., with the WL model RIBASIM for river basin simulation). But the question of how to establish environmental flow requirements is not yet satisfactorily solved (Marchand et al., 2002). It involves the recognition of all relevant and foreseeable on-site and off-site effects, i.e., in the river, along the river, and along the coast as far as this is influenced by the river-flow input, in order to achieve a comprehensive assessment of environmental flow requirements.

In a case study in Trinidad (Anonymous, 1998), it was found that it was not the sheer average quantity of water available to the river that was essential, but rather the whole hydrological regime, with droughts and flushes during the normal/natural seasonal cycles. It is again the case that it is the conditions rather than the sheer availability of "a resource" that are important. A case study in Bangladesh has considered relating changes in the flow regime to ecological effects by applying a classification of ecotopes (Marchand *et al.*, 2002). This is partly because they can be mapped relatively easily, but also because the relationship to flow regime and inundation frequency can be established relatively accurately. This allows predictive modeling under various discharge scenarios and comparison of the results for an assessment of management alternatives; in other words EIA. Finally, ecotopes allow easy communication through maps accompanied by photographs. Such "language" can be understood from the relatively illiterate to the Netherlands water management authorities, who use ecotopes for reasons of their communicative advantages.

As for monitoring in the context of water management planning, the European Union Water Framework Directive is a relevant recent development. It prescribes that all EU member states tune their surface water quality monitoring networks to European standards, which implies, among other things, (1) the distinction of catchments and sub-catchments; (2) the definition of quality standards for water courses and bodies according to eco-regional differences within these (sub-)catchments (see also Hughes and Larsen, 1988; Clarke *et al.*, 1991) as well as according to different functions of the water courses (e.g., primarily shipping, fishing); and (3) the monitoring of both physicochemical and biotic variables. As for the latter, the Netherlands authorities propose to also include a monitoring of ecotopes, since these encompass biotic and physicochemical variables in "whole systems," and because they can be regarded as constituting the relevant content of the combination of eco-regions/water systems in the context of habitat availability and quality. In fact, monitoring the main water courses and bodies of the Netherlands implies the monitoring of ecotopes, both their extent (by recurrent mapping and GIS analyses) and their quality (in terms of species richness established through field survey).

Flood risk management

A second field which requires that landscape ecologists apply their knowledge and experience to water management questions is related to the likely increase of flood risks and how to anticipate this increase. In the past, flood protection was the one and only answer; that was to build and heighten dikes and to regulate rivers. It was the world of the civil engineer and of a society silently supporting the engineers' approach by its faith in technology. Presently, however, society is often well aware of the negative side effects of many civil-engineering solutions. In the Netherlands, there has been massive societal opposition to further dike reinforcement, which can devastate the landscape with its characteristic cultural heritage. Vis *et al.* (2001; see also Hooijer *et al.*, 2002) argued that an unbridled, and hence normal, economic growth of 2% per year causes the damage potential in flood-prone areas to double about every 30 years, whether protected or not. This implies that flood risks (the product of flood probability and damage, or, alternatively, of flood hazard and vulnerability) will increase anyhow, whether we get more floods or not. The longer we wait, the worse things get. There seems, therefore, sufficient reason for a change of strategy to flood-risk management.

Two different strategies can be discerned (Klijn and Duel, 2001), one aiming at providing room for the river by excavating the floodplains and thus "rejuvenating" natural developments (Duel *et al.*, 2001), the other providing room in presently protected areas by dike relocation and/or the construction of bypasses (Vis *et al.*, 2001). These alternative strategies affect both the socioeconomy and the landscape equally strongly; they have direct negative impacts, but they also provide opportunities – for example, in the long run for "river restoration," by allowing the design of a corridor of floodplain areas where natural hydrological, morphological, and biological processes are freed and where now-isolated habitats are again connected. This can be regarded as an opportunity for spatial planning based on landscape-ecological principles. It must be a challenge for landscape planners and landscape architects to design the "cultural heritage of the future" at such large spatial scales as required for a sound flood-risk management (compare Vis *et al.*, 2001). I consider it essential that landscape ecologists participate in this design process, at least by providing information on what ecosystems can be expected to support (i.e., land evaluation), and perhaps even on what may be desired from them.

Summarizing, I maintain that professional landscape ecologists are urgently needed, primarily because information technologists without geographical and ecological knowledge produce mainly a "virtual reality." These technologists do not know what things look like in the field, they cannot judge input data, they make overlays without knowing what they are doing and without being able to judge the (intermediate) results. Finally, they use illogical colors (even the standard color schemes of some well-known GIS systems are awful) for their output maps, thus inhibiting communication rather than enhancing it. (I shall come back to communication later.) What is required is that a well-educated and experienced landscape ecologist judges and filters the information overload by distinguishing between the worthwhile/important and the worthless/unimportant. Acquaintance with functional, spatial, and temporal hierarchies may be very helpful in this context (Klijn, 1995).

Furthermore, only landscape ecologists are trained to see the relevant relationships between ecosystem/landscape components and between different locations at many different spatial scales. This is essential for setting up a sound EIA or integrated policy analysis. And only by sufficient experience can one judge the relative importance of such relationships within a larger context (the "whole"). This may sound like a plea for generalists, which it is of course, but I want to emphasize that landscape ecologists should also be aware of ensuring sufficient disciplinary depth; otherwise, they just tiptoe over things and may truly be regarded as "dilettantes" by the supportive disciplines. This requires education and experience as a generalist, but with a firm disciplinary basis in either ecology or physical geography (as my teachers A. P. A. Vink and I. S. Zonneveld maintained more than 25 years ago).

The role of the landscape ecologist: generalist amongst specialists, specialist amongst generalists

Thus, I gradually move toward the subject of disciplinary depth and pragmatic "holism." What, then, is the niche for landscape ecologists among specialists and real generalists such as "environmental scientists"? As for specialists, it is easy to think of examples: zoologists, geochemists, meteorologists, physiologists, etc. But what about this "environmental science"? This "transdiscipline" may not be well known outside the Netherlands, where we have experienced an evolution of environmental science. It began in the 1970s as an interdisciplinary approach to environmental problems encompassing the environmental sciences in the Anglo-Saxon tradition (see Bowler, 1992). In the Netherlands it was started by geographers such as A. P. A. Vink in Amsterdam and ecologists such as H. A. Udo de Haes in Leiden. Gradually it evolved into a problem-oriented discipline incorporating social sciences (human behavior, economy, management studies) and normative sciences such as philosophy (especially ethics) and planning, design, and engineering. During this process attempts were made to develop an individual theoretical framework, which was, not surprisingly, very ambitious, as may be seen from titles such as *Environmental Science Theory: Concepts and Methods in a One-World, Problem Oriented Paradigm* (De Groot, 1992). In more recent years, attempts to become more "scientifically respectable" have given rise either to a focus on very narrowly defined subjects, such as "life cycle analysis" or "industrial ecology" or to the splitting up of the single transdiscipline into social, physical, and policy-oriented environmental sciences. This evolution may be regarded as exemplary and may also befall landscape ecology if it were to expand, for example, toward "landscape science" as proposed by Vos and Klijn (2000) in *From Landscape Ecology to Landscape Science*. Though I feel with them in their concern about landscape degradation and societal alienation, I do not think a new "science," or a further extension of landscape ecology, is the answer. Instead, we need the commitment of concerned people, including scientists of many disciplines. It would not surprise me if this, in practice, would include many landscape ecologists.

Back to my subject: that is, the niche of the landscape ecologists. I think we should be aware of the societal context and normative context of landscape management and planning. This implies that we should read De Groot (1992), despite my comments about his ambitions, as the essence of this theory of environmental science is worthwile, and as the framework he presents is quite simple. Similarly, the *Framework of Analysis*, as proposed by WL/Delft Hydraulics (1993) for application in policy analyses, is also very simple. And again, so too is the essence of the theory of landscape ecology. In fact, all theories may be regarded as essentially simple, but it is very hard and it needs lots of practice to internalize their full scope and consequences and to act accordingly in everyday work. On the other hand, we should stick to our profession, which means that we should try to integrate the "environmental sciences" – in the Anglo-Saxon sense – but not attempt to expand our discipline toward becoming the one-and-only, all-encompassing "science-of-the-landscape" (in German: *Weltanschauungssysteme mit Totalanspruch*) (compare this approach to that of Naveh and Liebermann, 1994). Try to be like a family doctor, who can handle most illnesses by himself and knows about his patients, their character, their personal circumstances, etc., but who also knows when to refer to a lung specialist (meteorologist), a dermatologist (vegetation scientist), a cardiologist (geohydrologist), or a psychologist (social scientist), and who also knows the limits of his knowledge and expertise. You will be rewarded by thankful patients, but don't expect to win a Nobel prize! This is the niche (and the fate) of the landscape ecologist. Also, like the family doctor, the landscape ecologist may bridge the communication gap and the distance between the views of various reductionists/specialists, and between specialists and policy makers. As we know enough of all relevant disciplines, we can judge and translate into the language of ordinary people, a lord mayor or minister, or administrators. Lately, I have become convinced that this ability is extremely important. It does, however, conflict with the natural tendency of a young discipline, which is trying to become established and requires theory, to expand its own jargon. In my opinion this should be avoided. We can well do without it!

Issues for the future, with special attention to integrated water management

After these outpourings, some future-oriented remarks. I will restrict myself to questions related to current and future water-management problems. This implies that I will not be advocating science-for-science's-sake, as that can be covered by specialists. In my opinion, landscape ecology's prime purpose lies in the close connection to applications in landscape planning and environmental management (see also Opdam *et al.*, 2001).

Land *and* water

Landscape ecology usually addresses land systems and only seldomly water systems (i.e., the real aquatic systems). Indeed, there are large differences in approach between aquatic ecologists, who focus on functional relationships between biota, seldomly map, and look for short-term processes, and terrestrial (landscape) ecologists, who focus on the relationship between abiotic environment and vegetation, who do map, and who focus on longer timescales (succession, groundwater flow). Such specializations each use their own journals. Eco-hydrology rarely involves research into large water bodies and is part of landscape ecology (*Landscape Ecology*, *Wetland Management*). Eco-hydraulics only addresses rivers and streams (*River Research and Applications*). Aquatic ecology is divided again into freshwater and marine systems. For the practical management of catchment areas, and also in relation to coasts, I consider it undesirable that these "worlds" remain apart.

Resources *and* conditions

As already mentioned, water-resources management focuses to a large extent on the resource function of water: the sheer quantities of a certain quality level. This indicates an emphasis on "economic thinking." For the sake of landscape quality, landscape ecologists should emphasize the importance of water as an environmental condition. This may require a great deal of policy-oriented research, for example, into environmental flow requirements in the context of direct and indirect on-site and off-site effects (such as the Aral Sea situation), but also into the scenic and ethical functions of water bodies.

In view of uncertainties

Land-use planning and management planning have to anticipate changes which are difficult to forecast or which cannot be foreseen. Moreover, the response of ecosystems, and also of society, to certain management measures, is difficult to estimate. This requires that decision makers confront the long-term consequences of their decisions. One should think of scenario analysis, in which one may also take into account different world views, implying, among other things, different expectations as to the predictability and stability of ecosystems. Such an approach has been tried by Van Asselt *et al.* (2001) in an attempt to establish the robustness of different flood-risk management strategies for the Rhine and Meuse in view of possible events in the physical environment (such as a speeding up or a sudden delay of climatic warming) or in the socioeconomic environment (such as an economic crisis). For landscape ecologists it means that their predictive models for ecosystem response should be able to cope with such uncertainties and with different response rules. This requires a different approach to predictive modeling and is one which is very challenging indeed.

Whole-system behavior

In policy analysis and EIA, data are important, but maps, pictures, photographs, and views/feelings are at least as important. In that connection the appeal of particular concepts also plays a role. For example, "sustainability" may be a badly defined concept, but policy makers love it. Recently, in the Netherlands, in the Water Management Policy the concepts of resilience and (new!) robustness have come to the fore, again because of their appeal. I think it is worthwhile to try to operationalize such concepts, as they do, indeed, refer to whole-system behavior and, perhaps, can be turned into assessment criteria. After all, anyone who deals with EIA in practice is often unhappy about the criteria he is forced to work with – they just don't cover the essence of landscape quality, for example. When policy makers find these concepts appealing, we should try to exploit the situation. Moreover, it is an intellectual challenge to transcend the level of "just the ecosystem" and to explore how these concepts can be applied to landscapes.

Whole-system qualities

Not only whole-system behavior, but also whole-system qualities need attention in this era of reviving reductionism. There have been some provisional attempts to define "river health." These studies have been inspired by the increased attention paid by the scientific community to the study and discussion of "ecosystem health" (already there exists a division in journals focusing on this topic – the journal *Ecosystem Health* and a journal *Aquatic Ecosystem Health*). Similarly, concepts like landscape health or landscape integrity may be examined, even if only as an intellectual exercise and in the knowledge that they are merely metaphors. (I would not be surprised if they prove to be a cul-de-sac.) But, since these concepts appeal to policy makers, they may help gain attention for our case.

Participation in normative discussions

Meanwhile we have arrived at "our case," which demonstrates that I have my doubts about objective science. On the other hand, I do feel we should distinguish between landscape ecology as science and us as scientists, and our concern for the landscape and its degradation. This is also "us," but as members of society, and thanks to our profession, we are more aware and better informed. This does require that we participate in discussions about how to protect and manage our landscape and how to influence human activities that negatively affect these landscapes. In fact, this is inevitable for landscape ecologists who participate in physical planning and management. They must constantly make decisions on the basis of both their professional judgment and their world view. But participating in normative discussions goes further, as it requires that we be explicit about our opinions in view of our scientific knowledge.

Enhancing engagement: a different attitude toward communication

Being explicit about our opinions means becoming involved in public debate. This is an opportunity to raise awareness about landscape issues and to add also to the further education of those who we experience in Europe as the "lost generation," a generation alienated from their direct physical environment who have grown up in a world of virtual reality (TV, computer, etc.), but without adequate knowledge of the real world. Communication is therefore essential for the sake of enhancing engagement in the environment and the landscapes. This requires that we invest in knowledge on how to communicate better, not through websites, but by demonstrating things in the field. This must be sustained by good cartography – simple, self-evident maps, simple legends, few and logical colors, and by not diverting attention to the unnecessary things or requiring lengthy study. Equally important are simple texts that do not underestimate the intellect of the public. A recent experience with public-oriented publishing proved to be my most satisfying product so far (Klijn *et al.*, 2001), not least because of the reactions it received. Landscape ecology was, however, not even mentioned once in 59 pages.

References

Anonymous (1998). *Water Resources Management Strategy for Trinidad and Tobago. Annex 8: Ecology of wetlands.* The Government of Trinidad and Tobago, Ministry of Planning and Development.

Bowler, P. J. (1992). *The Environmental Sciences.* London: Fontana.

Claessen, F. A. M., Klijn, F., Witte, J. P. M., and Nienhuis, J. G. (1994). Ecosystem classification and hydro-ecological modelling for national water management. In *Ecosystem Classification for Environmental Management*, ed. F. Klijn. Dordrecht: Kluwer, pp. 199–222.

Claessen, F. A. M., Beugelink, G. P., Witte, J. P. M., and Klijn, F. (1996). Predicting species loss and gain caused by alterations in Dutch national water management. *European Water Pollution Control*, 6, 36–42.

Clarke, S. E., White, D., and Schaedel, A. L. (1991). Oregon, USA, ecological regions and subregions for water quality management. *Environmental Management*, 15, 847–856.

De Groot, W. T. (1992). *Environmental Science Theory: Concepts and Methods in a One-World, Problem-oriented Paradigm.* Amsterdam: Elsevier.

Duel, H., Baptist, M. J., and Penning, W. E. (2001). *Cyclic Floodplain Rejuvenation: a New Strategy Based on Floodplain Measures for both Flood Risk Management and Enhancement of the Biodiversity of the River Rhine.* NCR-publication 14-2001. Delft: Netherlands Centre for River Studies.

Hooijer, A., Klijn, F., Kwadijk, J., and Pedroli, B. (2002). *Towards Sustainable Flood Risk Management in the Rhine and Meuse River Basins: Main Results of the IRMA-SPONGE Research Program.* NCR-publication 18-2002. Delft: Netherlands Centre for River Studies.

Hughes, R. M., and Larsen, D. P. (1988). Ecoregions: an approach to surface water protection. *Journal of the Water Pollution Control Federation*, 60, 486–493.

Klijn, F. (1997). A hierarchical approach to ecosystems and its implications for ecological land classification; with examples of ecoregions, ecodistricts and ecoseries of the Netherlands. Ph.D. thesis, Leiden University.

Klijn, F. and Duel, H. (2001). Nature rehabilitation along Rhine River branches: dilemmas and strategies for the long term. In *River Restoration in Europe: Practical Approaches*, ed. H. J. Nijland and M. J. R. Cals. Proceedings of the Conference on River Restoration, 15–19 May 2000, Wageningen, the Netherlands. Lelystad: ECRR/RIZA rapport 2001.023, pp. 179–188.

Klijn, F. and Witte, J. P. M. (1999). Eco-hydrology: groundwater flow and site factors in plant ecology. *Hydrogeology Journal*, 7, 65–77.

Klijn, F., Silva, W., and Dijkman, J. P. M. (2001). *Room for the Rhine in the Netherlands: Summary of Research Results.* Arnhem: WL/ Delft and RIZA.

Klijn, J. A. (1995). *Hierarchical Concepts in Landscape Ecology and its Underlying Disciplines.* SC-DLO report 100. Wageningen:Winand Staring Centre.

Marchand, M., Penning, W. E., and Meijer, K. (2002). Environmental flow requirements as an aid for integrated management. In *Environmental Flows for River Systems.* 4th International Ecohydraulics Symposium, 3–8 March 2002, Cape Town.

McHarg, I. L. (1969). *Design with Nature.* New York, NY: Natural History Press.

Naveh, Z. and Lieberman, A. S. (1994). *Landscape Ecology: Theory and Application.* 2nd edn. New York, NY: Springer.

Opdam, P., Foppen, R., and Vos, C. (2001). Bridging the gap between ecology and spatial planning in landscape ecology. *Landscape Ecology*, 16, 767–777.

Silva, W., Klijn, F., and Dijkman, J. P. M. (2001). *Room for the Rhine Branches in the*

Netherlands. *What the Research has Taught Us.* Arnhem:WL/Delft and RIZA.

Van Asselt, M. B. A., Middelkoop, H., van 't Klooster, A. A., *et al.* (2001). *Development of Flood Management Strategies for the Rhine and Meuse Basins in the Context of Integrated River Management.* NCR-report 16-2001. Delft: Netherlands Centre for River Studies.

Van Ek, R., Witte, J. P. M., Runhaar, J., and Klijn, F. (2000). Ecological effects of water management in the Netherlands: the model DEMNAT. *Ecological Engineering*, 16, 127–141.

Vis, M., Klijn, F., and van Buuren, M. (2001). *Living with Floods: Resilience Strategies for Flood Risk Management and Multiple Land Use in the Lower Rhine River Basin. Executive Summary.* NCR-report 10-2001. Delft: Netherlands Centre for River Studies.

Vos, W. and Klijn, J. A. (2000). Trends in European landscape development: prospects for a sustainable future. In *From Landscape Ecology to Landscape Science*, ed. J. A. Klijn and W. Vos. Dordrecht: Kluwer, pp. 13–29.

WL/Delft Hydraulics (1993). *Methodology for Water Resources Planning.* WL-report T635. Delft:WL.

KATHRYN FREEMARK

19

Farmlands for farming and nature

Since the Second World War, there have been dramatic declines both in the diversity of farmland habitats available to wildlife (animals and plants) and in the quality of the remaining habitat elements. These changes have been brought about by agricultural intensification (i.e., striving for greater output per unit area) and development of the rural–urban fringe. Haphazard growth-management planning has resulted in residential and commercial sprawl that has converted farmlands, fragmented forestlands, increased infrastructure and transportation needs, consumed and compromised wildlife habitat, increased air pollution from more vehicles traveling more miles, and increased water pollution from the widespread use of on-site septic systems. Recent farming policies and technological developments in agricultural practices and their widespread adoption have produced external costs to the environment that are largely borne by non-farmers. In the United States and Canada, both the species richness and abundance of game and non-game wildlife have been adversely affected. Grassland birds, for example, have exhibited steeper and more consistent declines than any other group of birds monitored by the Breeding Bird Survey. In Europe, faunal and floral diversity have been shown to be more threatened on farmland than on almost any other habitat. Of the bird species associated with farmland in Europe, almost half are of conservation concern.

Loss and biotic impoverishment of farmland are concerns because humans depend on the presence and functioning of a diversity of species for services such as pollination, pest control, nutrient cycling, and recreation. Maintaining biodiversity retains subsets of species with similar capabilities, which can provide a functional redundancy that buffers against changes in the capacity or abundance of any one species. Since species must co-occur in space to provide redundancy and functional substitution, spatial patterns

Issues and Perspectives in Landscape Ecology, ed. John A. Wiens and Michael R. Moss. Published by Cambridge University Press.
© Cambridge University Press 2005.

in diversity are one important descriptor of biodiversity at any scale. Hence, studies of spatial pattern of species are useful for assessing risk to values derived from biodiversity and, ultimately, to formulating options to manage those risks. Spatial pattern is used as a surrogate measure of ecological integrity (i.e., the presence of all appropriate biotic and abiotic elements and occurrence of natural evolutionary and biogeographic processes at appropriate rates and spatio-temporal scales) because process is presumed to produce pattern. Process, however, is more costly and difficult to observe across the hierarchy, especially at the larger spatial extents relevant to biodiversity, such as birds that migrate long distances between breeding and over wintering areas.

Effects of farming

The following factors have been found to have adverse effects on patterns of species richness, abundance, survival, and reproduction of wildlife in farmland (especially birds, which have been the most studied), primarily in North America, Europe, and Australia (see also Fig. 19.1). These effects are so well documented in Europe that they have become a fixed element of debate

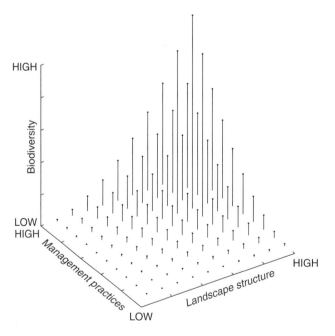

FIGURE 19.1
Model showing the increase in biodiversity as a function of improved landscape structure (composition and configuration) and better management practices. See text for details. Adpated from A. Evans in Pain and Pienkowski (1999: 347).

about agricultural reform. Recent work in Canada suggests that while both habitat and management practices affect wildlife, habitat effects tends to be more important.

Landscape composition

Crop   Loss of variety in crop types, especially more permanent cover (e.g., pasture, hay); increase in monocultures.

Non-crop   Loss of non-crop habitats, especially native habitats (upland/riparian woodlands, prairie, wetlands, streams), but also early successional habitats (e.g., old fields, shrublands) and semi-natural habitats adjoining fields (e.g., wooded fencerows, grassy margins).

Development   Loss of farmland to residential and commercial development; improvement and expansion of road networks.

Landscape configuration

Interspersion   Loss of habitat interspersion of crop : non-crop; more development (e.g., rural residential and roads) in close proximity to native habitats; polarization of farming systems or abandonment has resulted in (former) agricultural landscapes that are homogeneous at local, watershed, and (in some cases) regional scales.

Patch size   Decreasing size of native habitat patches; increasing size of crop fields; decreasing width of non-crop strip cover.

Patch shape/edge   Rectilinearization of fields, more abrupt crop : non-crop boundaries, increased perimeter : area ratio for remnant native habitat patches.

Isolation   Increased among native habitats due to decreased proximity as a result of habitat loss, and, to a lesser extent, loss of interconnecting habitat features (e.g., fencerows); barrier effects from intervening habitats (e.g., roads, urban development, intensive agriculture).

Management practices and use

Pesticides   Increase in the scale and quantity of use; indirect effects from loss of food resources such as insects and weed seeds are particularly important; also direct effects (e.g., poisonings); off-site movement degrades habitat quality (e.g., field margins, wetlands, streams).

Fertilizers   Increase in the scale and quantity of use of chemical fertilizers; decline in use of composted manure; decrease in quantity and diversity of food resources; loss of feeding and reproductive opportunities; earlier and more frequent cutting; off-site movement degrades habitat quality (e.g., field margins, wetlands, streams).

Passes   Increasing numbers of passes through fields from activities such as tilling, fertilizing, pesticide spraying.

Tile and other types of draining   Impacts wetlands through loss and lowering of water tables; also reduces within-field heterogeneity.

Stream channelization   Loss of in-stream and riparian habitat; loss of interconnecting habitat.

Rotation   Decline in use and complexity.

Inter-cropping   Less use, particularly under-sowing of cereals.

Grazing   Increase in stocking rate; grazing of woodlands; livestock access to streams and other wetlands.

Mechanization   Use of larger and heavier machines results in increase in field size, loss of adjoining habitat, soil compaction.

Irrigation   Use causes considerable disturbance losses to shy species; reduces habitat quality by speeding crop growth, salinization and lowering of the water table; contributes to loss of marginal habitats.

Crop improvements   Fast-growing, disease-resistant varieties reduce feeding and reproductive opportunities; earlier and more frequent cutting.

Crop seeding   Increase in rate reduces feeding and reproductive opportunities.

Crop timing   Autumn sowing reduces over-winter and spring food resources.

Abandonment   Loss of croplands and pastures, farmsteads, old buildings, and early successional habitats (e.g., old fields, shrublands).

Traffic density   ased volume from road improvements and exurban/suburban development increases wildlife roadkill and barrier effects.

Positive effects

The following agricultural practices have been found to benefit wildlife:

Conservation Reserve Program (CRP)   Provides grassland, which is particularly beneficial if in large blocks and relatively undisturbed (not mowed or grazed especially during the reproductive season).

Annual set-aside   Provides weedy stubble over winter and, in some cases, fallow; needs to be reframed/relaunched as conservation farmland rather than as a mechanism for reducing production surplus.
Conservation headlands   Outer 6 m of cereal fields grown as the rest of the field but without insecticides and herbicides that remove broad-leaved weeds beneficial to wildlife.
Organic/ecological agriculture   Higher carrying capacity of cropland for both species richness and abundance compared to conventionally (chemically) farmed croplands.

The new millennium

Landscape-scale ecological studies

More landscape-scale studies are needed in farmland to understand the effects of different landscape mosaics on spatio-temporal patterns in species distributions and demographics (e.g., reproductive success, dispersal, survival, metapopulation dynamics) as well as other ecological processes (e.g., ground/surface water quality and quantity, nutrient cycling). The long-term conservation of biodiversity is ultimately dependent on maintaining hospitable environments and viable populations within managed landscapes. Parks and reserves may be important core areas in these landscapes, but even the largest national forest or national park is not ecologically isolated from activities and conditions in the surrounding landscapes. Furthermore, the viability of species in reserves may often depend on inter-reserve migration through intervening habitats managed for agricultural (or forestry) production.

Policy and planning for alternative landscapes

Our challenge is to figure out how to better link ecological knowledge with the social sciences and humanities to gain greater diversity and depth of understanding in order to enlighten our efforts to conserve nature in farmland (and other human-dominated landscapes such as towns, cities, and managed forests). Phrased more simply, how do we integrate conservation with food production in farmland?

In Europe, extensification (i.e., producing less from a given area of land) using environmentally sensitive management systems is being recommended as a way to conserve and restore wildlife in farmland. Extensification is a compelling solution to the conservation crisis because extensive systems are more likely to be sustainable (as they indeed were for many centuries in parts of the world).

However, this will require farming within natural environmental constraints, rather than finding artificial means of supporting systems operating outside of these constraints. Achieving this most likely means that many farmers will be required to reduce production. As a consequence, they may suffer financially. Thus, for conservation to succeed, individual farmers will require policies and financial incentives that assist them in adopting different farming practices. More broadly, policies must fully "internalize" the environmental costs and benefits of agriculture into practices, markets, and policies. That is, when farmers, agribusiness, or policy makers make decisions, positive rewards for environmental benefits and penalties for environmental damage must be built in so that the environment is incorporated as part of the decision-making process.

New or improved growth-management strategies are needed to avoid development that wastes land, is expensive to service, and diverts private investment and public funds from maintaining and enhancing existing villages, towns, and cities to stem the flow of people to the countryside. In addition, federal, state or provincial, and municipal spending, taxation, and regulatory programs that encourage development sprawl need to be reformed to promote "smart" growth.

Beneficial actions need to be adopted over a wide scale; within-farm and other local changes will have minimal impact if carried out in isolation. Thus, we need to learn how to develop, evaluate, and implement land-use plans that are more comprehensive and hierarchical in space and time so as to be more effective in the proactive conservation of nature in farmland. Approaches will have to include ecological, socioeconomic, legal, cultural, ethical, and aesthetic considerations. To minimize and resolve conflicts, effective education, communication, and carefully designed mechanisms for planning, cooperation, and coordination are required. Articulating appropriate goals or targets for landscape and ecosystem management in collaboration with rural communities is a critical activity in the development and evaluation of alternative land-use scenarios for farmland. The linkage of models that capture key properties of ecological and socioeconomic systems observed in the field should become an increasingly important component of land-use decision making. A closer linkage with the arts could further enhance and facilitate the process of social choice through better formulation and communication of what the natural and social sciences attempt to explain.

Modeling the effects of global climate change

We have not yet figured out how to predict and plan for the effects of global climate change on farmland. To accomplish this, we need to integrate information on climate, landforms, landscape structure, and dynamics of species' distributions across a hierarchy of spatial and temporal scales. Comparative studies across gradients, regions, or larger geographic areas (e.g., countries, continents, the globe) will be particularly important in predicting the impacts of changes in landscape structure produced by global change and its associated human-driven land-use change. For example, the International Geosphere–Biosphere Programme is interested in the possible effects of changing the diversity within agricultural and forestry production systems on ecological complexity and function at the regional scale. Agricultural and forestry production systems that are more diverse and complex may be not only more sustainable, but also more conducive to the migration of species among nature reserves and hence lead to reduced rates of extinction as species cope with rapidly changing environmental regimes.

Quantitative measures of landscape structure derived from remote-sensing technology can provide appropriate metrics for monitoring regional ecological changes in response to factors such as global change. Potential effects of global change on biota may then be inferred from contemporary landscape studies. Use of spatially explicit models should help to focus related research, monitoring, and conservation activities in relation to global change. If landscape structure can be linked to population demographics, then spatially explicit models can be used to simulate impacts of global change on species. Spatially explicit, multispecies models also need to be developed to understand expected changes in biotic interactions at broad spatial and temporal scales.

Closing thoughts

Effective approaches for cross-boundary decision making and management (administratively and on the ground) need to be developed. Otherwise, the "tyranny of small decisions" will continue to prevail, with many local, relatively unimportant land-use decisions cumulatively resulting in profound, adverse landscape changes over greater extents. Our challenge is to create the sociocultural commitment and spatially integrated decision-making processes in which the rural character of farmlands can be sustained and farmers, other landowners, citizens, the development community, planners, and elected officials act as managers and stewards of the countryside, rather than just as consumers or producers for the market. Such a transition is beginning in Europe and possibly Australia but, for the most part, not in North America. Until attitudes change, agricultural and other land-use reforms intent on protecting and enhancing farmland will be unlikely. Without this, the ideals and international agreements forged in the United

Nations Conference on Environment and Development on sustainability, climate change, and the protection of biodiversity will continue to be undermined.

Selected references

Best, L. B., Bergin, T. M., and Freemark, K. E. (2001). Influence of landscape composition on bird use of rowcrop fields. *Journal of Wildlife Management*, 65, 442–449.

Bergin, T. M., Best, L. B., Freemark, K. E., and Koehler, K. J. (2000). Effects of landscape structure on nest predation in roadsides of a midwestern agroecosystem: a multiscale analysis. *Landscape Ecology*, 15, 131–143.

Daniels, T. (1999). *When City and Country Collide.* Washington, DC: Island Press.

Forman, R. T. T., Sperling, D., Bissonette, J. A., *et al.* (2002). *Road Ecology: Science and Solutions.* Washington, DC: Island Press.

Freemark, K. E. (1995). Assessing effects of agriculture on terrestrial wildlife: developing a hierarchical approach for the US EPA. *Landscape and Urban Planning*, 31, 99–115.

Freemark, K. E. and Kirk, D. A. (2001). Birds breeding on organic and conventional farms in Ontario: partitioning effects of habitat and practices on species composition and abundance. *Biological Conservation*, 101, 337–350.

Freemark, K., Bert, D., and Villard, M.-A. (2002a). Patch-, landscape-, and regional-scale effects on biota. In *Applying Landscape Ecology in Biological Conservation*, ed. K. J. Gutzwiller. New York, NY: Springer, pp. 58–83.

Freemark, K. E., Boutin, C., and Keddy, C. J. (2002b). Importance of farmland habitats for conservation of plant species. *Conservation Biology*, 16, 399–412.

Hulse, D. W., Eilers, J., Freemark, K., Hummon, C., and White, D. (2000). Planning alternative future landscapes in Oregon: evaluating effects on water quality and biodiversity. *Landscape Journal*, 19, 1–19.

Kareiva, P. M., Kingsolver, J. G., and Huey, R. B. (eds.) (1993). *Biotic Interactions and Global Change.* Sunderland, MA: Sinauer.

Kirk, D. A., Boutin, C., and Freemark, K. E. (2001). A multivariate analysis of bird species composition and abundance between crop types and seasons in southern Ontario, Canada. *Ecoscience*, 8, 173–184.

Montgomery, C. A., Pollak, R. A., Freemark, K., and White, D. (1999). Pricing biodiversity. *Journal of Environmental Economics and Management*, 38, 1–19.

Pain, D. J. and Pienkowski, M. W. (1997). *Farming and Birds in Europe.* New York, NY: Academic Press.

Santelmann, M., Freemark, K., White, D., *et al.* (2001). Applying ecological principles to land-use decision making in agricultural watersheds. In *Applying Ecological Principles to Land Management*, ed. V. H. Dale and R. A. Haeuber. New York, NY: Springer, pp. 226–252.

Saunders, D. A., Hobbs, R. J., and Ehrlich, P. R. (eds.) (1993). *Nature Conservation 3. The Reconstruction of Fragmented Ecosystems: Global and Regional Perspectives.* Chipping Norton, NSW: Surrey Beatty.

White, D., Preston, E. M., Freemark, K. E., and Kiester, A. R. (1999). A hierarchical framework for conserving biodiversity. In *Landscape Ecological Analysis: Issues and Applications*, ed. J. M. Klopatek and R. H. Gardner. New York, NY: Springer, pp. 127–153.

Wilson, E. O. (1998) *Consilience: The Unity of Knowledge.* New York, NY: Knopf.

20

Landscape ecology and forest management

Almost all activities associated with forest management affect the composition and structure of the landscapes in which they occur. For example, forest harvesting profoundly affects the composition, size, shape, and configuration of patches in the landscape matrix (Table 20.1). Even-age regeneration techniques such as clearcut harvesting have been applied in blocks of uniform size, shape, and distribution, and as strip cuts with alternating leave and cut strips or as progressive cutting of strips, or as patches with variable sizes, shapes, and distributions. In contrast to the coarse-grained pattern (Table 20.1) produced on the landscape by even-age management, uneven-aged regeneration techniques produce small openings in the canopy where individual trees or small groups of trees are periodically harvested.

Roads, another important landscape feature associated with forest management, are essential for a variety of activities including timber and wildlife management, recreation, and the management of fire, insects, and pathogens. Once in place, however, roads greatly alter the ecological character as well as the amount, type, and distribution of human activity on the landscape. At the landscape scale (Table 20.1), roads form a network and road density is closely correlated with the level of forest fragmentation, the amount of forest edge, and, conversely, the amount of forest interior available in the landscape (Forman and Alexander, 1988; Forman, 2000). In addition to maintained or improved roads that are often viewed as external to the forest, every managed forest has a network of unimproved haul roads and skid trails within the forest. In a study of the influence of haul roads and skid trails on plant composition and richness in forested landscapes of Upper Michigan, Buckley et al. (2003) found that these features comprised from 3% to 22% of the total area in managed forests. Soil compaction, soil moisture, solar radiation, and surface temperature are greater in skid trails and haul roads compared to the closed-canopy forest.

Issues and Perspectives in Landscape Ecology, ed. John A. Wiens and Michael R. Moss. Published by Cambridge University Press.
© Cambridge University Press 2005.

Table 20.1. Key concepts from landscape ecology and their application to managing natural resources

Ecological concept	Applications
Spatial scale	Landscapes consist of multiple and interacting ecosystems that are generally considered to occur at spatial scales of a few to many km². For management purposes, it is useful to think of landscapes as intermediate between local and regional scales.
Temporal scale	The concept of scale applies to both time and space. There is a general relationship between time and space, i.e., space-time principle, that suggests that more variable and shorter-term changes occur in smaller areas and less variable and longer-term changes occur in larger areas.
Patches and the landscape matrix	Patches are the basic spatial element of the landscape and the predominant land cover forms the landscape matrix. Land cover is generally used to define patches. Patches results from the interaction of the physical environment, natural and human disturbances.
Spatial and temporal heterogeneity	Heterogeneity or variation occurs in both time and space. Understanding heterogeneity is a core objective of landscape ecology. The degree of heterogeneity depends on the scale at which a system is viewed. Human activities may increase heterogeneity at some spatial scales, but decrease heterogeneity at other scales.
Landscape structure	Landscape structure is a measure of heterogeneity. The size, shape, and configuration of patches determine landscape structure. For management purposes, the size-class distribution of patches is useful for characterizing structure. Landscapes frequently contain many small and a few large patches. Large patches serve as connecting features in a landscape. The breaking up of large land areas into smaller parcels is a common feature of human land use.
Landscape grain	Grain refers to the coarseness in texture of the landscape, and mean and variance in grain size are measures of structure and heterogeneity. A fine-grained landscape is composed largely of small patches, while large patches dominate a coarse-grained landscape.
Landscape composition	Both natural features (e.g., vegetation, rivers, lakes) and human land use (e.g., agricultural land, urban and industrial land use, transportation systems) are generally used to define landscape composition.

Table 20.1 (cont.)

Ecological concept	Applications
Ecological context	Since landscapes consist of multiple and interacting ecosystems, the composition and function within a local ecosystem can be affected by other ecosystems. In addition to ecological context, social and economic context are important concepts in landscape ecology.
Hierarchical organization	This is another form of ecological context with local ecosystems embedded in larger landscape and regional ecosystems. At an operational level, management is generally conducted at local scales. When managing natural resources, it is important to consider the landscape and regional context (ecological, social, and economic) in which a local ecosystem exists.
Landscape change	Natural succession, natural and human disturbances all cause change in the composition and structure of landscapes. Deforestation, urbanization, and agricultural intensification are among the major causes of landscape change.

Other impacts of forest management on landscape composition and structure are common. Many fire-driven ecosystems are nearly monotypes of tree species and so diversity within the site may be low; however, the renewing effects of fire can create a spatial mosaic of community types, age classes, and forest structures that are highly diverse among sites (Heinselman, 1973). The combination of fire suppression and forest harvesting, however, has significantly changed the composition and structure of many forested landscapes throughout the world.

In addition to management activities, or, more generally, land-use activities, landscape patterns reflect the physical environment and natural disturbances such as wind and fire, as well as the interaction among these factors (Crow *et al.*, 1999). Regardless of the source of spatial variation, the type and number of patches, their size and shape, and their spatial arrangement strongly influence the benefits and the values that can be derived from a landscape.

There is a reciprocal relationship between landscape pattern and forest management as well – that is, landscape composition and structure strongly affect forest management. The ability to move from a pattern of dispersed harvesting to a pattern of aggregated harvesting, for example, is difficult when small, dispersed harvest units dominate the landscape matrix (Wallin *et al.*, 1994). Furthermore, small, widely dispersed patches of forest are more costly to harvest than large, aggregated patches. The opportunities for conducting intensive forestry operations (e.g., whole-tree harvesting, establishing plantations of fast-growing trees, or applying herbicides to control competing vegetation) are limited in landscapes where human population densities, defined in terms of people or houses per unit area, are high. Opportunities for intensive management of forests for timber are greatly diminished even when people and their housing densities are low but widely dispersed throughout a forested matrix. In the recently published Southern Forest Resource Assessment (USDA Forest Service, 2002), urban sprawl, not timber harvesting, was cited as the biggest threat to southern forests in the United States. Between 1992 and 2020, about 6% of the South's forests or about 4.8 million ha of forestland is projected to be lost to urban uses.

Adding a spatial element to multiple use

A landscape perspective is useful when applying the common management paradigm of multiple use (Crow, 2002). Foresters believe that multiple products and benefits can be derived from forests through the wise and careful application of scientifically based management practices. In the United States and elsewhere, such beliefs are codified into public policy (e.g., the Multiple-Use Sustained-Yield Act of 1960). In practice, however, the multiple-use paradigm has failed to provide an adequate framework for providing diverse resource benefits and values (Shands, 1988). As recognized in the language of the Multiple-Use Sustained-Yield Act of 1960, "some land will be used for less than all of the resources." That is to say, all multiple uses cannot and should not be practiced on every unit of land to the same degree or intensity; instead, managers need to utilize the different capabilities and potentials that exist within a landscape. Yet a formal framework for evaluating opportunities in time and space is rarely applied as part of forest planning and management (Crow and Gustafson, 1997).

Obviously some forest uses are in direct conflict, and when presented with this dilemma, forest managers tend to partition the land into different uses in order to meet specific management goals. When a wilderness area is designated, land is taken out of timber production. If a natural area is established, no trees will be harvested and it may be necessary to limit recreational use of the area in order to sustain the qualities for which the natural area was designated. Protective buffers are often placed around areas populated by rare or endangered species, resulting in numerous, small, but widely distributed management units that are difficult to administer and difficult to integrate with other land uses. Independently, each of these actions may be justified, but collectively the result is the compartmentalization of the land through a series of separate decisions instead of through comprehensive planning that is spatially and temporally explicit.

Multiple use works best when the land base is large and demands for outputs and benefits are small. Yet, in reality, just the opposite is true. On a global scale, the land base available for resource management is finite and the demands for both commercial products and intangible values are growing dramatically. The result is increasing conflict and seemingly intractable problems related to forest management (Shands, 1988). A spatial and temporal framework should be added to the multiple-use paradigm. Clearly, the application of any management system will benefit from evaluating the spatial and temporal context in which decisions are made and treatments occur, so that potential conflicts might better be minimized and so that unintended and undesirable cumulative impacts from multiple actions can be better anticipated.

Practicing the science of landscape ecology

A landscape perspective fosters a multi-scale approach to forest management (Table 20.1). Historically, foresters have managed at local spatial scales, i.e., the forest stand, and applied their treatments as if each stand was independent and existed in isolation of every other forest stand. An alternative approach to managing a forest is to first consider the broader landscape in which the management unit exists. It is important to recognize that ecosystems comprising a landscape interact by exchanging energy, materials, and organisms. The context in which an ecosystem exists can profoundly affect the content of that ecosystem. The hierarchical organization of ecological systems relates to both context and scale (Table 20.1). This concept, in which local ecosystems are viewed as being nested within larger ecosystems, enables managers to evaluate large-scale influences on conditions and processes at smaller spatial scales.

Franklin and Forman (1987) have demonstrated the importance of evaluating the spatial consequences of forest harvesting in the Douglas-fir region of the Pacific Northwest. They suggest a two-point guide for forest harvesting. First, harvesting should feature progressive or clustered harvest units instead of dispersed harvest units to reduce forest fragmentation. Approaches featuring progressive or clustered harvesting reduce the risks of disturbance associated with forest edges, and these spatial configurations also reduce the amount of maintained road systems necessary compared to more dispersed harvest patterns. The size of a cluster depends on management objectives and landscape characteristics. Retaining networks of corridors and small forest patches within the clustered harvest areas provides additional cover and edge for game species, reduces wind fetches and soil erosion, and enhances movement of species among forest patches (in this case, primeval forest). Large patches play especially important roles and they should be maintained in the landscape to facilitate flow and movement of materials and species, to enhance amenity values, and to provide critical habitat for interior species (Forman, 1995; Crow et al., 1999). To use the morphologic metaphor of an organism, large patches are the connecting tissue for landscapes.

The tools needed for applying a landscape perspective to forest management – aerial photography, satellite imagery, laser technology, airborne radar, geographic information systems (GIS), mathematical models – are available and, in some cases, already familiar to foresters (McCarter et al., 1998). Spatially explicit models that combine remote sensing with GIS offer great promise to land managers because they consider the arrangement of landscape elements in time and space. Furthermore, their visual and geographic nature facilitates the comparison of alternative management strategies and their associated landscape patterns (Gustafson, 1996, 1998; Gustafson and Crow, 1996, 1998). Ecosystem management of landscapes is accomplished using a combination of custodial management (e.g., wilderness, natural areas) and active management to produce a variety of benefits, including commodities. Spatial models provide the means for incorporating both custodial and active management into real landscapes to create a variety of uses and benefits.

Providing an array of benefits and values representing multiple social expectations will continue to be an important part of forest planning and management. More attention is needed to the spatial and temporal distributions of these allocations and more attention should be given to their cumulative impacts. These needs can best be met by complementing a stand approach to management with a landscape perspective. Landscape ecology confronts us with the realities of connections and of interdependencies that characterize our relationship with nature (Nassauer, 1997). A landscape perspective facilitates an integrated, holistic approach to resource management and conservation.

Final thoughts

Human activities are transforming landscapes to a greater extent and at a faster rate than at any time in human history. To deal with this transformation, new and improved collaborations are needed among scientists, planners, managers, and the public for developing land-use policies and for managing our natural resources. The science of landscape ecology attracts people from many different fields. And perhaps therein lies its strength – in bringing people from different disciplines together who have a common interest in the landscape in its broadest sense and who recognize the value of working collaboratively to solve problems that are beyond their individual capability.

References

Buckley, D. S., Crow, T. R., Nauertz, E. A., and Schulz, K. E. (2003). Influence of skid trails and haul roads on understory plant richness and composition in managed forest landscapes in Upper Michigan, USA. *Forest Ecology and Management*, 175, 509–520.

Crow, T. R. (2002). Putting multiple use and sustained yield into a landscape context. In *Integrating Landscape Ecology into Natural Resource Management,* ed. J. Liu and W. W. Taylor. Cambridge: Cambridge University Press, pp. 349–365.

Crow, T. R. and Gustafson, E. J. (1997). Ecosystem management: managing natural resources in time and space. In *Creating a Forestry for the 21st Century: The Science of Ecosystem Management*, ed. K. Kohm and J. F. Franklin. Washington, DC: Island Press, pp. 215–228.

Crow, T. R., Host, G. E., and Mladenoff, D. J. (1999). Ownership and ecosystem as sources of spatial heterogeneity in a forested landscape. *Landscape Ecology*, 14, 449–463.

Forman, R. T. T. (1995). *Land Mosaics: the Ecology of Landscapes and Regions*. Cambridge: Cambridge University Press.

Forman, R. T. T.(2000). Estimate of the area affected ecologically by the road system in the United States. *Conservation Biology*, 14, 31–35.

Forman, R. T. T. and Alexander, L. E. (1988). Roads and their ecological effects. *Annual Review of Ecology and Systematics*, 29, 207–231.

Franklin, J. F. and Forman, R. T. T. (1987). Creating landscape patterns by forest cutting: ecological consequences and principles. *Landscape Ecology*, 1, 5–18.

Gustafson, E. J. (1996). Expanding the scale of forest management: allocating timber harvests in time and space. *Forest Ecology and Management*, 87, 27–39.

Gustafson, E. J. (1998). Clustering timber harvests and the effects of dynamic forest management policy on forest fragmentation. *Ecosystems*, 1, 484–492.

Gustafson, E. J. and Crow, T. R. (1996). Simulating the effects of alternative forest management strategies on landscape structure. *Journal of Environmental Management*, 46, 77–94.

Gustafson, E. J. and Crow, T. R. (1998). Simulating spatial and temporal context of forest management using hypothetical landscapes. *Environmental Management*, 22, 777–787.

Heinselman, M. L. (1973). Fire and succession in the conifer forests of northern North America. In *Forest Succession: Concepts and Applications*, ed. D. C. West, H. H. Shugart, and D. B. Botkin. New York, NY: Springer, pp. 374–405.

McCarter, J. B., Wilson, J. S., Baker, P. J., Moffett, J. L., and Oliver, C. D. (1998). Landscape management through integration of existing tools and emerging technologies. *Journal of Forestry*, 96, 17–23.

Nassauer, J. I. (1997). Action across boundaries. In *Placing Nature, Culture and Landscape Ecology*, ed. J. I. Nassauer. Washington, DC: Island Press, pp. 65–169

Shands, W. E. (1988). Beyond multiple use: managing national forests for distinctive values. *American Forests*, 94, 14–15, 56–57.

USDA Forest Service (2002). *The Southern Forest Resource Assessment*. Asheville, NC: Southern Research Station.

Wallin, D. O., Swanson, F. J., and Marks, B. (1994). Landscape pattern response to changes in pattern generation rules: land-use legacies in forestry. *Ecological Applications*, 4, 569–580.

21

Landscape ecology and wildlife management

In his seminal book *Game Management* (1933: 128–129), Aldo Leopold set the stage for a marriage between landscape ecology and wildlife management:

> The game must usually be able to reach each of the essential types each day. The maximum population of any given piece of land depends, therefore, not only on its environmental types or composition, but also on the interspersion of these types in relation to the cruising radius of the species. Composition and interspersion are thus the two principal determinants of potential abundance on game range ... Management of game range is largely a matter of determining the environmental requirements and cruising radius of the possible species of game, and then manipulating the composition and interspersion of types on the land, so as to increase the density of its game population.

Although Leopold did not explicitly mention landscape ecology, he definitely introduced a landscape ecological perspective to wildlife management, at a time in history when ivory-billed woodpeckers (*Campephilus principalis*) still roamed swamp forests in Louisiana. Thirty years later radiotelemetry was made generally available, opening up a new era in wildlife biology. Now wildlife managers could see for themselves how the wildlife was moving around in the landscape. Some 70 years since Leopold's book, and 40 years since radiotelemetry was introduced, what is the state of the art? Have wildlife managers grasped the concepts of landscape ecology? Have landscape ecologists found wildlife management an interesting arena in which to play out their scientific endeavors? What are the future challenges facing landscape ecologists trying to solve practical matters of wildlife management?

The first issue of the *Journal of Wildlife Management*, published in 1937, stated that wildlife management embraces the practical ecology of all

 Issues and Perspectives in Landscape Ecology, ed. John A. Wiens and Michael R. Moss. Published by Cambridge University Press.

vertebrates and their plant and animal associates." Although many will argue that this definition has broadened over the years, I think it still captures the essence of what most people think wildlife is and what wildlife management is about. Leaning more toward game species, wildlife management differs from conservation biology (see With, this volume, Chapter 24) by putting more emphasis on vertebrate species with some sort of economic value. At the core of wildlife management lies the key ecological question: why are there too few of some species (e.g., grouse and deer) and too many of others (e.g., crows and raccoon dogs)? Too few and too many stress the practical, value-oriented idea that underpins the field as an applied scientific venture. To understand how and why wildlife numbers vary, wildlife management draws heavily on population ecology on the one side. Because it also deals with fairly large, mobile organisms and tries to understand how their numbers are affected by environmental variables and their spatial distribution, landscape ecology comes in as an essential counterpart. How has landscape ecology influenced the way wildlife research is conducted?

A landscape ecological perspective

Traditionally, wildlife managers start out with some simple questions about why there are too few or too many of a particular species. They proceed with censuses to get a more precise estimate of abundance, and they characterize the habitat to figure out whether this would give any clues as to what might explain the pattern of abundance. For instance, in Finland hunters have organized nationwide yearly line-transect censuses of forest grouse species since 1964 (Lindén and Rajala, 1981). Sites where birds were flushed were considered good habitat and the rest were considered less good or poor habitat. Comparing the numbers of birds flushed in different forest types using simple statistical inference enabled wildlife managers to come up with more precise preference indices. The message was straightforward: substituting poor habitats with good habitat would give more grouse. This procedure worked in some places but failed in others. Why? Because the spatial arrangement of the habitat patches matters (Kurki et al., 2000).

The reasons for the discrepancies between expected and observed responses of forest grouse to a simple substitution of good for poor habitat encompass a variety of ecological mechanisms. Here we are at the core of what landscape ecology is about: to explain the ecological effects of spatial variation. Two landscapes with similar habitat composition may vary considerably in terms of ecological processes, depending on how the habitat types are spatially arranged. In the case of forest grouse, the birds need feeding sites, mating sites (communal leks in the case of many species), nesting sites, and safe havens from predators. Most species have different seasonal diets. The capercaillie (*Tetrao urogallus*), for example, eats pine needles in winter and herbs and berries in summer. Small chicks are obligate insectivores the first weeks after hatching, whereas adults are vegetarians. During daytime birds rest at the ground in dense vegetation to avoid being detected by day-active raptors, whereas they roost in trees at night to avoid night-active mammalian predators searching for prey by smell. To stay alive and produce viable off-spring during its lifetime, a grouse needs a wide variety of different habitats within its ecological neighborhood.

At a first glance, close proximity of a variety of habitats may seem to be the perfect solution, but this may not always be the case. It has long been recognized that the dynamics of grouse and voles may be linked through what has become known as the "alternative prey model" (Angelstam *et al.*, 1984). In many northern regions voles fluctuate widely, with peak years occurring at three- to four-year intervals. In peak years, generalist predators like fox and marten rely on voles as staple food and produce large litters. In the following years, during the crash and low phases of the vole cycle, these predators shift their diet to grouse eggs, chicks, and adults. In some cases, the production of grouse in these years approaches zero, both due to a numerical (more predators) and a functional (different prey search) response in the generalist predator community. The landscape structure resulting from modern forestry leads to high densities of voles on clearcuts, which presumably increases the amplitude of the vole cycle. Because home ranges of the generalist predators encompass both clearcuts and forest patches, predation on grouse species extends from clearcuts into adjacent forest patches. Therefore, close proximity, or a fine-grained mosaic, of clearcuts and forests may in fact turn out to be far more negative for the grouse than a coarse-grained pattern (Rolstad and Wegge, 1989; Kurki *et al.*, 2000). Thanks to a landscape-ecological approach to wildlife studies, these issues, falling within the general subject of habitat fragmentation, have made their way into forestry policy plans today.

A landscape ecological perspective also has helped clarify the way we look at habitat selection in wildlife species. Although the idea of habitat selection as a hierarchical process was brought forward in the 1960s (Hildén, 1965), it was not until recently that this point was made explicit in wildlife studies (e.g., Swenson, 1993; Rolstad *et al.*, 2000). Imagine a dispersing bird looking for a place to live. First it has to decide where to establish a home range or territory, traveling perhaps tens or hundreds of kilometers. The spatial scale we are dealing with easily adds up to a million hectares. We are looking at complex landscape mosaics with spatially structured populations. Some areas are "sinks," being composed of surplus birds from "source" areas. Large areas may be totally uninhabitable. As we accomplish our study, how do we analyze the information at this spatial scale? Categorical map analysis using GIS techniques may be a good starting point.

When the bird has decided on a landscape in which to settle, it may choose a large home range that includes a few scattered patches of good habitat, or it may settle entirely within a large patch of suitable habitat. At this scale, in the range of thousands of hectares, it might be appropriate to compare the habitat composition of home ranges of a subsample of a population of the bird species. At a third scale, which may be in the range of tens to hundreds of hectares, the bird has to decide which parts of its home range it wants to use and which parts it will avoid. Here, it may be useful to approach the issue of habitat selection by comparing the frequency with which the bird is using the different habitat compartments. Alternatively, we might wish to conduct a point-data analysis using geostatistics, assuming that the habitat characteristics are spatially continuous. Finally, within a habitat compartment or patch (the scale usually termed microhabitat selection) the bird has choices as to where it wants to nest, where it wants to search for food, and where it wants to hide from predators. At this scale, detailed measures of habitat structure will be the method of habitat study. At the end of this hierarchy of scales we could add selection of food items, as a final choice within a preferred feeding site.

Clearly, habitat selection can be envisaged as a hierarchical spatial process, from choice of home range to choice of dietary item. Although the absolute scale, and to a certain degree the number of scale levels, may vary among organisms or landscape types, the principle of a hierarchy of scales generally applies. Isn't this obvious? Perhaps, but far too often we see that conclusions about habitat selection are drawn on the basis of analyses at an inappropriate scale, at an inappropriate organizational level, or with inappropriate methods. To extrapolate across scales, one asks whether the system would behave in the same way at other spatial or temporal scales or whether abrupt, nonlinear changes occur between domains of scales (see Mac Nally, this volume, Chapter 7). It is also important to distinguish clearly between levels of spatial scales and levels of biological organization. The first and second spatial scales above lie within the realm of population organization, whereas the three latter ones deal with the individual level of organization (King, 1997, this volume, Chapter 4). Extrapolating between scales and organizational levels is central to landscape ecology and ideally requires a close interplay between theoretical studies, experimental model systems (EMS), and long-term empirical field studies. As wildlife management has benefited from conceptual and theoretical developments in landscape ecology, so also landscape ecology will continue to benefit from empirical field studies of wildlife populations.

Pattern and process

Although I see shortcomings regarding scale and organizational-level issues, I would generally argue that we have made significant progress in adding a landscape ecological perspective to wildlife management. How come, then, that we still argue about whether old-growth forest is essential for species like spotted owl, capercaillie, northern goshawk and pine marten? Two basic problems are inherent to these studies. The first is related to how we define habitat heterogeneity, whereas the second deals with how successful we are in identifying the underlying ecological processes that are operating. In the end, both issues have bearings on the transition from micro- to macrohabitat scale, which in many cases coincides with the transition from individual to population level of organization.

First, how do we define a habitat patch? In boreal forests of northern Scandinavia and temperate conifer forests of the Pacific Northwest of North America, the task of delineating habitat patches comes fairly easy. New clearcuts in old forest tracts can be recognized on air photos and even on satellite images. But we need not go farther than to southern Scandinavia or northern California to realize that drawing sharp lines between forest stands is a daunting task even in the field. Put simply, when does a forest become old growth? Or when does a deciduous stand become coniferous? In most cases the delineation of habitat patches is a subjective issue. If we asked a professional forester and a non-governmental environmentalist to identify the remnant old-growth forest in a tract, we can be pretty sure they would come up with quite different maps. The forester would presumably rely heavily on tree height, stand volume, and growth rate, whereas the environmentalist would put more emphasis on tree age and the amount of coarse woody debris. The environmentalist might perhaps use "indicator species" to define old-growth forest. Some of these difficulties may be reduced by more careful transformations of microhabitat characteristics to pixel-based GIS images. If simplified maps are to be used, details of the microhabitat characteristics should be made explicit.

As mentioned earlier, point-data analysis might be a way of circumventing dubious map categories in cases were the habitat patchiness gets fuzzy. A whole suite of geostatistical techniques has been developed over the past years, and many of these are now being applied to ecological studies. This approach makes fewer assumptions about the spatial configuration of the system, and there are no explicit boundaries. Consequently, real discontinuities that might have ecological relevance are not as easily recognized as with the categorical map techniques. Thus, these two methods of characterizing landscape patterns should be perceived as complementary approaches (Gustafson, 1998). Point-data analyses can provide useful insight into the scale of patchiness, and thereby be used as a statistical tool guiding the appropriate scale to construct categorical maps.

The second and perhaps more fundamental problem facing landscape-ecological studies of wildlife is to identify the ecological processes that are operating. For example, in southern Scandinavia young spruce plantations seem to be preferred feeding habitat for black woodpeckers (*Dryocopus martius*). This is because the clearcuts feature rotten stumps with colonies of carpenter ants, the staple food source of this woodpecker (Rolstad *et al.*, 1998). Old-growth stands with snags and large woody debris, which also provide ample colonies of carpenter ants, do not exist because the forests have been logged by selective cutting for centuries. In northern Scandinavia, snow often covers the stumps on clearcuts, but snags and logs still occur in old-forest stands due to less intensive logging. In this setting, the old forest provides feeding sites for the woodpeckers whereas the stumps on clearcuts are inaccessible due to heavy snow (Rolstad and Rolstad, 2000). In southern Scandinavia black woodpecker numbers seem to increase with increasing amounts of clearcut and young plantation in the landscape, whereas in northern, snow-rich regions, populations appear to decline for the same reason. Like the capercaillie or spotted owl, these birds do not die of a heart attack when they see a clearcut. They starve, get killed, or compete with other species. If possible, analyses at macrohabitat (or landscape) scales should be accompanied by an evaluation of the underlying reasons why a habitat patch is favorable or why a larger tract is a "source" landscape. Put another way, descriptions of pattern should be accompanied by an understanding of the ecological process. This is perhaps the most compelling challenge within landscape ecology. Whereas landscape ecologists have done pretty well in describing patterns, they have been kind of slow in grasping the underlying ecological processes.

EMS and PVA

The best recipe for unraveling the underlying ecological processes is to conduct good field research over appropriate temporal and spatial scales. But what do we do when it appears that collecting the appropriate field data is not feasible? It might be that the species we are interested in is too rare or its home ranges are too large. Or we simply do not have enough money or field assistants to conduct a comprehensive field study. Two shortcut approaches, theoretical simulation models and experimental model systems (EMS), may come in handy. These methods are intended to substitute for "real data" to gain insight into the ecological processes that interest us.

Assume that a study of a "real system" has given us some hints about the ecological processes that may explain an observed pattern. To gain reliable knowledge about the underlying mechanisms, we have the option of designing experiments that efficiently discriminate between alternative hypotheses regarding the cause–effect relationship. Due to the logistic problems that often encumber large-scale studies of wildlife species, we might decide to "scale down" the system and select a more tractable setting that is amenable to experimental manipulation. Experimental model systems (EMS) have long been accepted as an efficient scientific tool within applied fields like medicine or engineering, where "real systems" are intractable due to practical or moral issues. Although ecologists also have used EMS to study population and community dynamics (Wiens et al., 1993), the general application of this procedure has at best been modest, especially within landscape-ecological studies of wildlife (Matter and Mannan, 1989). The reason for this might be that wildlife biologists have been reluctant to accept that "artificial" model systems can substitute for hardcore data from the natural world. Although one should be cautious when extrapolating across spatial scales, landscape ecologists and wildlife biologists should be more willing to explore the various possibilities that lie within the realm of this approach, thereby gaining better knowledge about pattern–process linkages within their real-world systems (e.g., Schmidt et al., 2001; Ims, this volume, Chapter 8).

Finally, I will briefly touch upon an even more abstract approach to gain knowledge from landscape-ecological wildlife studies, which, perhaps as a result of the explosive growth in computer capacity, has been more widely applied than EMS – pure theoretical models. The use of demographic models in wildlife biology has been thoroughly reviewed by Beissinger and Westphal (1998) (see also Verboom and Wamelink, this volume, Chapter 9). I therefore restrict myself here to a few comments. A popular application of demographic models is to make decisions for managing populations of threatened or endangered species. This suite of models is termed Population Viability Analysis (PVA). Metapopulation and source–sink models may fall into this category. When applied to individuals in landscape mosaics we call them individual-based, spatially explicit simulation models (e.g., Letcher et al., 1998). Although increasingly popular, the most profound limitation of these models is that they have immense data requirements. Such detailed data sets may not exist, and even though we might have a fairly good empirical foundation, the time and resources needed to construct and validate the model often restrict its application. For instance, everyone would agree that knowledge about the dispersal ability of a species is crucial for understanding its long-term spatial dynamics. In very few cases do we actually have these data to put into our models.

Inspiration or perspiration?

Why is it that we rarely see wildlife studies firmly based upon and backed up by the whole suite of scientific approaches, from theoretical models through down-scaled empirical models to real-world studies? I think the reason is fairly straightforward, as described by Aarssen (1997) in a general comment about progress in ecology:

> The "centrifugal force" in ecology that keeps theory and data apart is largely a consequence of human nature of some to be more preoccupied with ideas than with facts, and vice versa. It is a chronic symptom of our limited minds that science progresses by a series of small steps made by both theoreticians and empiricists, often working in isolation. The coming together of theory and data certainly contributes to progress and is cause for celebration, but history has produced relatively few great integrators and it is pointless to ask for this to change.

We all, more or less, live within our narrow sphere of financial support systems, struggling in everyday life to keep our labs and graduate students "alive." Whether we like it or not, this automatically restrains us from sharing our grant funds with colleagues occupying "competing territories."

I therefore close this essay by pleading for a pluralistic approach to explore new "territories." I have picked upon concepts, methods, and techniques that are at our disposal, and I have tried to pinpoint areas that might prove fruitful to pursue in future studies. Quoting a recent book review, "Landscape ecology is a novel way of understanding the world because it integrates facts and ideas from a multitude of sources to produce new insights" (McIntyre, 2002). In a nutshell, it all comes down to keeping our minds open. I know this does not come easy in a world where technical papers in high-ranking journals are all that count. It is very tempting to stick to the field we already know and keep on fine-tuning the techniques we already are good at. In a thought-provoking paper, "A guide to increased creativity in research: inspiration or perspiration?," Loehle (1990) urges us to explore new approaches to stimulate our creative achievements. Aldo Leopold had the gift and guts to expand into new fields, starting out as forester, continuing as wildlife biologist, ending up as philosopher with the *Sand County Almanac* (Leopold, 1949). Today, no one would blame him for that. Today, no one would deny that Leopold also was a genius proponent for landscape ecology. So let's get inspired by his writing in 1939: "The basic skill of the wildlife manager is to diagnose the landscape, to discern and predict trends in its biotic community, and to modify them where necessary in the interest of conservation."

References

Aarssen, L. W. (1997). On the progress of ecology. *Oikos*, 80, 177–178.

Angelstam, P., Lindström, E., and Widén, P. (1984). Role of predation in short-term population fluctuations of some birds and mammals in Fennoscandia. *Oecologia*, 62, 199–208.

Beissinger, S. R. and Westphal, M. I. (1998). On the use of demographic models of population viability in endangered species management. *Journal of Wildlife Management*, 62, 821–841.

Gustafson, E. J. (1998). Quantifying landscape spatial pattern: what is the state of the art? *Ecosystems*, 1, 143–156.

Hildén, O. (1965). Habitat selection in birds. *Annales Zoologici Fennici*, 2, 53–75.

King, A. W. (1997). Hierarchy theory: a guide to system structure for wildlife biologists. In *Wildlife and Landscape Ecology: Effects of Pattern and Scale*, ed. J. A. Bissonette. New York, NY: Springer, pp. 185–212.

Kurki, S., Nikula, A., Helle, P., and Lindén, H. (2000). Landscape fragmentation and forest composition effects on grouse breeding success in boreal forests. *Ecology*, 81, 1985–1997.

Leopold, A. (1933). *Game Management*. New York, NY: Charles Scribner's Sons.

Leopold, A.(1939). Academic and professional training in wildlife work. *Journal of Wildlife Management*, 3, 156–161

Leopold, A.(1949). *A Sand County Almanac and Sketches Here and There*. New York, NY: Oxford University Press.

Letcher, B. H., Priddy, J. A., Walters, J. R., and Crowder, L. B. (1998). An individual-based, spatially-explicit simulation model of the population dynamics of the endangered red-cockaded woodpecker, *Picoides borealis*. *Biological Conservation*, 86, 1–14.

Lindén, H. and Rajala, P. (1981). Fluctuations and long-term trends in the relative densities of tetraonid populations in Finland, 1964–1977. *Finnish Game Research*, 39, 13–34.

Loehle, C. (1990). A guide to increased creativity in research: inspiration or perspiration? *BioScience*, 40, 123–129.

Matter, W. J. and Mannan, R. W. (1989). More on gaining reliable knowledge: a comment. *Journal of Wildlife Management*, 53, 1172–1176.

McIntyre, N. E. (2002). Landscape ecology explained. *Ecology*, 83, 301.

Rolstad, J. and Rolstad, E. (2000). Influence of large snow depths on black woodpecker *Dryocopus martius* foraging behavior. *Ornis Fennica*, 77, 65–70.

Rolstad, J. and Wegge, P. (1989). Capercaillie *Tetrao urogallus* populations and modern forestry: a case for landscape ecological studies. *Finnish Game Research*, 46, 43–52.

Rolstad, J., Majewski, P., and Rolstad, E. (1998). Black woodpecker use of habitats and feeding substrates in a managed Scandinavian forest. *Journal of Wildlife Management*, 62, 11–23.

Rolstad, J., Løken, B., and Rolstad, E. (2000). Habitat selection as a hierachical spatial process: the green woodpecker at the northern edge of its distribution range. *Oecologia*, 124, 116–129.

Schmidt, K. A., Goheen, J. R., and Naumann, R. (2001). Incidental nest predation in songbirds: behavioral indicators detect ecological scales and processes. *Ecology*, 82, 2937–2947.

Swenson, J. E. (1993). The importance of alder to hazel grouse in Fennoscandian boreal forest: evidence from four levels of scale. *Ecography*, 16, 37–46.

Wiens, J. A., Stenseth, N. C., Van Horne, B., and Ims, R. A. (1993). Ecological mechanisms and landscape ecology. *Oikos*, 66, 369–380.

Restoration ecology and landscape ecology

The recent history of the world has been one of a dramatic increase in the incidence of human-induced disturbances as humans utilize an increasing proportion of the earth's surface in some way or another and appropriate an increasing amount of the earth's productive capacity and natural resources (Vitousek *et al.*, 1997). Human modification has led in many cases to increasing degradation of ecosystem components, resulting in a decline in the value of the ecosystem, either for production or for conservation purposes. This has been met with an increasing recognition that measures need to be taken to halt or reverse this degradation, and hence the importance of restoration or repair of damaged ecosystems is increasing (Dobson *et al.*, 1997; Hobbs, 1999).

Restoration ecology is the science behind attempts to repair damaged ecosystems. Here I provide a brief outline of recent developments in the field of restoration ecology, and highlight where I think a strong synergy exists between restoration ecology and landscape ecology. The material presented in this chapter is based in part on Hobbs and Norton (1996), Hobbs (1999), and McIntyre and Hobbs (1999, 2000)

What is restoration ecology?

The term "ecological restoration" covers a wide range of activities involved with the repair of damaged or degraded ecosystems. An array of terms has been used to describe these activities including restoration, rehabilitation, reclamation, reconstruction, and reallocation. Generally, restoration is used to describe the complete reassembly of a degraded system to its undegraded state, while rehabilitation describes efforts to develop some sort of functional protective or productive system on a degraded site. In addition, some authors also use the term "reallocation" to describe the transfer of a site from one land use to a more productive or otherwise beneficial use.

Issues and Perspectives in Landscape Ecology, ed. John A. Wiens and Michael R. Moss. Published by Cambridge University Press.

Unfortunately, a stable terminology has been slow to develop and the above terms are frequently used interchangeably and differently by different authors. Here I will follow Hobbs and Norton (1996) and use the term *restoration* to refer broadly to activities which aim to repair damaged systems.

Ecological restoration is usually carried out for one of the following reasons:

1 To restore highly disturbed, but localized sites, such as mine sites. Restoration often entails amelioration of the physical and chemical characteristics of the substrate and ensuring the return of vegetation cover.

2 To improve productive capability in degraded production lands. Degradation of productive land is increasing worldwide, leading to reduced agricultural, range, and forest production. Restoration in these cases aims to return the system to a sustainable level of productivity, e.g., by reversing or ameliorating soil erosion or salinization problems in agricultural or range lands.

3 To enhance nature conservation values in protected landscapes. Conservation lands worldwide are being reduced in value by various forms of human-induced disturbance, including the effects of introduced stock, invasive species (plant, animal, and pathogen), pollution, and fragmentation. In these cases, restoration aims to reverse the impacts of these degrading forces, for example, by removing an introduced herbivore from a protected landscape. In many areas, there is also a recognized need to increase the areas of particular ecosystem types; for instance, attempts are being made to increase the area of native woodlands in the United Kingdom in order to reverse past trends of decline and to increase the conservation value of the landscape (Ferris-Kaan, 1995).

4 To restore ecological processes over broad landscape-scale or regional areas. In addition to the need for restoration efforts within conservation lands, there is also a need to ensure that human activities in the broader landscape do not adversely affect ecosystem processes. There is an increasing recognition that protected areas alone will not conserve biodiversity in the long term, and that production and protection lands are linked by landscape-scale processes and flows (e.g., hydrology, movement of biota). Methods of integrating conservation and productive use are thus required, as for instance in the Biosphere reserve and core–buffer–matrix models (Hobbs, 1993; Noss and Cooperrider, 1994; Morton *et al.*, 1995). Restoration in this case entails (1) returning conservation value to portions of the productive landscape, preferably through an integration of production and conservation values; and/or

(2) ensuring that land uses within a region do not have adverse impacts on the region's ecological processes.

Ecological restoration thus occurs along a continuum from the rebuilding of totally devastated sites to the limited management of relatively unmodified sites (Hobbs and Hopkins, 1990). The specific goals of restoration and the techniques used will obviously differ between these different cases. In general terms, however, restoration aims to return the degraded system to some form of cover which is protective, productive, aesthetically pleasing, or valuable in a conservation sense (Hobbs and Norton, 1996). A further tacit aim is to develop a system which is sustainable in the long term.

Within these broad general aims, more specific goals are required to guide the restoration process. Ecosystem characteristics which may be considered when considering restoration goals include (from Hobbs and Norton, 1996):

1 Composition: species present and their relative abundances
2 Structure: vertical arrangement of vegetation and soil components (living and dead)
3 Pattern: horizontal arrangement of system components
4 Heterogeneity: a complex variable made up of components 1–3
5 Function: performance of basic ecological processes (energy, water, nutrient transfers)
6 Species interactions: includes pollination, seed dispersal, etc.
7 Dynamics and resilience: succession and state-transition processes, recovery from disturbance

This set of characteristics is complex, and often individual components are considered as primary goals. For instance, restoration of a mine site may aim to replace the complement of plant species present prior to disturbance, while other situations may have the restoration of particular ecosystem functions as a primary aim (e.g., bioremediation of eutrophication in lakes, or the manipulation of vegetation cover to modify water use).

Unfortunately, restoration goals are often poorly defined, or stated in general terms relating to the return of the system to some pre-existing condition. The definition of the characteristics of this condition has proved problematic, since it assumes a static situation. Ecologists increasingly consider that natural systems are dynamic, that they may exhibit alternative (meta-)stable states, and that the definition of what is the "natural" ecosystem in any given area may be difficult (Sprugel, 1991). Indeed, the concept of "naturalness" has itself been the subject of much recent debate, especially in relation to landscapes with long histories of human habitation.

Landscape-scale restoration

Most of the information and methodologies on ecological restoration center on individual sites. This is reflected in the discussion above. However, site-based restoration has to be placed in a broader context and is often insufficient on its own to deal with large-scale restoration problems (Hobbs and Norton, 1996; Hobbs and Harris, 2001). Landscape- or regional-scale processes are often either responsible for ecosystem degradation at particular sites, or alternatively have to be restored to achieve restoration goals. Hence, restoration is often needed both within particular sites and at a broader landscape scale.

How are we, then, to go about restoration at a landscape scale? What are the relevant aims? What landscape characteristics can we modify to reach these aims, and do we know enough to be able to confidently make recommendations on priorities and techniques?

There are several steps in the development of a program of landscape-scale restoration, which can be outlined as follows:

1   Assess whether there is a problem which requires attention: for instance,

    (a) changes in biotic assemblages (e.g., species loss or decline, invasion)
    (b) changes in landscape flows (e.g., species movement, water and/or nutrient fluxes)
    (c) changes in aesthetic or amenity value (e.g., decline in favored landscape types)

2   Determine the causes of the perceived problem: for instance,

    (a) removal and fragmentation of native vegetation
    (b) changes in pattern and abundance of vegetation/landscape types
    (c) cessation of historic management regimes

3   Determine realistic goals for restoration: for instance,

    (a) retention of existing biota and prevention of further loss
    (b) slowing or reversal of land or water degradation processes
    (c) maintenance or improvement of productive potential
    (d) integrated solutions tackling multiple goals

4   Develop cost-effective planning and management tools for achieving agreed goals:

    (a) determining priorities for action in different landscape types and conditions
    (b) spatially explicit solutions

(c) acceptance and "ownership" by managers and landholders
(d) an adaptive approach which allows course corrections when necessary

This short list hides a wealth of detail, uncertainty, and science yet to be done. For instance, the initial assessment of whether there is a problem or not requires the availability of a set of readily measurable indicators of landscape "condition" or "health." This ties in with recent attempts to use the concept of ecosystem health as an effective means of discussing the state of ecosystems (Costanza et al., 1992; Cairns et al., 1993; Shrader-Frechette, 1994). Central elements of ecosystem health are the system's vigor (or activity, production), organization (or the diversity and number of interactions between system components), and resilience (the system's capacity to maintain structure and function in the presence of stress) (Rapport et al., 1998). Attempts have also been made to produce readily measurable indices of ecosystem health for a number of different ecosystems, although there is still debate over whether these are useful or not. In the same way, there have been recent attempts to develop a set of measures of landscape condition (Aronson and Le Floc'h, 1996).

Aronson and Le Floc'h (1996) present three groups of what they term "vital landscape attributes" which aim to encapsulate landscape structure and biotic composition, functional interactions among ecosystems, and degree, type, and causes of landscape fragmentation and degradation. While their list of 16 attributes provides a useful start for thinking about these issues, it fails in its attempt to provide a practical assessment of whether a particular landscape is in need of restoration and, if so, what actions need to be taken. Steps towards this are being developed, at least for landscape flows, in the Landscape Function Analysis approach developed for Australian rangelands (see Ludwig et al., 1997).

Once a problem has been perceived, the correct diagnosis of its cause and prescription of an effective treatment is by no means simple. The assumption underlying landscape ecology is that landscape processes are in some way related to landscape patterns. Hence, by determining the relationship between pattern and process, one is better able to predict what will happen to the processes in which one is interested (biotic movement, metapopulation dynamics, system flows, etc.) if the pattern of the landscape is altered in particular ways. Thus, we are becoming increasingly confident that we can, for instance, predict the degree of connectivity in a landscape from the proportion of the landscape in different cover types. As proportion of a particular cover types decreases, a threshold value is reached at which connectivity rapidly decreases (Pearson et al., 1996; Wiens, 1997; With, 1997). Similarly, as landscapes become more fragmented, a greater proportion of the biota drops out, and again there may be thresholds or breakpoints where relatively large numbers of species drop out. Hobbs and Harris (2001) have argued that there may be different types of thresholds at the landscape scale, with some being biotically driven (in the case of connectivity-related processes) and others being abiotically driven (in the case of physical changes such as altered hydrology). The possibility of the existence of different types of threshold means that clear identification of the primary driving forces is essential before restoration is attempted. There will be little point in trying to deal with biotic issues before treating abiotic problems.

A number of other important questions have to be asked in terms of restoration. First, does the threshold work the same way on the way up as it did on the way down, or is there a hysteresis effect? In other words, in a landscape in which habitat area is being increased, will species return to the system at the same rate as they dropped out when habitat was being lost? Second, what happens when pattern and process are not tightly linked? For instance, studies in central Europe have illustrated the important role of traditional management involving seasonal movement of sheep between pastures in dispersing seeds around the landscape (Bakker *et al.*, 1996; Fischer *et al.*, 1996; Poschlod *et al.*, 1996). The long-term viability of some plant species may be threatened by the cessation of this process, and restoration in this case will not involve any modification of landscape pattern; rather, it will entail the reinstatement of a management-mediated process of sheep movement. Hence, correct assessment of the problem and its cause and remedy require careful examination of the system and its components rather than generalized statements of prevailing dogma.

From assessment to action

Given the considerations above, how does one then go about determining how to conduct restoration at a landscape scale? Here, I relate what we have been thinking about in the context of rural Australia, where landscape fragmentation and habitat modification have caused numerous and extensive problems of land degradation and biodiversity decline. We have been examining the question of what remedial measures can be taken to prevent further loss of species and assemblages in these altered landscapes. A set of general principles, derived from island biogeography theory, suggest that bigger patches are better than small patches, connected patches are better than unconnected, and so on. For fragments in agricultural landscapes, such principles can be translated into the need to retain existing patches (especially large ones) and existing connections, and to revegetate in such a way as to provide larger patches and more connections (Hobbs,

1993). Ryan (2000) indicates clearly the lack of evidence to date that carrying out such revegetation will actually do anything useful, although some examples cited by him and Barrett and Davidson (2000) provide some hopeful signs that revegetation and regeneration do, in fact, result in conservation benefits.

Nevertheless, important questions still remain concerning what sort of landscape-level management and revegetation is appropriate for different landscapes. If we can accept that priority actions involve firstly the protection of existing fragments, secondly their effective management, and thirdly restoration and revegetation, where do we go from there? Which are the priority areas to retain? Should we concentrate on retaining the existing fragments or on revegetation, and relatively how many resources (financial, manpower, etc.) should go into each? How much revegetation is required, and in what configuration? When should we concentrate on providing corridors versus additional habitat? If we are to make a significant impact in terms of conserving remaining fragments and associated fauna, these questions need to be addressed in a strategic way.

McIntyre and Hobbs (1999) have examined these questions in terms of the range of human impacts on landscapes. They recognized two gradients of human impact on ecosystems: destruction and modification. These can both be conceptualized as a continuum and each is associated with the effects of disturbance resulting from human activities. Such disturbances tend to result in alteration of the ecosystem and irreversible loss of species, and can take the form of novel types of disturbance or changes to the natural disturbance regime. They can result in the destruction and modification of habitats as described below. Habitat destruction results in loss of all structural features of the vegetation and loss of the majority of species, as occurs during vegetation clearance. McIntyre and Hobbs (1999, 2000) identified four broad types of landscapes (Table 22.1), with intact and relictual landscapes at the extremes, and two intermediate states, variegated and fragmented. In variegated landscapes, the habitat still forms the matrix, whereas in fragmented landscapes, the matrix comprises "destroyed habitat."

Each of the four levels described in Table 22.1 is associated with a particular degree of habitat destruction, and the categories are not entirely arbitrary. For instance, the distinction between variegated and fragmented landscapes reflects suggestions discussed earlier that landscapes in which habitats persist on more than 60% of the area are operationally not fragmented, since they consist of a continuous cluster of habitat. This broad division can be regarded as a "first cut," and the provision of names for each category is for convenience rather than to set up a rigid classification. Further investigation is required to test these categories and to examine the need for further subcategories. For

Table 22.1. Four landscape states defined by the degree of habitat destruction. Characteristic connectivity (Pearson *et al.*, 1996), and degree and patterns of modification associated with each state, are also given.

Landscape type	Degree of destruction of habitat (% remaining)	Connectivity of remaining habitat	Degree of modification of remaining habitat	Pattern of modification of remaining habitat
Intact	Little or none (>90%)	High	Generally low	Mosaic with gradients
Variegated	Moderate (60–90%)	Generally high but lower for species sensitive to habitat modification	Low to high	Mosaic which may have both gradients and abrupt boundaries
Fragmented	High (10–60%)	Generally low but varies with mobility of species and arrangement on landscape	Low to high	Gradients within fragments less evident
Relictual	Extreme (<10%)	None	Generally highly modified	Generally uniform

instance, functionally different types of "fragmented" landscapes could be recognized.

Habitat modification alters the condition of the remaining habitat and can occur in any of the situations illustrated in Table 22.1. Modification acts to create a layer of variation in the landscape over and above the straightforward spatial patterning caused by vegetation destruction. There is a tendency for habitats to become progressively more modified with increasing levels of destruction, owing to the progressively greater proportion of edge in remaining habitats.

We are exploring the proposition that the framework in Table 22.1 can assist in deciding where on the landscape to allocate greater and lesser efforts toward different management actions (McIntyre and Hobbs, 2000). Three types of action could be applied to habitats for their conservation management:

1 **Maintain** the existing condition of habitats by removing and controlling threatening processes. It is generally much easier to avoid the effects of degradation than it is to reverse them.
2 **Improve** the condition of habitats by reducing or removing threatening processes. More active management may be needed to initiate a reversal of condition (e.g., removal of exotic species, reintroduction of native species) in highly modified habitats.
3 **Reconstruct** habitats where their total extent has been reduced below viable size using replanting and reintroduction techniques. As this is so difficult and expensive, it is a last-resort action that is most relevant to fragmented and relictual landscapes. We have to recognize that restoration will not come close to restoring habitats to their unmodified state, and this reinforces the wisdom of maintaining existing ecosystems as a priority.

The next stage is to link these activities to specific landscape components (matrix, connecting areas, buffer areas, fragments) in which they would be most effective, and to determine priorities for management action in different landscape types. A general approach might be to build on strengths of the remaining habitat by filling in gaps and increasing landscape connectivity, increasing the availability of resources by rehabilitating degraded areas, and expanding habitat by revegetating to create larger blocks and restore poorly represented habitats.

The first priority is the maintenance of elements which are currently in good condition. This will be predominantly the vegetated matrix in intact and variegated landscapes and the remnants which remain in good condition in fragmented landscapes. There may well be no remnants left in good condition in relictual landscapes. Maintenance will involve ensuring the continuation of population, community, and ecosystem processes which result in the persistence of the species and communities present in the landscape. Note that maintaining fragments in good condition in a fragmented system may also require activities in the matrix to control landscape processes, such as hydrology.

The second priority is the improvement of elements that have been modified in some way. In variegated landscapes, buffer areas and corridors may be a priority, while in fragmented systems, improving the surrounding matrix to reduce threatening processes will be a priority, as indicated above. In relict landscapes, improving the condition of fragments will be essential for their continued persistence. Improvement may involve simply dealing with threatening processes such as stock grazing or feral predators, or may involve active management to restore ecosystem processes, improve soil structure, encourage regeneration of plant species, or reintroduce flora or fauna species formerly present there (Hobbs and Yates, 1997).

Reconstruction is likely to be necessary only in fragmented and relict areas. Primary goals of reconstruction will be to provide buffer areas around fragments, to increase connectivity with corridors, and to provide additional habitat (Hobbs, 1993). While some basic principles of habitat reconstruction have been put forward, the benefits of such activities have rarely been quantified. Questions remain about which characteristics of "natural" habitat are the most important to try to incorporate into reconstruction, and what landscape configurations are likely to be most effective.

In order to answer such questions, it becomes very important to clearly specify what the conservation goals are for the area. Lambeck (1997) has recently contended that more efficient solutions to conservation problems can be developed if we take a strategic approach rather than a generalized one. This involves developing a clear set of conservation objectives rather than relying on vague statements of intent. One set of objectives relates to the achievement of a comprehensive, adequate, and representative set of reserves or protected area networks. Another, complementary set of objectives relate to the adequacy of the existing remnant vegetation (not only reserves). Lambeck has suggested that the process of setting conservation objectives in any given area can be simplified by identifying a set of key or "focal" species. This approach approximates to a multi-species indicator/umbrella species approach.

To identify focal species, Lambeck (1997) recognized three distinct sets of species, each of which was likely to be limited or threatened by particular characteristics of the landscape. These were:

1   Area- or habitat-limited species: species whose numbers are limited by the availability of large enough patches of suitable habitat

2 Movement-limited species: species whose numbers are limited by the degree to which they can move between habitat patches

3 Management-limited species: species whose numbers are limited by processes such as predation, disturbance, fire, and the like, which can be manipulated within particular sites

In Lambeck's approach, design of landscape reconstructions is based on the requirements of the most sensitive species in each of these categories. For instance, if you can identify which species have the requirement for the largest areas of habitat, you can start assessing the adequacy of the current landscape for that species, and hence all other species with less demanding habitat requirements, and can also start making recommendations on where and how much habitat reconstruction needs to be undertaken.

Conservation objectives of an area can be discussed in terms of which species and communities are at risk, what the likely source of that risk is, and how prepared society is to address the risk. The focal species approach put forward by Lambeck (1997) could profitably be combined with the framework for categorizing landscapes suggested by McIntyre and Hobbs if the relative incidence of species in different categories could be linked to landscape configuration. Perhaps a useful approach is the development of a set of principles/guidelines to guide activities in a general way; i.e., to decide the relative efforts needed in remnant protection or revegetation. More detailed guidelines then become necessary in relation to goals for particular sets of species; i.e., to decide on the relative need for corridors versus provision of enlarged habitat patches. Lambeck (1997) has indicated how the identification of focal species and a rapid assessment of their habitat requirements can result in the production of quantitative guidance as to how much vegetation is needed, and in what configuration. The further development of this work involves being able to make spatially explicit recommendations as to where revegetation should occur. This is the essential outcome if real solutions are to be developed and implemented.

Conclusion

This chapter has explored the interface between landscape ecology and restoration ecology. There is a pressing need for interaction between the two fields, and the opportunity for synergy is obvious. Both are relatively new sciences, and both are tackling important problems currently facing humanity. And yet few scientists from either field make much effort to foster interaction. While there are obvious barriers and disincentives to interaction with other fields, the science of landscape ecology by its very nature needs to make linkages across a range of disciplines. I encourage all landscape ecologists to be involved, not just in the description and analysis of landscape change and decline, but also in the active development of effective strategies for the restoration of the world's degraded landscapes.

References

Aronson, J. and Le Floc'h, E. (1996). Vital landscape attributes: missing tools for restoration ecology. *Restoration Ecology*, 4, 377–387.

Bakker, J. P., Poschlod, P., Strykstra, R. J., Bekker, R. M., and Thompson, K. (1996). Seed banks and seed dispersal: important topics in restoration ecology. *Acta Botanica Neerlandica*, 45, 461–490.

Barrett, G. and Davidson, I. (2000). Community monitoring of woodland habitats: the Birds on Farms Survey. In *Temperate Eucalypt Woodlands in Australia: Biology, Conservation, Management and Restoration*, ed. R. J. Hobbs and C. J. Yates. Chipping Norton, NSW: Surrey Beatty, pp. 382–399.

Cairns, J. J., McCormick, P. V., and Niederlehner, B. R. (1993). A proposed framework for developing indicators of ecosystem health. *Hydrobiologia*, 263, 1–44.

Costanza, R., Norton, B. G., and Haskell, B. D. (1992). *Ecosystem Health: New Goals for Environmental Management*. Washington, DC: Island Press.

Dobson, A. P., Bradshaw, A. D., and Baker, A. J. M. (1997). Hopes for the future: restoration ecology and conservation biology. *Science*, 277, 515–522.

Ferris-Kaan, R. (ed.) (1995). *The Ecology of Woodland Creation*. Chichester: Wiley.

Fischer, S. F., Poschlod, P., and Beinlich, B. (1996). Experimental studies on the dispersal of plants and animals on sheep in calcareous grasslands. *Journal of Applied Ecology*, 33, 1206–1222.

Hobbs, R. J. (1993). Can revegetation assist in the conservation of biodiversity in agricultural areas? *Pacific Conservation Biology*, 1, 29–38.

Hobbs, R. J. (1999). Restoration of disturbed ecosystems. In *Ecosystems of the World* 16, ed. L. Walker. Amsterdam: Elsevier, pp. 673–687.

Hobbs, R. F. and Harris, J. A. (2001). Restoration ecology: repairing the earth's ecosystems in the new millennium. *Restoration Ecology*, 9, 239–246.

Hobbs, R. J. and Hopkins, A. J. M. (1990). From frontier to fragments: European impact on Australia's vegetation. *Proceedings of the Ecological Society of Australia*, 16, 93–114.

Hobbs, R. J. and Norton, D. A. (1996). Towards a conceptual framework for restoration ecology. *Restoration Ecology*, 4, 93–110.

Hobbs, R. J. and Yates, C. J. (1997). Moving from the general to the specific: remnant management in rural Australia. In *Frontiers in Ecology: Building the Links*, ed. N. Klomp and I. Lunt. Amsterdam: Elsevier, pp. 131–142.

Lambeck, R. J. (1997). Focal species: a multi-species umbrella for nature conservation. *Conservation Biology*, 11, 849–856.

Ludwig, J., Tongway, D., Freudenberger, D., Noble, J., and Hodgkinson, K. (eds.) (1997). *Landscape Ecology, Function and Management: Principles from Australia's Rangelands*. Melbourne: CSIRO.

McIntyre, S. and Hobbs, R. J. (1999). A framework for conceptualizing human impacts on landscapes and its relevance to management and research. *Conservation Biology*, 13, 1282–1292.

McIntyre, S. and Hobbs, R. J. (2000). Human impacts on landscapes: matrix condition and management priorities. In *Nature Conservation 5: Nature Conservation in Production Environments*, ed. J. Craig, D. A. Saunders and N. Mitchell. Chipping Norton, NSW: Surrey Beatty, pp. 301–307.

Morton, S. R., Stafford Smith, D. M., Friedel, M. H., Griffin, G. F., and Pickup, G. (1995). The stewardship of arid Australia: ecology and landscape management. *Journal of Environmental Management*, 43, 195–217.

Noss, R. F. and Cooperrider, A. Y. (1994). *Saving Nature's Legacy: Protecting and*

Restoring Biodiversity. Washington, DC: Island Press.

Pearson, S. M., Turner, M. G., Gardner, R. H., and O'Neill, R. V. (1996). An organism-based perspective of habitat fragmentation. In *Biodiversity in Managed Landscapes: Theory and Practice*, ed. R. C. Szaro and D. W. Johnston. New York, NY: Oxford University Press, pp. 77–95.

Poschlod, P., Bakker, J., Bonn, S., and Fischer, S. (1996). Dispersal of plants in fragmented landscapes. In *Species Survival in Fragmented Landscapes, vol. 35*, ed. J. Settele, C. Margules, P. Poschlod, and K. Henle. Dordrecht: Kluwer, pp. 123–127.

Rapport, D. J., Costanza, R. and McMichael, A. J. (1998). Assessing ecosystem health. *Trends in Ecology and Evolution*, 13, 397–402.

Ryan, P. (2000). The use of revegetated areas by vertebrate fauna in Australia: a review. In *Temperate Eucalypt Woodlands in Australia: Biology, Conservation, Management and Restoration*, ed. R. J. Hobbs and C. J. Yates. Chipping Norton, NSW: Surrey Beatty, pp. 318–335.

Shrader-Frechette, K. S. (1994). Ecosystem health: a new paradigm for ecological assessment. *Trends in Ecology and Evolution*, 9, 456–457.

Sprugel, D. G. (1991). Disturbance, equilibrium, and environmental variability: what is "natural" vegetation in a changing environment? *Biological Conservation*, 58, 1–18.

Vitousek, P. M., Mooney, H. A., Lubchenco, J., and Melillo, J. (1997). Human domination of Earth's ecosystems. *Science*, 277, 494–499.

Wiens, J. A. (1997). Metapopulation dynamics and landscape ecology. In *Metapopulation Biology: Ecology, Genetics, and Evolution*, ed. I. A. Hanski and M.E. Gilpin. New York, NY: Academic Press, pp. 43–62.

With, K. A. (1997). The theory of conservation biology. *Conservation Biology*, 11, 1436–1440.

23

Conservation planning at the landscape scale

A major challenge for the science of ecology, to make it relevant, is to build a bridge between the local scale of reductionist science and the landscape scale of planning and decision making. This is, of course, the task that landscape ecology has set for itself. Planning for biodiversity conservation is a practice that illustrates the opportunities, as well as the risks and challenges, in bringing ecological science to bear on problems in the real world of human activities.

The objective of conservation planning is to balance production and other forms of exploitation with the conservation of biodiversity in a way that allows for the realization of the evolutionary potential of as many life forms as possible. To help achieve this objective, some areas within regions (countries, biomes, landscapes, etc.) should be primarily managed for the protection of biodiversity. I will call these biodiversity priority areas. Priority areas will not encompass all biodiversity nor will they sustain the biodiversity they encompass over time if they are managed in isolation from the surrounding matrix of other natural, semi-natural, and production lands. However, biodiversity priority areas should form the core of biodiversity conservation plans.

Not many existing protected areas (current biodiversity priority areas) were selected with an explicit biodiversity goal in mind. Some were chosen for their outstanding natural beauty and others because they protected rare species or wilderness values. Most were chosen because there were few competing land uses (Pressey, 1994). With a handful of notable exceptions (see MacKinnon and MacKinnon, 1986), protected-area selection has been opportunistic and ad hoc. As a result, much of the biodiversity most in need of protection has not been protected and now there is a strong bias favoring species associated with areas with the least potential for alternative exploitative uses (Pressey and Tully, 1994). A more systematic and rational approach would be to measure the contribution every area in a region (or landscape) makes to an agreed

 Issues and Perspectives in Landscape Ecology, ed. John A. Wiens and Michael R. Moss. Published by Cambridge University Press.
© Cambridge University Press 2005.

biodiversity goal, identify those areas with high contributions, and manage them as biodiversity priority areas.

Measuring and mapping biodiversity and setting biodiversity goals are sources of contention among ecologists and conservation practitioners. Both draw on ecological knowledge gained from scientific studies at local scales and extrapolate that knowledge in an attempt to generalize it to regional scales. There are real dangers in doing this because ecological knowledge is incomplete. There never seems to be enough information on hand for ecologists to be certain about the relative merits of different courses of action. For example, if biodiversity is described as forest types, what level of resolution is correct? Since the level of resolution can go from one class (the whole forest) right down to the number of spatial units clustered to form the forest types – the grid cells, catchments, or any other polygons – how far along that continuum should we go? How many different types should be mapped and used in practical conservation planning and management? Once that decision is made, how should targets for each type be set? If a percentage – say 10% for argument's sake – which 10%? Forest types, like all classifications, are spatially heterogeneous and protecting a proportion of them is no guarantee that they have been adequately represented in protected areas. In addition, goal setting is seen as dangerous because by implication, once a goal is achieved – say, 10% of each forest type in a region is under protection – the remainder might be considered available for any exploitative use regardless of the impacts of that use. The persistence of biodiversity priority areas, which are connected in space and time by ecological processes to the whole landscape in which they are embedded, depends on appropriate management outside those priority areas, as well as within them.

However, land-use planning and decision making will proceed regardless. If we say nothing because we believe our knowledge is inadequate, we will have no input to decisions concerning the fate of biodiversity and the use of natural resources. Because the need is urgent in the face of continuing land-use change and because biodiversity protection competes with legitimate, alternative uses of biodiversity, methods for identifying priority areas have to be explicit, efficient, cost-effective, and flexible. They also have to make the most effective use of existing knowledge to measure and map biodiversity and to set goals, acknowledging that it will always be necessary to re-examine priorities as knowledge accumulates. In order to identify biodiversity priority areas it is necessary to do three things. First, there must be an acceptable way of measuring biodiversity and mapping its spatial distribution. Second, there must be a way of determining an acceptable level of representation, i.e., setting the goal. Third, having set that goal, there must be a cost-effective and socially acceptable way of allocating limited resources to secure it. These three requirements are discussed below and considered in more detail in Margules and Pressey (2000) as well as in Margules *et al.* (2002), Sarkar and Margules (2002), and other papers in that same special issue of the *Journal of Biosciences* (volume 27, Supplement 2).

Measuring and mapping biodiversity

Biological systems are organized hierarchically from the molecular to the ecosystem level. Logical classes such as populations, species, assemblages, and ecosystems are heterogeneous, which means that all members of each class can be distinguished from one another. The complete description of a class requires the inclusion of all members. The variety of viable biological configurations at all levels is extremely large, currently unknown, and probably unmeasurable. Yet this is biodiversity, and sustaining such complexity is the goal of biodiversity protection. Unfortunately, it is not practical to enumerate all of the species of an area, let alone the logical classes at lower levels, such as populations and individuals.

For the foreseeable future it will be necessary to accept this incomplete knowledge and adopt methods for making the most of what we do know. One implication is that surrogate or partial measures of biodiversity must be used. Some people advocate the use of particular taxa as surrogates, while others favor higher-order surrogates such as habitat types or environmental classes. We have to be honest with ourselves here and admit that there is no known surrogate in the true sense of the word, i.e., one that stands for all of biodiversity in all situations. Intuitively, to me at least, it seems unlikely that we will ever find one. Therefore, and returning to the over-arching goal in the introduction – the realization of the evolutionary potential of as many life forms as possible – we should accept that we can only use partial measures of biodiversity and agree that these partial measures should focus on expressing the range of natural variation across regions and landscapes in order to see that biodiversity priority areas capture that variation. While it may be desirable to plan for biodiversity protection using the more precise measures of species, especially rare, threatened, or endemic species, taxa subsets such as plants, birds, butterflies, etc. represent only a tiny proportion of all of biodiversity. More heterogeneous levels of biological organization have the practical advantage that information on the distribution of, say, assemblages or habitat types is more widely available or more easily acquired. These levels may also integrate more of the functional processes that are important for maintaining both ecosystem processes and the viability of populations (McKenzie *et al.*, 1989). But most importantly, with limited knowledge and limited resources, they allow for the possibility that a set of priority areas within a region might sample that range of natural variation and therefore maximize the likelihood that the evolutionary potential of as many life forms as possible is realized.

Planning is essentially a matter of comparison, and it is not valid to compare two or more areas unless the same kind of information with the same level of detail is available for all areas. Thus, obtaining spatially consistent data is a planning requisite. Museum and herbarium data are notoriously biased, having been collected for a different purpose (systematics) and therefore from locations where collectors expected to find what they were looking for or, worse, which were conveniently accessible (Margules and Austin, 1994). Plot the field locations of many collections and you will find that they map the road network.

A range of analytical procedures is available for reducing spatial bias. Numerical clustering and ordination can be used to detect general patterns in large complex data sets. Empirical models such as BIOCLIM (Hutchinson et al., 1996) and DOMAIN (Carpenter et al., 1993) and statistical models (e.g., Margules and Austin, 1994; Austin and Meyers, 1996) can be used to estimate wider spatial distribution patterns from the point records that field collections represent. These methods are not substitutes for new knowledge, which should always be sought wherever and whenever possible, but they facilitate the current planning process by making the most of existing data.

There is no single best partial measure of biodiversity. The choice, in any given situation, will depend on what data are available and what resources and facilities there are for data analysis and the collection of new data. In parts of Europe and North America it may be possible to use taxa subsets with some confidence because the field records of taxa are a true representation of the distribution patterns of those taxa, although this still leaves the problem that any set of taxa represents only a tiny portion of biodiversity. In many other parts of the world, only information on higher-level measures is available at comparable levels of detail across regions. It seems likely that combinations of measures will be most practicable in most situations. In a recent countrywide conservation planning project in Papua New Guinea (Faith et al., 2001a, 2001b), environmental domains generated from climate, landform, geology (Nix, 1982; Hutchinson et al., 1996), vegetation types mapped from aerial photographs (Hammermaster and Saunders, 1995), known locations of rare and threatened taxa, and areas of vertebrate endemism were all used as biodiversity surrogates.

Biodiversity goals

Just as there is no best way to decide which measures of biodiversity to use, determining how much biodiversity is enough, setting the level of representation, is an unresolved, and probably unresolvable, problem. Realizing the evolutionary potential of as many species as possible is an appropriate over-arching goal, but in order to judge the success or failure of a conservation plan it is necessary to set more explicit goals. Setting such goals is difficult because we know that protecting all biodiversity means excluding all areas from alternative uses, a goal that is not very helpful because it cannot be achieved. Recently, the international community, individual nations, and jurisdictions within nations have been concerned with quantifying conservation goals and setting targets. Conservation International and lUCN have campaigned for a minimum of 10% of all forest types to be represented in forest protected-area networks. While a number of countries have committed to this goal, some have exceeded it. The Australian target for forests is 15% of the extent of pre-1750 (European settlement) forest ecosystems (Commonwealth of Australia, 1997). There is no reason why targets of this kind should be the same for all forest types or ecosystems. Localized habitats such as mound springs in central Australia might require 90% or 100% protection to ensure persistence. More widespread habitats such as mopane woodlands of southern Africa, for example, might require only 10% or 15% protection.

The setting of targets has both advantages and disadvantages. On the one hand, any biodiversity target is arbitrary, perhaps guided but certainly not defined by science. Achieving an arbitrary target is unlikely to satisfy the broader objective of biodiversity protection. On the other hand, a target is a clear goal against which achievement can be assessed and it is probably necessary to have one (or more) if societies are to agree on conservation objectives and make progress toward them. Setting targets for conservation planning should therefore be seen in the same light as target setting in other areas of human endeavor: as a means to an end rather than an end in itself. As knowledge accumulates and as social, economic, and political conditions change, biodiversity goals should be revisited and plans revised.

Biodiversity conservation planning

Systematic planning methods which aim for cost-effectiveness and social acceptability are currently under development and are now being implemented in Australia, southern Africa, Papua New Guinea, and parts of Europe. Two features, in particular, characterize these methods: complementarity as a measure of conservation value, and the incorporation of constraints, including opportunity-cost trade-offs.

Complementarity

The contribution that any one area within a region makes to the agreed conservation goal is its complementarity value: that is, the contribution it makes to the full regional complement of biodiversity measures (for example, species, forest types). This can be thought of as the marginal gain in bio-diversity that the addition of a new area makes to an existing set of areas. Complementarity explicitly addresses the need for biodiversity priority areas to represent the range of natural variation across regions because areas with highest complementarity will be most different from one another.

An important property of complementarity is that its value may change as the entire set of areas is enlarged. This is because some of the species in a particular area may already be represented by the inclusion of other areas. This stands in contrast to the more traditional measures of conservation value, such as the number of species or the number of rare or endemic species. Those values are fixed. Further, complementarity is quite different from species richness. Areas with few species can have a very high complementarity value if those species do not occur anywhere else or in only a few other places. Gaps in the coverage of biodiversity by existing priority areas are at least as likely to be in species-poor areas as in species-rich areas.

Opportunity-cost trade-offs and other constraints

To gain credibility and, therefore, stand some chance of being imple-mented, a conservation plan must achieve a conservation goal in a cost-effective way that is socially and politically acceptable. This means minimizing forgone opportunities for production, explicitly avoiding, where possible, areas already intensively used and densely populated, and building on any existing protected-area network or other previous conservation plan.

Area selection methods that employ complementarity are inherently flexible and able to accommodate, up to a point, competing demands on biodiversity. This is because there are many possible combinations of areas that can achieve a conservation goal (Pressey et al., 1993). It's just that some solutions have a greater cost (in area of land, forgone production opportunities, etc.) attached to them than do others. Early proponents of the use of complementarity saw the advantages of this flexibility and envisaged the application of cost constraints (Margules et al., 1988; Nicholls and Margules, 1993). Pressey (1998) has shown how area selection methods using complementarity and incorporating compet-ing land-use demands can be effective tools in negotiating land-use plans. Faith and Walker (1996) developed methods for trading off opportunity costs with biodiversity gain and implemented these, and other constraints, in their TARGET software (Faith and Nicholls, 1996) in a countrywide biodiversity planning study in Papua New Guinea (Faith et al., 2001a, 2001b). It is now possible to measure the opportunity costs of achieving a biodiversity goal. It is also possible to measure the biodiversity cost (in biodiversity surrogate units) of meeting a production goal, where that goal requires land allocation.

Conclusions

Conservation planning is developing rapidly but many important questions remain unanswered. Three challenges for the immediate future are as follows. First, we must improve the measurement of biodiversity so that it is both more precise and at a consistent level of detail across regions. In part, this is happening as incremental scientific advances in the description of biodiversity occur and as field collections are built up. But more focus is needed, in particular on tests of the ability of different surrogates to predict more of biodiversity.

Second, we must incorporate some measure of the probability of persistence of the various biodiversity surrogates we use in conservation planning, based, perhaps, on ideas of population viability and landscape connectivity. Faith *et al.* (2001c), have proposed a somewhat different approach. They suggest using the probability of persistence based on tenure to measure complementarity, in which case priority areas become those that, if converted to other uses, have the greatest impact on the probability of persistence of most biodiversity in the region. All these possibilities need to be explored and tested.

Finally, and probably most importantly, we must participate in real conservation planning processes, which incorporate explicit social and economic goals as well as biodiversity goals, even if we think we don't know as much as we would like to. If we do this we will see that all knowledge is incomplete, not just ecological knowledge. People working in other fields routinely try to make the most of what they do know to do the best job they can, given that one certainty in life is that change will occur. Conservation planners, in common with all other kinds of planners, must fully expect to revisit their goals and their plans as knowledge accumulates and as social and economic conditions change.

Acknowledgments

Many colleagues have contributed to the ideas expressed here. I hope they all appear in the references and, in any case, they know who they are. Liz Poon commented critically on the typescript and I thank her for that.

References

Austin, M. P. and Meyers, J. A. (1996). Current approaches to modelling the environmental niche of eucalypts: implications for management of forest biodiversity. *Forest Ecology and Management*, 85, 95–106.

Carpenter, G., Gillison, A. N., and Winter, J. (1993). DOMAIN: a flexible modelling procedure for mapping potential distributions of plants and animals. *Biodiversity and Conservation*, 2, 667–680.

Commonwealth of Australia (1997). *Nationally Agreed Criteria for the Establishment of a Comprehensive, Adequate and Representative Reserve System for Forests in Australia*. Canberra: Australian Government Publishing Service.

Faith, D. P. and Nicholls, A. O. (eds.) (1996). *BioRap Vol. 3. Tools for Assessing Biodiversity Priority Areas*. Canberra: The Australian BioRap Consortium.

Faith, D. P. and Walker, P. A. (1996). Integrating conservation and development: effective trade-offs between biodiversity and cost in the selection of protected areas. *Biodiversity and Conservation*, 5, 431–446.

Faith, D. P., Margules, C. R., Walker, P. A., Stein, J., and Natera, G. (2001a). Practical application of biodiversity surrogates and percentage targets for conservation in Papua New Guinea. *Pacific Conservation Biology*, 6, 289–303.

Faith, D. P., Margules, C. R., and Walker, P. A. (2001b). A biodiversity conservation plan for Papua New Guinea based on biodiversity trade-offs analysis. *Pacific Conservation Biology*, 6, 304–324.

Faith, D. P., Walker, P. A., and Margules, C. R. (2001c). Some future prospects for systematic conservation planning in Papua New Guinea – and for biodiversity planning in general. *Pacific Conservation Biology*, 6, 325–343.

Hammermaster, E. T. and Saunders, J. C. (1995). *Forest Resources and Vegetation Mapping of Papua New Guinea*. PNGRIS Publication 4. Canberra: AusAID.

Hutchinson, M. F., Belbin, L., Nicholls, A. O., Nix, H. A., McMahon, L. P., and Ord, K. D. (1996). *BioRap Vol. 2. Spatial Modelling Tools*. Canberra: The Australian BioRap Consortium.

MacKinnon, J. and MacKinnon, K. (1986). *Review of the Protected Area System in the Indo-Malayan Realm*. Gland, Switzerland: IUCN/UNEP.

Margules, C. R. and Austin, M. P. (1994). Biological models for monitoring species decline: the construction and use of data bases. *Philosophical Transactions of the Royal Society of London B*, 344, 69–75.

Margules, C. R. and Pressey, R. L. (2000). Systematic conservation planning. *Nature*, 405, 243–253.

Margules, C. R., Redhead, T. D., Hutchinson, M. F., and Faith, D. P. (1995). *Guidelines for using the BioRap Methodology and Tools*. Canberra: CSIRO and the World Bank.

Margules, C. R., Nicholls, A. O., and Pressey, R. L. (1988). Selecting networks of reserves to maximise biological diversity. *Biological Conservation*, 43, 63–76.

Margules, C. R., Pressey, R. L., and Williams, P. H. (2002). Representing biodiversity: data and procedures for identifying priority areas for conservation. *Journal of Biosciences*, 27 (Suppl. 2), 309–326.

McKenzie, N. L., Belbin, L., Margules, C. R., and Keighery, G. J. (1989). Selecting representative reserve systems in remote areas: a case study in the Nullarbor region, Australia. *Biological Conservation*, 50, 239–261.

Nicholls, A. O. and Margules, C. R. (1993). An upgraded reserve selection algorithm. *Biological Conservation*, 64, 165–169.

Nix, H. A. (1982). Environmental determinants of biogeography and evolution in Terra Australis. In *Evolution of the Flora and Fauna of Arid Australia*, ed. W. R. Barker and P. J. M. Greenslade. Adelaide: Peacock Press, pp. 47–66.

Pressey, R. L. (1994). Ad hoc reservations: forward or backward steps in developing representative reserve systems? *Conservation Biology*, 8, 662–668.

(1998). Algorithms, politics and timber: an example of the role of science in a public political negotiation process over new conservation areas in production forests. In *Ecology for Everyone: Communicating Ecology to Scientists, the Public and the Politicians*, ed. R. Willis and R. Hobbs. Sydney: Surrey Beatty, pp. 73–87.

Pressey, R. L., Humphries, C. J., Margules, C. R., Vane-Wright, R. I., and Williams, P. H. (1993). Beyond opportunism: key principles for systematic reserve selection. *Trends in Ecology and Evolution*, 8, 124–128.

Pressey, R. L. and Tully, S. L. (1994). The cost of ad hoc reservations: a case study in western New South Wales. *Australian Journal of Ecology*, 19, 375–384.

Sarkar, S. and Margules, C. (2002). Operationalizing biodiversity for conservation planning. *Journal of Biosciences*, 27 (Suppl. 2), 299–308.

24

Landscape conservation: a new paradigm for the conservation of biodiversity

We are in the midst of one of the greatest ecological disasters ever to befall this planet. Species are vanishing worldwide at a rate rivaling the mass extinction events chronicled in the geological record, a rate which exceeds the "normal" or expected rate of extinction by several orders of magnitude (Wilson, 1988). Unlike previous mass extinctions, however, this one has been precipitated by a single species, *Homo sapiens*. It is no coincidence that the global biodiversity crisis occurs at a time when landscapes are being transformed at a rate unprecedented in human history. Humans have transformed up to 50% of the land surface on the planet, such that no landscape (or "aquascape") remains untouched by the direct or indirect effects of human activities (Vitousek *et al.*, 1997). Habitat destruction, in the form of outright loss, degradation, and fragmentation of habitat, is the leading cause of the current extinction crisis (Wilcove *et al.*, 1998). Humans are the primary drivers of landscape change, and thus the current ecological crisis is really a cultural one (Naveh, 1995; Nassauer, this volume, Chapter 27). An understanding of the factors affecting land-use decisions, which involve cultural, political, and socioeconomic dimensions, must be integrated with the ecological consequences of landscape transformation if a full rendering of the biodiversity crisis is to be had and the crisis averted. This will require a holistic approach that transcends disciplines.

Conservation biology and landscape ecology are each touted as being emergent, holistic, problem-solving disciplines that transcend the traditional boundaries between science and policy, theory and practice, society and nature. While the historical and philosophical roots of both disciplines date back centuries, conservation biology and landscape ecology were formalized as scientific disciplines relatively recently, in the early 1980s. On the surface, conservation biology and landscape ecology appear to address both sides of the biodiversity crisis. Landscape ecology originated as the study of

Issues and Perspectives in Landscape Ecology, ed. John A. Wiens and Michael R. Moss. Published by Cambridge University Press.
© Cambridge University Press 2005

the ways in which human systems affect land-use decisions and from a need to direct landscape planning at a regional scale (Turner *et al.*, 2001). Conservation biology is often defined as "the science of scarcity and diversity" and is concerned with halting and reversing the alarming loss of biodiversity (Soulé, 1986). Clearly, conservation strategies will have to be implemented within the context of human-dominated landscapes.

Landscape ecology and conservation biology should thus be able to tackle the major land-use and conservation issues that are at the core of the global biodiversity crisis. Why, then, is landscape ecology perceived to have failed in its "obligation" (Hobbs, 1997) to provide the concepts and techniques to tackle these issues? If landscape transformation is acknowledged to be the primary driving force behind the recent mass extinctions, then why does the perception exist among conservation biologists that landscape ecology has little to offer in this regard (Hobbs, 1997)?

A mission for landscape ecology

Landscape ecology has long suffered from an "identity crisis" (Hobbs, 1994). While this is perhaps expected of any discipline in its adolescence, conservation biology was able to articulate a mission and statement of purpose from infancy. In part, this was due to the fact that it was conceived in response to a crisis, but also because conservation biologists were required to explain early on how their new discipline differed from existing fields such as wildlife biology. The response was that none of the resource management fields, which generally focused on the management of economically important species, was comprehensive enough to deal with the global biodiversity crisis (Edwards, 1989; Jensen and Krausman, 1993; Bunnell and Dupuis, 1995). Conservation biology also promised to provide a theoretical foundation required for developing the scientific framework and guiding principles necessary for the management of complex systems (Simberloff, 1988; With, 1997a).

In contrast, landscape ecology has not been expressly "crisis-driven" or "mission-oriented" in either its origin or subsequent development. Thus, it lacked the early focus and disciplinary cohesion that guided the development of conservation biology. A true synthesis of the disparate scientific and design professions that make up the nexus that is landscape ecology has been slow to emerge as a result of the discipline evolving independently, in different directions, on different continents (Wiens, 1997). Little wonder, then, that landscape ecology was viewed as lacking a comprehensive scientific framework for the analysis, planning, and management of landscapes. The development of this scientific framework was one of the goals of the 1998 mission statement of the International Association for Landscape Ecology (IALE,

1998). Several recent texts highlight landscape ecological principles for resource and land management (e.g., Dale and Haeuber, 2001; Liu and Taylor, 2002).

Although the synthesis must come from within, it also needs to be developed externally by establishing stronger linkages with other disciplines that would benefit from the application of landscape ecological principles. Landscape ecologists must effectively communicate to researchers and practitioners outside the discipline what landscape ecology is all about, what is unique about it, and what it has to offer above and beyond approaches developed in other resource-management disciplines. In the present context, this involves examining how landscape ecology can contribute to the resolution of the biodiversity crisis, by demonstrating how landscape ecology can be applied to problems in land use and conservation.

How can landscape ecology contribute to conservation biology?

Landscape ecology can contribute to the resolution or mitigation of the biodiversity crisis in a number of ways.

The adoption of a landscape perspective in conservation biology

There is a growing consensus that the landscape is the relevant scale at which to manage biodiversity (e.g., Noss, 1983; Salwasser, 1991; Petit *et al.*, 1995; Gutzwiller, 2002; Margules, this volume, Chapter 23). Conservation strategies need to be implemented at broad scales if they are to be effective. This follows from the recent shift in management focus away from individual species and toward entire ecosystems, which necessitates a broader-scale perspective (see below). In addition, nature reserves cannot be viewed in isolation of their landscape context. Human land-use activities in the surrounding matrix affect processes occurring within the reserve, and thus the ultimate success of the reserve in protecting biodiversity depends upon managing the entire landscape (Wiens, 1996; Jongman, this volume, Chapter 31).

Facilitating the shift from species to systems management in conservation

Conservation biology is undergoing a paradigm shift from single-species management to ecosystem management. Ecosystem management emphasizes the importance of maintaining the functional relationships among components of the system, and not just the components themselves

(Christensen *et al.*, 1996). This emphasis on functional relationships ultimately requires an understanding of how landscape structure affects the flows of energy, matter, or individuals across heterogeneous land mosaics. Landscape ecology focuses on how spatial patterns affect ecological flows (Turner, 1989). Although the description and analysis of landscape structure dominated much of the early research activity in landscape ecology (e.g., Turner and Gardner, 1991), there is now more emphasis being placed on the study of landscape function, particularly in regard to issues of flows among boundaries (e.g., Hansen and di Castri, 1992; Wiens *et al.,* 1993) and overall landscape connectivity.

Providing a landscape mosaic perspective in assessing connectivity

Connectivity is a dominant theme in both landscape ecology and conservation biology. In conservation biology, connectivity is an essential component of ecosystem integrity, reserve design, and metapopulation dynamics (Noss, 1991). While the importance of maintaining the functional connectivity of systems is often recognized, this is often interpreted literally to mean maintaining structural connectivity (e.g., actual physical linkages among system components). For example, habitat corridors have been suggested as an obvious means of connecting isolated reserves or habitat patches. Corridors have become a controversial issue in conservation biology, however (Hobbs, 1992; Simberloff *et al.*, 1992; Mann and Plummer, 1995). There is limited empirical evidence regarding the efficacy of corridors and the costs may outweigh the benefits if corridors also facilitate the spread of disease or predators (e.g., Simberloff and Cox, 1987; Hess, 1994). Structural connectivity is thus no guarantee of functional connectivity.

Because landscape ecology focuses on ecological flows across landscapes, it has provided a new paradigm for thinking about landscape connectivity. Landscapes are not viewed simply as patches embedded within an inhospitable matrix, but as integrated mosaics of different habitat types, land uses, and other structural features that may facilitate or impede movement to varying degrees across the landscape (Wiens, 1997; With, 1999). The landscape-mosaic approach emphasizes the importance of defining connectivity from the perspective of the species or process of interest (e.g., Taylor *et al.*, 1993; With *et al.*, 1997). In other words, connectivity is an emergent property of landscapes, resulting from an interaction between the scale at which the process or species operates and the scale of the landscape pattern. For example, species may possess different perceptions as to whether a given landscape is connected depending upon their ability or willingness to cross gaps of unsuitable habitat (Dale *et al.*, 1994; With, 1999). Dispersal or gap-crossing abilities dictate the scales at which organisms interact with landscape pattern, and the gap or patch structure of a landscape is a function of the scales of disturbance or habitat destruction, whether natural or anthropogenic.

How can we quantify connectivity or predict when landscapes become disconnected? A number of approaches for quantifying landscape connectivity have been developed (Tischendorf and Fahrig, 2000a, 2000b; Urban and Keitt, 2001). For example, applications of percolation theory, in the form of neutral landscape models, were developed within the discipline of landscape ecology and have provided a means of modeling ecological flows across structured landscapes (Gardner *et al.*, 1987; Gardner and O'Neill, 1991). Neutral landscape models have been used to quantify when landscapes become disconnected, and thus when the functional integrity of systems may become compromised (With, 1997b; With and King, 1997; With, 2002). Landscape connectivity is predicted to be disrupted abruptly, as a threshold phenomenon, which may have dire consequences for biodiversity. Critical thresholds in landscape connectivity may not coincide with ecological thresholds, such as in dispersal success or population persistence, however (e.g., With and Crist, 1995; With and King, 1999a, 1999b). Nevertheless, landscape thresholds may precipitate other ecological thresholds, setting off a "threshold cascade." Evidence for this has been found in the relationship between landscape thresholds and thresholds in the search efficiency of biocontrol agents (biocontrol thresholds; With *et al.*, 2002). This has implications for the field of conservation biological control, which seeks to manage landscapes so as to enhance the efficacy of natural enemies in controlling pest outbreaks (Barbosa, 1998). Predicting thresholds in the ecological consequences of habitat loss and fragmentation has thus been identified as a major unsolved problem facing conservation biologists (Pulliam and Dunning, 1997).

Developing a general landscape ecological theory

Although conservation biology is viewed as having a strong theoretical framework, there has been very little theory developed specifically for conservation (With, 1997a). Conservation biology has borrowed heavily from the theoretical foundations of its parent disciplines (population genetics, population and community ecology; Simberloff, 1988). Because this theory was not developed with conservation applications in mind, however, it may contain restrictive assumptions that ultimately limit its utility for management or result in its misuse if such constraints are ignored. Some conservation biologists therefore discredit the use of theory in conservation, failing to recognize that the problem lies not so much with the theory itself as with the misapplication of theory (Doak and Mills, 1994). Furthermore, much of the ecological theory that is used in conservation biology is patch-based (e.g., metapopulation theory, theory of island biogeography), which ignores the spatial heterogeneity of real landscapes and thus offers little insight into how scenarios of land-use change might affect population persistence in managed landscapes. Geographical Information Systems (GIS) have become powerful tools in both landscape ecology and conservation biology. For example, population simulation models linked with landscape maps in a GIS can be used to evaluate extinction risk for species under different land-management plans or scenarios of land-use change (e.g., Dunning *et al.*, 1995). Such "spatially realistic models" tend to be site- or species-specific, however, and thus are not able to provide a general landscape theory.

Although landscape ecology has been criticized for lacking a theoretical foundation (Wiens, 1992), landscape ecologists have at least been able to build upon general systems theory which has given rise to hierarchy theory (Allen and Starr, 1982; O'Neill *et al.*, 1986; O'Neill, this volume, Chapter 3). This could be a useful framework for the management of complex integrated systems now targeted in conservation, particularly in contributing to an understanding of the extent to which phenomena at a given scale are simultaneously the product of processes operating at finer scales and system constraints at broader scales. In addition, there is an urgent need for a theoretical framework for assessing the impacts of landscape transformation on biodiversity. Neutral landscape models, coupled with computer simulation models of dispersal, gene flow, population dynamics, or species interactions, provide one example of how a general landscape theory might be developed (With and Crist, 1995; With, 1997b; With and King, 1999b, 2001; With *et al.*, 2002).

Using landscape design principles to guide conservation efforts

Reserve design is still primarily governed by principles derived (supposedly) from the theory of island biogeography – e.g., the debate over the advantages of "single large or several small" (SLOSS) reserves. As discussed previously, reserve systems must be developed within the context of human land-use activities. This is illustrated, for example, by UNESCO's Man and the Biosphere reserve model, in which strictly protected core areas are surrounded by buffer zones and transitional zones that allow varying degrees of research, restoration, resource extraction, recreation, and human settlement. Regional reserve networks take this concept a step further by adopting a landscape perspective that emphasizes the importance of maintaining functional connectivity (or at least structural connectivity) by the creation of broad corridors to facilitate animal movement among reserves (Noss, 1983).

Deciding where to establish reserves is another problem in landscape reserve design, which has been addressed using gap analysis to identify current gaps in the protection of biodiversity at a regional level (Scott *et al.*, 1993). Overlays of existing reserves with the distribution of species across the landscape may reveal "hotspots" of species diversity that are currently unprotected and thus vulnerable to future landscape development and human depredations. Gap analysis also provides a means of prioritizing conservation efforts and directing land acquisition and future land-use activities. What it fails to take into account is whether such areas are actually capable of supporting viable populations of these species. Species richness may be high on a landscape because the landscape is productive and therefore capable of sustaining viable populations of many species. Alternatively, high species richness may arise from the juxtaposition of various habitat types or land uses (i.e., high habitat diversity). Populations may not be viable (self-sustaining) within some or even most of these different habitats, yet persist there owing to immigration from elsewhere. Gap analysis does not discriminate between these two alternatives (Maurer, 1999).

Finally, the mitigation of land-use activities for the conservation or restoration of biodiversity can only be achieved through careful landscape planning and management (Hobbs, this volume, Chapter 22; Margules, this volume, Chapter 23). Landscape ecologists need to become more involved as active partners in the development of conservation strategies to ensure that these will be based on sound land-management and design principles.

Landscape conservation: the new paradigm?

The landscape approach to conservation involves much more than the adoption of a broader-scale, regional perspective in species or ecosystem management. One of the hallmarks or distinguishing characteristics of landscape ecology is its emphasis on how spatial pattern affects ecological processes. Subsequently, landscape ecology can be profitably applied at any scale. For example, connectivity must be assessed and managed across a range of scales, from the spatial patterning of resources or habitat required to fulfill an individual's minimum area requirements, to populations within a metapopulation, to reserves in a regional network. Landscape ecology also explicitly addresses the importance of landscape context and recognizes the mosaic nature of landscape structure. It thus affords a new perspective on connectivity and for understanding how landscape structure affects ecological processes, as well as the consequences of human land-use activities on the structural and functional integrity of terrestrial and aquatic ecosystems. Although theory development has not been a particularly vigorous activity in landscape ecology, the synthesis of neutral landscape models, based on percolation theory with ecological theory, may help contribute to a general landscape theory. This is required if a predictive science of the ecological consequences of landscape transformation is to emerge. Landscape ecology possesses the design principles necessary for effective land management and planning, and thus should play an active role in directing land-use activities and reserve design so as to benefit conservation and restoration efforts. The goal for the future should be to establish "landscape conservation" as the new paradigm for the conservation of biodiversity – not for the conservation of landscapes per se, but for conservation that is founded on landscape ecological principles (Gutzwiller, 2002).

Acknowledgments

I thank John Wiens for inviting me to contribute to this volume, thereby giving me the opportunity to explore how landscape ecological principles can contribute to the conservation of biodiversity. My research on applications of landscape ecology for the conservation of biodiversity has been supported by past grants from the National Science Foundation, and most recently by a STAR grant from the Environmental Protection Agency (R829090).

References

Allen, T. F. H. and Starr, T. B. (1982). *Hierarchy: Perspectives for Ecological Complexity*. Chicago, IL: University of Chicago Press.

Barbosa, P. (1998). *Conservation Biological Control*. San Diego, CA: Academic Press.

Bunnell, F. L. and Dupuis L. A. (1995). Conservation biology's literature revisited: wine or vinaigrette? *Wildlife Society Bulletin*, 23, 56–62.

Christensen, N. L., Bartuska, A., Brown, J. H., *et al.* (1996). The report of the Ecological Society of America committee on the scientific basis for ecosystem management. *Ecological Applications*, 6, 665–691.

Dale, V. H. and Haeuber R. A. (2001). *Applying Ecological Principles to Land Management*. New York, NY: Springer.

Dale, V. H., Pearson, S. M., Offerman, H. L., and O'Neill, R. V. (1994). Relating patterns of land-use change to faunal biodiversity in the central Amazon. *Conservation Biology*, 8, 1027–1036.

Doak, D. F. and Mills, L. S. (1994). A useful role for theory in conservation. *Ecology*, 75, 615–626.

Dunning, J. B., Stewart, D. J., Danielson, B. J., *et al.* (1995). Spatially explicit population models: current forms and future uses. *Ecological Applications*, 5, 3–11.

Edwards, T. C. Jr. 1989. The Wildlife Society and the Society for Conservation Biology: strange but unwilling bedfellows. *Wildlife Society Bulletin*, 17, 340–343.

Gardner, R. H. and O'Neill, R. V. (1991). Pattern, process, and predictability: the use of neutral models for landscape analysis. In *Quantitative Methods in Landscape Ecology*, ed. M. G. Turner and R. H. Gardner. New York, NY: Springer, pp. 289–307.

Gardner, R. H., Milne, B. T., Turner M. G., and O'Neill, R. V. (1987). Neutral models for the analysis of broad-scale landscape pattern. *Landscape Ecology*, 1, 19–28.

Gutzwiller, K. J. (ed.) (2002). *Applying Landscape Ecology in Biological Conservation*. New York, NY: Springer.

Hansen, A. J. and di Castri, F. (eds.) (1992). *Landscape Boundaries: Consequences for Biotic Diversity and Ecological Flows*. New York: Springer.

Hess, G. R. (1994). Conservation corridors and contagious disease: a cautionary note. *Conservation Biology*, 8, 256–262.

Hobbs, R. J. (1992). The role of corridors in conservation: solution or bandwagon? *Trends in Ecology and Evolution*, 7, 389–392.

Hobbs, R. (1994). Landscape ecology and conservation: moving from description to application. *Pacific Conservation Biology*, 1, 170–176.

Hobbs, R. (1997). Future landscapes and the future of landscape ecology. *Landscape and Urban Planning*, 37, 1–9.

IALE (1998). IALE mission statement. *IALE Bulletin*, 16, 1.(http://www.wsl.ch/land/lale/bulletin.php)

Jensen, M. N. and Krausman, P. R. (1993). Conservation biology's literature: new wine or just a new bottle? *Wildlife Society Bulletin*, 21, 199–203.

Liu, J. and Taylor, W. W. (2002). *Integrating Landscape Ecology into Natural Resource Management*. Cambridge: Cambridge University Press.

Mann, C. C. and Plummer, M. L. (1995). Are wildlife corridors the right path? *Science*, 270, 1428–1430.

Maurer, B. A. (1999). *Untangling Ecological Complexity: The Macroscopic Perspective*. Chicago, IL: University of Chicago Press.

Naveh, Z. (1995). Interactions of landscapes and cultures. *Landscape and Urban Planning*, 32, 43–54.

Noss, R. (1983). A regional landscape approach to maintain diversity. *BioScience*, 33, 700–706.

Noss, R. F. (1991). Landscape connectivity: different functions at different scales. In *Landscape Linkages and Biodiversity*, ed. W. Hudson. Washington, DC: Island Press, pp. 27–39.

O'Neill, R. V., DeAngeles, D. L., Waide, J. B., and Allen, T. F. H. (1986). *A Hierarchical Concept of Ecosystems*. Princeton, NJ: Princeton University Press.

Petit, L. J., Petit, D. R., and Martin, T. E. (1995). Landscape-level management of migratory birds: looking past the trees to see the forest. *Wildlife Society Bulletin*, 23, 420–429.

Pulliam, H. R. and Dunning, J. B. (1997). Demographic processes: population dynamics on heterogeneous landscapes. In *Principles of Conservation Biology*, 2nd edn, ed. G. K. Meffe and C. R. Carroll. Sunderland, MA: Sinauer, pp. 203–232.

Salwasser, H. (1991). New perspectives for sustaining diversity in US national forest ecosystems. *Conservation Biology*, 5, 567–569.

Scott, J. M., Davis, F., Csutin, B. *et al*. (1993). Gap analysis: a geographic approach to protection of biological diversity. *Wildlife Monographs*, 123.

Simberloff, D. (1988). The contribution of population and community biology to conservation science. *Annual Review of Ecology and Systematics*, 19, 473–511.

Simberloff, D. and Cox, J. (1987). Consequences and costs of conservation corridors. *Conservation Biology*, 1, 63–71.

Simberloff, D., Farr, J. A., Cox, J., and Mehlman, D. W. (1992). Movement corridors: conservation bargains or poor investments? *Conservation Biology*, 6, 493–504.

Soulé, M. E. (ed.) (1986). *Conservation Biology: the Science of Scarcity and Diversity*. Sunderland, MA: Sinauer.

Taylor, P. D., Fahrig, L., Henein, K. and Merriam, G. (1993). Connectivity is a vital element of landscape structure. *Oikos*, 68, 571–573.

Tischendorf, L. and Fahrig, L. (2000a). On the usage and measurement of landscape connectivity. *Oikos*, 90, 7–19.

Tischendorf, L. and Fahrig, L. (2000b). How should we measure landscape connectivity? *Landscape Ecology*, 15, 633–641.

Turner, M. G. (1989). Landscape ecology: the effect of pattern on process. *Annual Review of Ecology and Systematics*, 20, 171–197.

Turner, M. G. and Gardner, R. H. (eds.) (1991). *Quantitative Methods in Landscape Ecology*. New York, NY: Springer.

Turner, M. G., Gardner, R. H., and O'Neill, R. V. (2001). *Landscape Ecology in Theory and Practice: Pattern and Process*. New York, NY: Springer.

Urban, D. and Keitt, T. (2001). Landscape connectivity: a graph-theoretic perspective. *Ecology*, 82, 1205–1218.

Vitousek, P. M., Mooney, H. A., Lubchenco, J., and Melillo, J. M. (1997). Human domination of Earth's ecosystems. *Science*, 277, 494–499.

Wiens, J. A. (1992). What is landscape ecology, really? *Landscape Ecology*, 7, 149–150.

Wiens, J. A. (1996). Wildlife in patchy environments: metapopulations, mosaics, and management. In *Metapopulations and Conservation*, ed. D. R. McCullough. Washington, DC: Island Press, pp. 53–84.

Wiens, J. A. (1997). Metapopulation dynamics and landscape ecology. In *Metapopulation Biology: Ecology, Genetics, and Evolution*, ed. I. A. Hanski and M. E. Gilpin. San Diego, CA: Academic Press, pp. 43–62.

Wiens, J. A., Stenseth, N. C., Van Horne, B., and Ims, R. A. (1993). Ecological mechanisms and landscape ecology. *Oikos*, 66, 369–380.

Wilcove, D. S., Rothstein, D., Dubow, J., Phillips, A., and Lossos, E. (1998). Assessing the relative importance of habitat destruction, alien species, pollution, over-exploitation, and disease. *BioScience*, 48, 607–616.

Wilson, E. O. (1988). *Biodiversity*. Washington, DC: National Academy Press.

With, K. A. (1997a). The theory of conservation biology. *Conservation Biology*, 11, 1436–1440.

With, K. A. (1997b). The application of neutral landscape models in conservation biology. *Conservation Biology*, 11, 1069–1080.

With, K. A. (1999). Is landscape connectivity necessary and sufficient for wildlife management? In *Forest Fragmentation: Wildlife and Management Implications*, ed. J. A. Rochelle, L. A Lehmann, and J. Wisniewski. Leiden, the Netherlands: Brill, pp. 97–115.

With, K. A. (2002). Using percolation theory to assess landscape connectivity and effects of habitat fragmentation. In *Applying Landscape Ecology in Biological Conservation*, ed. K. J. Gutzwiller. New York, NY: Springer, pp. 105–130.

With, K. A. and Crist, T. O. (1995). Critical thresholds in species' responses to landscape structure. *Ecology*, 76, 2446–2459.

With, K. A. and King, A. W. (1997). The use and misuse of neutral landscape models in ecology. *Oikos*, 79, 219–229.

With, K. A. and King, A. W. (1999a). Dispersal success on fractal landscapes: a consequence of lacunarity thresholds. *Landscape Ecology*, 14, 73–82.

With, K. A. and King, A. W. (1999b). Extinction thresholds for species in fractal landscapes. *Conservation Biology*, 13, 314–326.

With, K. A. and King, A. W. (2001). Analysis of landscape sources and sinks: the effect of spatial pattern on avian demography. *Biological Conservation*, 100, 75–88.

With K. A., Gardner, R. H., and Turner, M. G. (1997). Landscape connectivity and population distributions in heterogeneous environments. *Oikos*, 78, 151–169.

With, K. A., Pavuk, D. M., Worchuck, J. L., Oates, R. K., and Fisher, J. L. (2002). Threshold effects of landscape structure on biological control in agroecosystems. *Ecological Applications*, 12, 52–65.

25

The "why?" and the "so what?" of riverine landscapes

Seeking to penetrate "the untranslatable dark," the astronomer and poet Rebecca Elson (2001) observed that "explanation is not understanding." This assertion was expanded by Ingrid Fiske (2001) in a review of the book: "Understanding comes through vigilant attention to the sensual world, through fidelity to the spirit and to the way our personal world interacts with the explanatory world of science."

Accordingly, when studying riverine systems, a key question is to know what is the relevance of the "explanatory world" of landscape ecology to understand these systems. In other words, how to answer at the same time questions such as "why?" (the explanation) and "so what?" (the significance). O'Neill and Smith (2002) remind us that hierarchy theory provides a framework for that: the explanation is related to the next lower hierarchical level, and the significance to the next higher level of organization of the systems under study.

Perhaps more than others, riverine landscapes illustrate the need to address this distinction between explanation and significance. And perhaps more than in other landscapes, this distinction relates to the two realities of landscapes: they are at the same time natural and cultural. I'd like to illustrate this on the basis of two hypotheses:

1 The hierarchical organization of riverine landscapes can be simplified to include two main levels – natural and cultural – the second level being higher than the first.
2 The interacting structures and processes that characterize riverine landscapes can be explained at the lower natural hierarchical level, but they must be understood at the higher cultural hierarchical level.

 Issues and Perspectives in Landscape Ecology, ed. John A. Wiens and Michael R. Moss. Published by Cambridge University Press.
© Cambridge University Press 2005

Defining riverine landscapes

According to the dictionary, the adjective riverine means "of or on a river or river bank." Herein, I use the term riverine landscape to indicate a holistic perspective of patterns and processes linking a river and its banks, or its riparian areas, within a fluvial system (Ward, 1998).

Researchers sometimes define and delineate riparian areas differently because a large array of life-history strategies and successional patterns determine their functional attributes via community composition, along with their environmental setting. Naiman and Décamps (1997) suggest that "the riparian zone encompasses the stream channel between the low and high watermarks and that portion of the terrestrial landscape from the high watermark toward the upland where vegetation may be influenced by elevated water tables or flooding and by the ability of the soils to hold water."

Riparian areas are unique environments because of their position in riverine landscapes (Malanson, 1993). Ecologists view them as aquatic–terrestrial transition zones (Junk *et al.*, 1989) or as interfaces between aquatic and terrestrial zones (Naiman and Décamps, 1997). Riparian habitats are created by lateral flood pulses of varying intensity, duration, and frequency, develop on alternatively erosional and depositional landforms, and are maintained by linked hydrological, geomorphologic, and biological processes. They run along stream networks over important linear distances, varying in width from simple narrow ribbons to complex, enlarged, and diversified alluvial forests.

Depending on drainage conditions, regional hydrography outlines patterns of riparian areas within an overall matrix. These patterns differ in sinuosity, degree of fragmentation, width, and area/perimeter ratio. They vary according to the scale on which they are perceived, from corridor-like elements within the surrounding matrix to transversal gradients away from the nearby river. Moreover, riparian areas have aesthetic and recreational values as well as social and economic values. For example, they are places for livestock grazing and forest harvest, they maintain water quality and bank stability, and they provide environmental services such as enhancing diversity of habitats and of species.

At the same time, riverine landscapes are among the most dominated by human societies, those where the interaction of nature and culture is most developed. As such, they are affected by, and affect, human perception, cognition, and values of landscapes (Nassauer, 1995). Their sustainability depends primarily on attention and care by people, which demands that ecological functions are clearly signaled to societies (Nassauer, 1992). Such a cultural understanding is fundamental to safeguard ecological health, as well as to imagine possible action plans for conservation, restoration, or creation of new riverine landscapes.

Explaining shifting habitat mosaics

The originality of riverine landscapes comes from their dynamics, which depend largely on the hydrological regimes of neighboring streams. Floods regularly reshape the banks, creating shifting mosaics of overlapping plant communities. These mosaics depict the various stages that form riparian plant successions from pioneer to mature communities.

Mature floodplain forests are remarkable elements of riverine landscapes. They may be kilometers wide along the lower reaches of large rivers where annual water fluctuations of 4–5 m are not rare, and may be as much as 10–15 m in the Amazon. Along such areas, relict point bars, levees, and channels often result in a ridge and swale topography. In natural conditions, this broken spatial structure of floodplain forests affects processes which can themselves have positive feedback as well. Such an apparent conflict between structure and process is at the origin of the renewal of floodplain forests after disturbance. Explaining this renewal requires a landscape perspective that involves hydrology, geomorphology, and ecology. It also requires one to link studies on processes to those on patterns.

Flooding is at the root of the formation of many landforms on floodplains through the processes of erosion and deposition. Point bars appear to be the key landforms in the establishment of regeneration of floodplain forests. They shelter increased numbers of species, although proliferation of one species can occur in certain years. Thus the development of forests proceeds through intense primary successions at meander points, leading to a sequential successional forest. At the same time, lateral erosion at the outer curves of the meanders leads to the formation of mosaic and transitional forests (Salo *et al.*, 1986). A sharp contrast distinguishes forest dynamics in the active zone from those in the rest of the floodplain.

This contrast is illustrated by many European and North American riverine landscapes: forests within the active band are regenerated through allogenic processes such as hydrologic events, whereas forests outside the active band are regenerated through autogenic processes such as competition and gap dynamics. A comparison of the different profiles of floodplain forests suggests that allogenic and autogenic types of regeneration are probably more intermingled than has been generally reported (Décamps, 1996). Firstly, lateral erosion and channel changes may repeatedly disturb all types of forest, resulting in mosaics of closed forest patches which differ in age, structure, and turnover time. Secondly, interactions between species may affect the response of floodplain forests to hydrological disturbances at any level along floodplain profiles. As a consequence, plant succession in floodplains results from complex interactions between stochastic processes, life-history traits, and inhibitory and facilitative effects. Investigating these interactions is necessary in order to explain the shifting habitat mosaics that characterize floodplain forests. Concentrating on mechanisms which link water, land-forms, and species in different landscape settings is also necessary to predict the effects on floodplain forests caused by manipulating flow.

Coming back to the title of this essay, this is where "why" questions are not entirely separated from "so what" questions. Rather, understanding is progres-sively built upon explanations as illustrated by the hierarchical classification of streams in space and time. The framework provided by Frissell *et al.* (1986) sustains a systematic approach for explaining and understanding the natural variability of riverine landscapes (Fig. 25.1). Their approach assumes a habitat-centered view of ecological systems. It assumes also that the structure and dynamics of stream habitats are determined by the surrounding catchment. In such a framework, different spatiotemporal scales define various stream systems and habitat subsystems. For example, riverine landscapes develop in floodplain and reach system levels, encompassing distances from 10^2 to 10^3 m and time periods from 10^1 to 10^3 yr. They may be affected by processes such as aggradation or degradation associated with large sediment-storing structures, bank ero-sion, and riparian plant succession. This allows an integrated and holistic view of riverine landscapes that may guide researchers and managers in conceiving protocols for conservation and restoration (Stanford *et al.*, 1996).

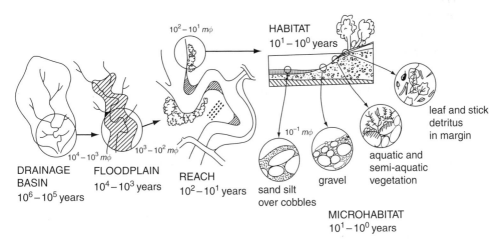

FIGURE 25.1
Hierarchical organization of a stream and its habitat subsystems (Pinay et al., 1990; adapted from Frissell et al., 1986).

Understanding the spirit of the place

"Of one thing at least I am certain: that not to take myth seriously in the life of an ostensibly 'disenchanted' culture like our own is actually to impoverish our understanding of our shared world" (Schama, 1995). Few landscapes are so constructed by imagination, so impregnated with the spirit of the place as riverine landscapes, whether these are the Mississippi (Twain, 1883), the Nile (Schama, 1995), or the Danube (Burlaud, 2001).

Every river on earth, every reach of river, has its identity, distinct from every other river, every other reach. This identity – or spirit of the place – comes from natural distinctive features, as well as from a cultural reading that continuously renews these features (Fig. 25.2). As a result, no two riverine landscapes are the same. This poses critical questions for landscape architects, designers, and managers in general. How to conjure up the spirit of the place? How to revive an old one? How to invent the future of a landscape on the basis of its present potential?

Fascinating in this respect are the emergence of "ecosymbols" from a relationship between humans and their terrestrial area (Berque, 1995), the inventive analysis applied by Lassus (1998) to create a new spirit of the place along the river Charente at Rochefort in France, or the design and planning developed in the Mediterranean context by Makhzoumi and Pungetti (1999). Fascinating also is the cultural sustainability advocated by Nassauer (1997) on the basis of aesthetic expectations that rest upon ecological health. Such approaches are necessary to understand riverine landscape (Décamps, 2001).

FIGURE 25.2
A valley in Lebanon: the spirit of the place in a Mediterranean landscape. Original drawing by Jala Makhzoumi.

An historical approach to the effect of human societies on riverine land-scapes is also necessary. In the Mediterranean area, centuries of land and water use have created unique landscapes (Vita-Finzi, 1969). Around 4000–3000 years BP some Cretian landscapes appeared already as mosaics of cultivated fields, orchards, and semi-natural exploited woodlands. In fact, the first significant deforestations, about 8000 BP, increased with the expansion of human populations, breeding and agricultural practices. They slowed down after the fall of the Roman Empire in the fifth century and restarted during the medieval times, with ups during periods of high natality and downs during periods of high mortality due, for example, to the bubonic plague in the fourteenth century. The power of the naval forces of Spain and Portugal between the fifteenth and sixteenth centuries was built on a regular deforestation of the Iberian Peninsula, particularly along the coasts and main rivers. At the present time, after extensive cuttings during the nineteenth and twentieth centuries, the Mediterranean forest is recovering along the European seacoast.

Besides land use, water use has always been a major concern in the Mediterranean area. Survival of people, livestock, and cultures has depended on water collection and storage capacity. There are still many remains of a surprising savoir-faire that culminated in Roman times with the construction of dams, aqueducts, and various devices for water transfer. Such remains obviously contribute to the identity of Mediterranean riverine landscapes, and to their understanding.

Improving our forecasting ability

A main issue for the coming decades is to improve our forecasting ability about riverine landscapes. Changes in climate, land and water use, human populations, technologies, and economic activity are affecting riverine landscapes everywhere in the world (Naiman, 1996). We need to anticipate the consequences of these changes if we are to deal with them (Clark *et al.*, 2001). This requires one to better explain and understand the dynamics of riverine landscapes.

Monitoring at the regional scale is a first requisite to explain changes in riverine landscapes. This means that we need to organize and sustain data networks over large catchments, for example to get an adequately distributed knowledge of precipitation, stream heights, and discharges. Similarly, we need to get a spatialized knowledge of the effects of dams and irrigation, as well as of habitat loss. Remote sensing and large-scale experiments are useful tools in this context, helping to identify the "slow variables" that constrain successional change (Carpenter *et al.*, 1999).

Interdisciplinary exchange is a second requisite to understand changes in riverine landscapes (Décamps, 2000). Our knowledge of riverine landscapes is indeed fragmented between approaches developed independently within the natural and the social sciences, the humanities, or the arts. Now, changes in these landscapes simultaneously create environmental, social, cultural, aesthetic, and economic issues. A common theoretical foundation is clearly needed to articulate these various disciplinary perspectives. The transdisciplinary systems approach recently proposed by Tress and Tress (2001) holds promise as it unites a landscape as a spatial entity, a mental entity, a temporal dimension, a nexus of nature and culture, and a complex system. To use the words of these authors, it is time to capitalize on plurality in landscape research. It is time to get involved in a holistic conception of our landscapes (Naveh, 2001).

In addition to explaining and understanding riverine landscapes, there is an urgent need for landscape researchers to be connected to the processes of conservation and restoration. Environmental problems facing riverine landscapes will be solved only through a dialogue between the various approaches of landscape research (including landscape ecology), policy makers, managers, and the general public. This dialogue is the third requisite I'd like to mention – a requisite for action. It does not mean that landscape ecologists must conserve and restore by themselves; it means that they should find their role in a decision-making process. They have a lot to offer stakeholders in estimating uncertainties, developing possible scenarios, and communicating the potential consequences of extreme events. However, they have to make their own distinctive contribution to solutions, in concert with the perspectives of the other approaches or interest groups (Risser, 1999; Wiens, 1999).

Placing landscape ecology

Riverine landscapes cannot be understood without an interdisciplinary approach linking natural and human sciences. However, there is an unwise and a wise use of such an approach. An unwise use could be to subordinate one disciplinary approach to another; a wise use is to reciprocally recognize the uniqueness of every disciplinary approach.

Landscape ecology is essential to explain how spatial configuration interacts with ecological processes in riverine landscapes. These interactions are particularly unstable, requiring the study of spatial and temporal heterogeneity at a variety of scales and the use of concepts and methods coming from the fields of geography and ecology. Thus a landscape ecology of riverine landscapes may appear as a hybrid discipline. Far from being a weakness, this character reinforces its ability to support the creation of new riverine systems by landscape architects. It is important to realize that this support, although necessary, is not enough: it is necessary because every landscape is a reality determined by laws of nature; it is not enough because every landscape refers also to subjective myths and symbols and culturally determined perceptions.

Landscapes are at the same time natural *and* cultural. This is why creating and anticipating the dynamics of new riverine landscapes require a concert of approaches and perspectives. A landscape ecology of rivers will be all the better if it finds its place in such a concert.

Acknowledgments

I am grateful to Jala Makhzoumi, Robert J. Naiman, and Bärbel and Gunther Tress for helpful discussions and comments. Jala Makhzoumi kindly drew the sketch of a Mediterranean riverine landscape in Lebanon (Fig. 25.2).

References

Berque, A. (1995). *Les Raisons du Paysage de la Chine Antique aux Environnements de Synthèse*. Paris: Hazan.

Burlaud, P. (2001). *Danube Rhapsodie: Images, Mythes et Représentations*. Paris: Grasset/Le Monde.

Carpenter S. R., Brock, W., and Hanson, P. (1999). Ecological and social dynamics in simple models of ecosystem management. *Conservation Ecology*, 3, 4 (online). www.consecol.org/vol3/iss2/art4.

Clark, J. S., Carpenter, S. R., Barber, M., *et al.* (2001). Ecological forecasts: an emerging imperative. *Science*, 293, 657–660.

Décamps, H. (1996). The renewal of floodplain forests along rivers: a landscape perspective. *Verhandlungen Internationale Vereinigung Limnologie*, 26, 35–59.

Décamps, H. (2000). Demanding more of landscape research (and researchers). *Landscape and Urban Planning*, 47, 103–109.

Décamps, H. (2001). How a riparian landscape finds form and comes alive. *Landscape and Urban Planning*, 57, 169–175.

Elson, R. (2001). *A Responsibility to Awe*. Manchester: Carcanet.

Fiske, I. (2001). The poetic mystery of dark matter. *Nature*, 414, 845–846.

Frissell, C. A., Liss, W. J., Warren, C. E., and Hurley, M. D. (1986). A hierarchical framework for stream habitat classification: viewing streams in a watershed context. *Environmental Management*, 10, 199–214.

Junk, W. J., Bayley, P. B., and Sparks, R. E. (1989). The flood pulse concept in river-floodplain systems. *Canadian Special Publication of Fisheries and Aquatic Sciences*, 106, 110–127.

Lassus, B. (1998). *The Landscape Approach*. Philadelphia, PA: University of Pennsylvania Press.

Makhzoumi, J. and Pungetti, G. (1999). *Ecological Landscape Design and Planning*. London: Spon Press.

Malanson, G. P. (1993). *Riparian Landscapes*. Cambridge: Cambridge University Press.

Naiman, R. J. (1996). Water, society, and landscape ecology. *Landscape Ecology*, 11, 193–197.

Naiman, R. J. and Décamps, H. (1997). The ecology of interfaces: riparian zones. *Annual Review of Ecology and Systematics*, 28, 621–658.

Nassauer, J. I. (1992). The appearance of ecological systems as a matter of policy. *Landscape Ecology*, 6, 239–250.

Nassauer, J.I. (1995). Culture and changing landscape structure. *Landscape Ecology*, 9, 229–237.

Nassauer, J. I. (1997). Cultural sustainability: aligning aesthetics and ecology. In *Placing Nature: Culture and Landscape Ecology,* ed. J. I. Nassauer. Washington, DC: Island Press, pp. 65–83.

Naveh, Z. (2001). Ten major premises for a holistic conception of multifunctional landscapes. *Landscape and Urban Planning*, 57, 269–284.

O'Neill, R. and Smith, M. (2002). Scale and hierarchy theory. In *Learning Landscape Ecology: a Practical Guide to Concepts and Techniques*, ed. S. E. Gergel and M. G. Turner. New York, NY: Springer, pp. 3–8.

Pinay, G., Décamps, H., Chauvet, E., and Fustec, E. (1990). Functions of ecotones in fluvial systems. In *The Ecology and Management of Aquatic-Terrestrial Ecotones*, ed. R. J. Naiman and H. Décamps. Paris: UNESCO and Parthenon, pp. 141–169.

Risser, P. G. (1999). Landscape ecology: does the science only need to change at the margin? In *Landscape Ecological Analysis: Issues and Applications*, ed. J. M. Klopatek and R. H. Gardner. New York, NY: Springer, pp. 3–10.

Salo, J., Kalliola, R., Häkkinen, I., *et al.* (1986). River dynamics and the diversity of Amazon lowland forest. *Nature*, 322, 254–258.

Schama, S. (1995). *Landscape and Memory*. London: Fontana.

Stanford, J. A., Ward, J. V., Liss, W. J., *et al.* (1996). A general protocol for restoration of regulated rivers. *Regulated Rivers*, 12, 391–413.

Tress, B. and Tress, G. (2001). Capitalising on multiplicity: a transdisciplinary systems approach to landscape research. *Landscape and Urban Planning*, 57, 143–157.

Twain, M. (1883). *Life on the Mississippi*. Boston, MA: Osgood.

Vita-Finzi, C. (1969). *The Mediterranean Valleys*. Cambridge: Cambridge University Press.

Ward, J. V. (1998). Riverine landscape: biodiversity patterns, disturbance regimes, and aquatic conservation. *Biological Conservation*, 83, 269–278.

Wiens, J. A. (1999). The science and practice of landscape ecology. In *Landscape Ecological Analysis: Issues and Applications*, ed. J. M. Klopatek and R. H. Gardner. New York, NY: Springer, pp. 371–383.

PART VI

Cultural perspectives and landscape planning

26

The nature of lowland rivers: a search for river identity

Rivers have, more than almost any other unanimated object, an animated gesture, some-thing resembling character   Macaulay (1838) describing the Rhône (see Schama, 1995)

River rehabilitation, on what scientific basis?

In doing research on a river, by discovering more and more of its secrets, the observer will come nearer and nearer to its identity. Every river ecologist has his or her favorite river mainly because of its character. This individual character or identity, however, is difficult to translate into scientific terms. Since we are educated to mistrust our subjectivity in science, personal impressions are generally kept for artists and general conversation. Can river identity be approached in a more objective way by making use of objective personal impressions?

In integrated river management in western Europe, scientific, technological, and political developments have led to an understanding that the immense social chances and constraints related to river management should be approached in a systematic and interactive way. A clear delineation of rehabilitation targets for nature should enhance unbiased public and scientific discussion of these opportunities and constraints. The aim of this essay is to explore the scientific dimension of river rehabilitation and to survey the possibility of using personal impressions as an instrument to approach river identity. The main focus is on lowland rivers, as illustrated by the Meuse.

River rehabilitation commonly aims at increased biodiversity or improved connectivity. Biodiversity, as such, has no meaning unless it is related to a coherent network of habitats. In fact, the indicators biodiversity and connectivity together make up the identity of a river reach. The ecological potential of

Issues and Perspectives in Landscape Ecology, ed. John A. Wiens and Michael R. Moss. Published by Cambridge University Press. © Cambridge University Press 2005.

a river can be used as a guideline to achieve this identity in a more expressive way. The ecological potential of river and floodplain ecosystems generally deviates from a historical reference point because river regulation, land use, diminished water quality, water-quantity management, and even climatic change have fundamentally transformed the boundary conditions for ecological development in lowland rivers in temperate climate zones in the last few centuries – especially so in the twentieth century. Just the rehabilitation of natural values known from any historical or earlier reference situation is therefore impossible. A purely historical reference point for nature rehabilitation is seldom adequate. And because every river is unique in its natural and social setting (see Schama, 1995; von Königslöw, 1995) the adoption of a geographical reference point, that is, a virtually untouched river with comparable characteristics elsewhere, can never define a perfect example.

In discussions of river restoration, an interesting change is gradually taking place in the way river ecologists and hydrologists are consulted. Traditionally, ecologists have been mainly engaged in the safeguarding of any remaining or threatened natural values and in the prediction of negative environmental impacts on these values. This has greatly enhanced nature conservation. In recent years the question is often posed in a different way: what targets should be used for nature's rehabilitation (Pedroli *et al.*, 2002)? Landscape ecology is challenged to give a new scientific basis for river rehabilitation: how can the identity of a river be defined in a way that can guide its rehabilitation? In the following discussion this is illustrated by the River Meuse.

The Meuse, a river rich in history

Impressions of the Lorraine Meuse

The River Meuse is a lowland river flowing from northeastern France through Belgium and the Netherlands to the North Sea (Fig. 26.1). It is a beautiful river, connecting the age-old cultural landscapes of the Lorraine plains and Ardennes hills, through the urban conglomerations of Liege and Maastricht, to the Rhine delta in the Netherlands. Part of the upper reach of the Meuse is still in quite a natural state. Commercial and recreational navigation take place here at a low intensity and mainly on constructed channels which date back to Napoleonic times. This leaves the original course to natural processes and ecosystem development, and to kayaks and sport fishermen.

When you enter the original course of the Meuse, somewhere in Lorraine, just downstream of a weir, the first few kilometers are often characterized by rapids where a kayak will touch the pebbles. Small islands form in the bed, some of them covered with annual plant species, others with one-year-old

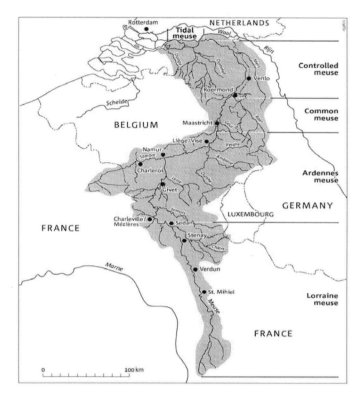

FIGURE 26.1
Catchment of the Meuse.

willows or poplars. Other islands are characterized by huge ruins of old willows, undermined by the rapidly flowing river water. The riverbed lies mostly between meadows with croplands, and sometimes wooded hills, farther away. Inner curves have gently sloping, sandy, or even clayey banks; outer curves have steep to straight walls. The latter harbour numerous sand martins (*Riparia riparia*), while the former are generally accessible for cattle that often stand halfway into the water, staring at you thoughtfully. When you look down into the clear water, you see waving water plants, and numerous small fish flee away astonishingly quickly. You smell the water and flowering herbs and grasses. Over the river you often see birds of prey. In early August you can see black kites (*Milvus migrans*) gathering for migration, sometimes 40 of them spiralling upwards majestically above wheat fields and gliding away southward. In this river section you can suddenly find yourself in a small channel passing by a weir, with kingfishers (*Alcedo atthis*) cruising under overhanging trees. I learnt that such channels often end at a sawmill, which means carrying your kayak over the sawmill factory premises back to the river. Hydropower is not used any more.

Below the mill, again a stretch of rapids begins, this time beside a village where children play in the water. Farther downstream, the river flows calmly toward the junction with the navigation channel. Here a wealth of water plants colonizes the nearly standing water, which is enriched by ground-water. The water is bordered by a margin of trees, mostly alder (*Alnus* sp.) and willow (*Salix* sp.). From the river, remote hills can be seen, covered by wheat fields and woodland, and sometimes a village, a castle, or a monastery.

For other sections of the river, considerably different characterizations can be given. For the sake of brevity I refer to the summary table (Table 26.1).

Impressions of the controlled Meuse

The controlled Meuse is a river section in the Netherlands, dominated by wide cloudy skies and black-and-white cows in green meadows. A few large locks and dams are present; groynes and bank protection allow for a reliable navigation route for ships of up to 2000 tons. Sailing down this stretch, the banks are mostly low and uniform rip-rapped edges of the meadows are lined with sparse poplar (*Populus* sp.) cultivars. Only a few places remain with natural vegetation or apparent erosion/sedimentation processes. Sharp bends have been straightened. Nearly 95% of the time water levels are around regulated levels, and thus agricultural use is possible nearly everywhere on the floodplain. Few alluvial ecosystems remain, but some valuable hedgerow landscapes are present on the floodplains. Water quality is far from optimal and underwater visibility is poor. Salmon and trout are absent, although formerly abundant. Some sand pits are in use as recreational lakes. It is mainly common birds that can be observed, while in the migration season migratory birds rest on and along the river. Villages and some castles face toward the water, whereas in the towns the river cannot be felt as a dominant presence.

Historical notes

From prehistoric times several civilizations left traces in the upper Meuse catchment, although they are not functionally linked with the Meuse itself. From Roman times on, however, the Meuse valley has played a distinct role in history, beginning with the river's important transportation function. In the early Middle Ages, monks from Ireland and Scotland founded mon-asteries in the area. In the churches of the Meuse region, especially in Liege, some fine examples of Christian art and classical thinking from AD 1000 have been preserved. In the following period, Gothic developments concentrated more toward the west and the Meuse region suffered in many wars fought

Table 26.1. Characteristics of the Meuse sections

Characteristic	Lorraine Meuse	Ardennes Meuse	Common Meuse	Controlled Meuse	Tidal Meuse
Average precipitation	800–900 mm	980 mm (max 1400)	775 mm	740 mm	740 mm
Geomorphology	wide valley, clay and sand	narrow valley, gravel wide	incised valley, gravel	wide valley, sand	narrow valley, clay
Drainage basin	narrow	wide	medium wide	medium wide	narrow–medium wide
Soil	calcareous, permeable	impermeable rock, near Liège gravel	gravel and sand	gravel and sand	sand and clay
Anthropogenic adjustments	navigable derivation canals	Meuse is canalized	Meuse cuts deep due to canalization	Meuse is canalized, artificial lakes	Meuse is canalized, artificial diversion
Side branches	present	hardly present	hardly present	some	not present
Islands	present	decreasing	not present	not present	not present
Floodplain	present	hardly present	within limits	within limits	within limits
Riffles	present	hardly present	present	hardly present	hardly present
Main tributaries	Mouzon, Vair	Semois, Sambre, Lesse, Ourthe, and others	Jeker, Geul	Roer, Swalm, Niers, Donge, Dieze	Dommel
Navigation with max. cargo	not on Meuse, 350 tons on Canal de l'Est	1,350–2,000 tons	not on Meuse itself	2,000 tons	2,000 tons
Power generation	low head; nuclear plant	low head; nuclear plant	—	low head; coal plant	—
Population density	low (Verdun, Sedan, Charleville–Mézières)	low in south, high in north (Charleroi, Namur, Liège, Visé)	high (Maastricht)	high (Roermond, Venlo)	low near river
Industry	metal, paper/cardboard, foodstuffs	heavy & metallurgic industry, fertilizer, soda	chemical industry	conventional power plants	—

Characteristic	Lorraine Meuse	Ardennes Meuse	Common Meuse	Controlled Meuse	Tidal Meuse
Mining	(sand)	gravel	gravel/sand	gravel/sand	—
Sport fishery	very frequent	frequent	frequent	frequent	frequent
Recreation	increasing	boating	increasing	on lakes	little
Main land use	agriculture/forestry	forestry	agriculture	agriculture	agriculture
Current natural values	high	low	medium	moderate	moderate
Potential natural values	high	medium	high	medium	medium

between French and Germanic invaders. Many churches along the Meuse were fortified in these times; these are still evident. Jeanne d'Arc (fifteenth century) is a famous heroine from this region. In the Renaissance, a revival of artistic creativity can be observed; for example, the sixteenth–century sculptures of Ligier Richier of St. Mihiel. In the seventeenth century the area was again a battlefield in recurrent wars. Almost all the towns along the Meuse, but especially St. Mihiel, Verdun, Stenay, Sedan, Charleville-Mézières, and Givet, were fortified and played significant roles in these battles – especially the First World War on the battlefields around Verdun. The region has now gradually recovered. The dominant economic activities are now agriculture (wheat and cattle) and forestry. Transport on the river still functions, although at a modest level (350 tons maximum).

The controlled Meuse also has its history, but one much less pronounced than the upper Meuse. Roman remains are found at several places. But this region apparently functioned in the shadow of developments along the upper Meuse and the lower Rhine. Agriculture dominated, and still dominates, the land use, with the church as an important landlord. Navigation has always been a function of the river, especially connecting the upper and middle Meuse with the Rhine. Until the twentieth century fisheries were important, both for eel and for salmon and trout. Clay extraction supported brick factories, with their typical tall chimneys, all along this stretch of the Meuse.

How to appreciate "river identity"

From the description of the Lorraine Meuse it is evident that both the traditional and the more recent values (e.g., for recreation) have become integrated to a considerable extent. Could this image be used as an example for the controlled Meuse? The latter currently serves mainly as a discharge channel for water and cargo, allowing for economic development along its banks. Recent flood events, however, have proved that the Meuse is still a living river, at times generating considerable damage to newly built houses, enterprises, and infrastructure. Currently, new guidelines are being sought for river management and restoration.

Comparison between the two river sections described prompts questions concerning the concept of river identity, since the Lorraine Meuse could readily be seen as the ideal reference for the controlled Meuse. These two sections are, however, only comparable to a certain degree: the identity of the river is multidimensional. To base target images for nature rehabilitation on this multidimensionality, decision makers and politicians will, however, require a reduction in scope.

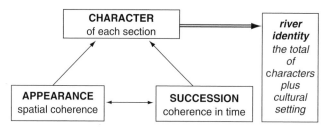

FIGURE 26.2
Appearance, succession, and character of a river as stages in identification.

The observations as described together give a firm, yet imprecise, personal impression of the river. I presume this is the way it works for every researcher. What makes it worthwhile is how to make use of it. To put the observations in order I propose a gradual approach to the river's identity, one leading through appearance, succession, and character as described below (Fig. 26.2).

Appearance: spatial coherence

Interestingly, a river cannot be described only from one point of view. It becomes an image as soon as the observer has combined, in his mind, observations of many sites which belong together. Young islands with willow seedlings are inseparable from the eroded banks at the next bend, and pools and riffles downstream of a weir belong to the same system as the quiet standing water in the backswamps. Some parts may be sandy, others clayey or gravely, some banks steep and others gently sloping, some flowers red and others yellow. But together they characterize the same section. These are the phenomena as they appear physically, but we have to bring about spatial coherence ourselves. Just as an individual tree produces a richer image, in ourselves, when we observe it from several sides, so too is the river's image richer and multifaceted when filled with diverse observations in a spatially coherent framework.

Succession: coherence in time

Another dimension is coherence in time. From items like plastic bags or straw in the tree branches, we can deduce that periods of high discharge also occur. The age of the seedlings on gravel islands tells us of flooding events in the past. Observing rivers at high-discharge stages creates a strong impression. But so does observing the river during all seasons, or even during one day. It strongly enriches our experience of a river section. At this stage, the question also arises whether the source of a river represents its past – since the water originates from the source – or its future – since water from the source will flow into the future. This dimension, in the arrangement of observations, yields an image that is constantly in motion. The same upper Meuse exhibits many different faces during the day, the seasons, the years. Phenomena observed are continually in transition, like the water itself. It requires an active effort of thought to build up a conscious image of this unsteady but nonetheless characteristic picture.

Character: the combination of appearance and succession.

The character of the river can be seen as a combination of aspects of appearance and features of succession, brought together in one's mind. For every section of the river this character is different, resulting in different processes, and in the plants and animals present; an upper course, middle course, and lower course can be differentiated. This is reflected in plants and animals, in the behavior of the water and in the river's banks and floodplains. The character is known about a river to anyone who knows it well. The inhabitants of the region know the difference between the Lorraine Meuse and the Ardennes Meuse. It can even be communicated between us, without requiring its quantitative characteristics such as discharge rates, length, gradient, etc.

At this stage, it helps to identify the character of river sections by using conceptual summaries. In general, in the upper course of a river, I could speak of a "powerful play of dissolving processes," whereas in the lower course a "steady enrichment of life" takes place. The character of the middle course can be generalized as a "flowing by-pass": transport processes play a dominant role. Of course, these conceptions are not exclusive and they are – depending on discharge stage and scale of detail – relevant in all sections, but they may inspire the composition of a target image of specific river sections as a whole. The Lorraine Meuse, in this sense, has more the character of a middle course than of an upper course; still the mineral-rich groundwater being fed to the river in this section is clearly an upper-course element.

River identity

Why is the Meuse a different river from, for example, the Marne? In both rivers very comparable physical phenomena can be observed, comparable processes play a role, and a comparable character may be attributed to the sections identified. But still these rivers differ completely from each other. Just as no landscape is identical to another landscape, every river has its own identity. It is in the specific composition of the character of the sections that the identity of a river is defined. The Meuse's biography is characterized by flowing a long distance on the gently sloping plains of northeastern France, then crossing the Ardennes, flowing out in the lowlands and into the delta near Rotterdam. The Marne has its source in the same area as the Meuse, but flows through the gentle Champagne hills toward the Paris Basin, and is in fact a tributary to the Seine, which in an estuary merges with the sea.

Moreover, it is also the cultural appreciation of the river that determines, to a large extent, the identity of a river (Antrop, 2000). Whether the river has this influence on society, or society on the river, is an unsolved question (Schama, 1995). The fact is that the Champagne and its Gothic cathedrals give the Marne a completely different expression than the Meuse with its meadows and fortified medieval churches. At the confluence of the Marne and Seine, Paris had a huge influence on the use of the river, giving it a special status for the transport of grains and wine. The lower course of the Meuse is dominated by Liege and Maastricht, and farther down Rotterdam, but ongoing traffic was always hampered by the gravel shallows downstream of Maastricht. Moreover, the river Meuse flows through three European states: France, Belgium, and the Netherlands. By tradition, each of these countries has a specific river-management style, which did not enhance a coherent development of the river as a whole.

Humans are inseparably associated with river landscapes. Thus, to find target images for river rehabilitation, we must find those images that are realistic and which refer to natural physical processes and to the variation of those processes in time, and also to the changes society has brought about and which, in most instances, are irreversible. Even if reversed, completely different situations would result because of the changes in political boundaries. The following section gives an example of implementing the approach outlined above.

The Meuse, artery of nature?

Application

The following is a description of an attempt to identify the type of nature that can develop under certain conditions in the floodplain of the Meuse. I will concentrate on sections in the Netherlands (see Postma *et al.*, 1995), because that is where the call for nature rehabilitation is the strongest – and is, in fact, most needed from an ecological point of view. In this example, the type of natural elements are expressed in areas (hectares) of "ecotopes," defined as spatial ecological units with uniform morphodynamic and hydrodynamic characteristics and a vegetation structure that either has resulted from land use (e.g., grazing or pasture) or is in a natural state.

Geomorphologic and hydrological processes determine the development of ecological units such as ecotopes. The Meuse in the Netherlands can be divided into three main sections (see Table 26.1 and Fig. 26.1), each with more-or-less uniform geomorphologic characteristics. For each of the river sections, a first estimate of recognizable ecotopes was derived from topographic maps produced in 1850 at a scale of 1 : 50000. Although physical processes in Dutch rivers have changed radically during the past century as a result of human interaction, the analysis of historical patterns gives a good deal of insight into river dynamics under varying conditions. These elements help clarify coherence in time, that is, the potential for succession. Also, images from the upper Meuse are of help, confirming the dimension of appearance, that is, spatial coherence. This information, on the historical situation and on recent features, was then combined and analyzed in a qualitative way ("expert judgment") to identify which geomorphologic and hydrological processes and which ecotopes still have the intrinsic ecological potential to develop. This intrinsic ecological potential corresponds with the "character" of the section, as described above. This is expressed quantitatively in ecotope distribution. Imagine, then, that with the exception of levees, no societal functions were supported by the river – no navigation, agriculture, or infrastructure in the floodplain. What, then, would be the resulting character of the river, the resulting ecosystems? This is referred to as the reference model (Figs. 26.3, 26.4).

These sections of the river, as described, are middle-course sections, with water flow and sediment transport as characteristics. It appears, for example, that large parts of the active floodplains of the Meuse will turn into floodplain forests if a natural development under current conditions were allowed. Forests, however, tend to raise water levels because they hinder rapid runoff. To prevent altering flood design levels, forest development on active floodplains should not be allowed on a large scale unless there is a compensatory increase in hydraulic resistance, achieved by restoring secondary channels, for example. The ecological potential to develop under existing prescribed conditions of acceptable flood risk (that is, along Dutch dike-protected rivers not exceeding the 1/1250 design flood) is referred to as the rehabilitation target model (Figs. 26.3, 26.4).

For a more realistic picture of the river landscape under existing requirements for flood safety and major infrastructure works, some restrictions and conditions were defined. For example, floodplain levels should be lowered compared from their present silted-up situation, and the proportion of forests should be locally decreased. This results in a realistic restoration

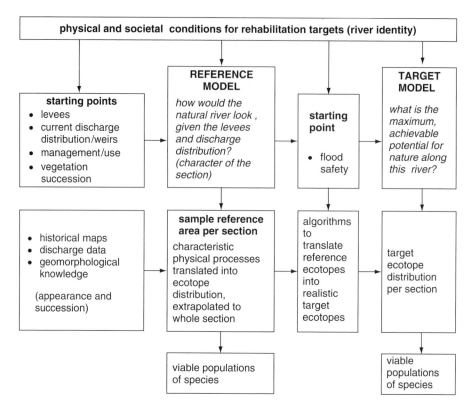

FIGURE 26.3
Approach of reference and target models for nature rehabilitation. After Pedroli et al. (1996).

objective, defined in terms of ecotope distribution at the scale of the river section.

Perspective: scenario analysis

Considering ecotope distribution, and given the rehabilitation target model, the effects of planning alternatives, such as reducing agricultural production in favor of semi-natural grazing, can be compared with distribution under the target model.

Given a certain configuration and distribution of ecotopes, based on the intrinsic ecological potential of the particular river section, it is possible to apply a simple habitat-evaluation procedure for selected plant and animal species. Based on the predicted ecotopes, the potential carrying capacity for characteristic river-related species has been estimated (Postma *et al.*, 1995). Not only would the total area of ecotopes (cover types or habitats) then determine the return of species, but the distribution of ecotopes over the

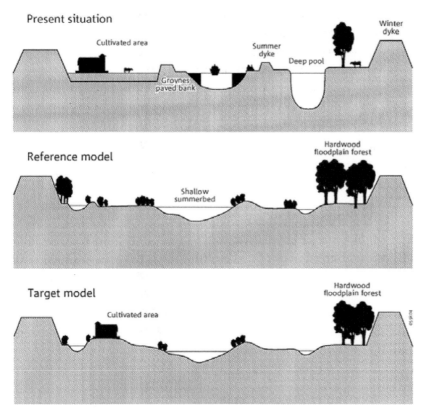

FIGURE 26.4
Cross-sections of the river Meuse: present situation, reference model, and rehabilitation target model. From Postma et al. (1995).

length of the river floodplain must meet their ecological network requirements. Foppen and Reijnen (1998) have developed an instrument to analyze the sustainability of species populations at different spatial patterns of ecotopes. It appears that spatial patterns can determine, to a large extent, the viability of species populations.

Conclusion

If the historical physical processes of a river floodplain cannot be restored, and a historical reference for the river and for its natural condition is not available, then an alternative reference should be chosen, one that is based on historical or natural river dynamics. A phenomenological approach would be adequate here (see Pickles, 1985), one which concentrates on the potential development of possible ecosystems under specific societal conditions. The scientific dimension of river rehabilitation is, however, not restricted to the prediction of the effects of the proposed measures. The processes used to arrive at a certain objectivity of river identity can, however, be treated in a scientific manner. This allows for consideration of personal impressions of phenomena observed, if these have been consciously structured. This then also allows for public involvement, since the definition of river identity in specific cases, and therefore of reference and rehabilitation target models, can generate fruitful discussion.

In the example referred to above, no attempt is made to reconstruct the vegetation and ecosystem types considered typical for that particular river. The major guideline is to allow the river to create its own ecosystems, starting from the river dynamics currently present, or attainable under current conditions of river regulation upstream, and within given environmental-quality ranges. This confidence in intrinsic ecological potential allows for a combination of efforts with third parties. In the case of the Meuse, this could be with gravel, sand, and clay mining operations that would also give projects a sound economic and financial basis. Little effort has been put into predicting the exact results of river rehabilitation in terms of numbers of plant and animal species or individuals. The focus is more on creating sound physical boundary conditions for ecosystem development, as expressed in terms of ecotope distribution.

Let me return, finally, to Macaulay's "animated gesture" of the river. Goethe, in his scientific work, tried to remain consciously connected with directly observed phenomena in nature when seeking to discover the intrinsic "gesture" (*Urbild*) or response of phenomena (Bortoft, 1996; Bockemühl, 1997). It is a challenge to follow this example in landscape ecology in those issues relating to the rehabilitation of nature. Under any circumstances it means regularly, and faithfully, returning to personal observation in the field.

Acknowledgments

This essay could not have been written without the enthusiastic support of Roeland A. Bom in observing and interpreting field phenomena along the rivers Marne and Meuse. Annejet Rümke gave valuable comments on an earlier draft of this essay.

References

Antrop, M. (2000). Where are the Genii Loci? In *Landscape : Our Home / Lebensraum Landschaft*, ed. B. Pedroli. Zeist: Indigo, pp. 29–35.

Bockemühl, J. (1997). Aspekte der Selbsterfahrung im phänomenologischen Zugang zur Natur der Pflanzen, Gesteine, Tiere und der Landschaft. In *Phänomenologie der Natur*, ed. G. Böhme and G. Schiemann. Frankfurt am Main: Suhrkamp, pp. 149–189.

Bortoft, H. (1996). *The Wholeness of Nature: Goethe's Way Toward a Science of Conscious Participation in Nature.* New York, NY: Lindisfarne.

Foppen, R. P. B. and Reijnen, R. (1998). Ecological networks in riparian systems: examples for Dutch floodplain rivers. In *New Concepts for Sustainable Management of River Basins*, ed. P. H. Nienhuis, R. S. E. W. Leuven, and A. M. J. Ragas Leiden: Backhuys, pp. 132–139.

Pedroli, B., De Blust, G., van Looy, K., and Van Rooij, S. (2002). Setting targets in strategies for river restoration. *Landscape Ecology*, 17, 5–18.

Pedroli, G. B. M., Postma, R., Kerkhofs, M. J. J., and Rademakers, J. G. M. (1996). Welke natuur hoort bij de rivier? *Landschap*, 13, 97–113.

Pickles, J. (1985). *Phenomenology, Science and Geography*. Cambridge: Cambridge University Press.

Postma, R., Kerkhofs, M. J. J., Pedroli, G. B. M., and Rademakers, J. G. M. (1995). *Een stroom natuur, Natuurstreefbeelden voor Rijn en Maas*. Ministerie van Verkeer en Waterstaat, project Watersysteemverkenningen, RIZA nota 95.060. Arnhem: RIZA. (In Dutch, summary in English.)

Schama, S. (1995). *Landscape and Memory*. London: HarperCollins.

von Königslöw, J. (1995). *Flüsse Mitteleuropas: Zehn Biographien*. Stuttgart: Urachhaus.

27

Using cultural knowledge to make new landscape patterns

Human interactions with ecological systems are typically described as impacts. Thinking of culture not only as the source of impacts but also as the source of clues to what motivates human behavior may help us integrate human effects into landscape ecological research and action. We can simulate and model the landscape ecological effects not only of current trends but also of distinctly different futures. Motivations may be difficult to change, but the particular behaviors that disturb, pollute, and consume landscapes may be malleable to the extent that human needs, including cultural preferences and desires, continue to be met (Bailly *et al.*, 2000).

For example, two very different land-use behaviors, sprawl and urban habitat restoration, may be motivated by similar needs. Both sprawl, the large-lot development pattern that has spread from metropolitan farmland to scenic rangeland and wildlands, and habitat restoration of abandoned urban industrial sites may fulfill the desire to live close to nature (Strong, 1965; Grove and Cresswell, 1983; Nelessen, 1994; Hough, 1995; Nassauer, 1995; Romme, 1997; Nasar, 1998). Sprawl disturbs habitats, pollutes water and air, and consumes agricultural land. Urban habitat restoration establishes small patches that may have aggregative effects across the larger landscape matrix (Collinge, 1996; Corry and Nassauer, 2002). If we understand the desire to live close to nature as part of what motivates people to choose to live on large lots far from traditional centers of cities and towns, we can propose different ways to meet the same perceived need. We can ask ourselves: what are ecologically beneficial substitutes for ecologically destructive behavior? What new landscape patterns would be improvements compared with present landscape patterns if they continue into the future?

People are not inherently averse to improvement. In fact, most intentional landscape change, from the eighteenth-century enclosure movement in England to the post-war rise of suburbia in the United States, has been

 Issues and Perspectives in Landscape Ecology, ed. John A. Wiens and Michael R. Moss. Published by Cambridge University Press.

understood and advocated as improvement. In contrast, people generally are unwilling to deny their desires and needs. Telling people to do less of what they are doing can sound more like piety than a plan. New landscape patterns that are immediately recognizable as improvements will be seen as real alternatives to present landscape trends.

Recognition is not automatic. It requires that what is new match what people know they value, and culture provides the clues to recognition. What people value is not surprising. People want to feel safe, they want to feel healthy, they want to take care of their children, they want to be proud of where they live, they want to get along with their neighbors, and they want to make a living (Nelessen, 1994; Nasar, 1998). Perhaps because these values are so common, they are sometimes unexamined. How particular cultures make these values concrete in the landscape is the source of our clues for designing new landscape patterns. We need to understand what people recognize when they look at the newest subdivision. What do people see that they want when they look at "Mountain Creek," or "Brookfield Farms," or any other development named to evoke the image of home that we desire? We can respect people's values at the same time as we invent new ways to fulfill them. The new should be familiar, looking like nature and home, at the same time as it is fundamentally new in the way it embodies ecological function (Nassauer, 1997).

The initial precepts of landscape ecology suggest how we can approach inventing new landscapes that accommodate human needs and also embody ecological function. Landscape ecology includes human behavior in ecosystems; it attends to inhabited as well as pristine ecosystems; it studies ecological function across landscapes at multiple scales including scales of everyday human experience; and it is interdisciplinary (Risser *et al.*, 1984). Fulfilling these precepts requires landscape ecologists to continue to experiment with our ways of working.

The best cultural indicators of landscape ecological quality may not be readily available numbers, like the economic and demographic data we have gathered for decades. The science and scholarship practised by environmental psychologists, cultural geographers, design behaviorists, and environmental historians have examined causes rather than only trends in human landscape perception and behavior. Knowledge about causes allows us to realign trends toward normative goals. For example, we know that people seek landscapes that afford perceived opportunities for the display of pride to others (e.g., Lowenthal and Prince, 1965; Nassauer, 1988; Gobster, 1997; Nasar, 1998; Westphal, 1999), rest and psychological restoration (e.g., Kaplan, 1995), safety (e.g., Schroeder and Anderson, 1984; Nasar, 1993; Bailly *et al.*, 2000; Ness and Low, 2000), information and locomotion (e.g., Lynch, 1960; Gibson, 1979;

Kaplan and Kaplan, 1982; Golledge and Stimson, 1987), prospect and refuge (Appleton 1975), and closeness to nature (e.g., Grove and Cresswell, 1983; Kaplan and Kaplan, 1989; Gobster and Westphal, 1998; Gobster, 2001). Humans will seek landscapes that are designed and planned to protect and enhance ecological function if they also provide these apparent opportunities.

To propose new landscapes, we need to be able to make good judgments and good hypotheses about how they will function ecologically. We need ecological data and models that describe subdivisions and cities as well as forest patterns and reserves. Landscape ecology has given us the strongest basis for judging the ecological function of settled landscapes to date, but we know our understanding is dramatically incomplete (Peck, 1998). By working together in our shared medium, the landscape, landscape ecologists of several disciplines can propose new landscapes for experimentation and for action.

One example of this kind of new landscape is a model 120-ha subdivision for the city of Cambridge, Minnesota (Nassauer *et al.*, 1997). The city of 5700 within the expanding commutershed of Minneapolis–St. Paul, a metropolis of 2.5 million, wanted this model to inform its negotiations with developers who see a burgeoning market for new homes. Using our best understanding of the evolving ecological principles in landscape ecology and seeking the critique and insights of our ecology and hydrology colleagues, we proposed a form of subdivision that was both familiar and radically new (Fig. 27.1). To developers and homebuyers, the unusual ecological function of this new subdivision would likely be of little immediate value. However, the familiar cultural cues that are apparent in the landscape pattern would be of immediate value. We designed the landscape to be a source of pride, to look well cared for, to create a sense of ownership, to look safe, to be legible, to afford prospect and refuge, to create a feeling of closeness to nature. We also designed it to include affordable housing, to be accessible by public transportation, to provide public access to high-amenity landscape features, and to minimize infrastructure costs. Improving surface- and groundwater quality and increasing habitat quality, connectivity, and extent were our leading goals, but not the leading goals for homebuyers or the developer. We designed with ecological goals and cultural means.

Compared with a large-lot subdivision designed under a typical ordinance intended to maintain rural character with 10-acre (4-ha) lots (Fig. 27.2), this plan provides more than 15 times as many homes on the same area at lower net costs to taxpayers. It keeps all homes close to nature and keeps the most high-amenity landscape, the lakeshore, open to public access. Compared with the typical plan, this plan creates greater connectivity and habitat patch size and restores some lake edge habitats. By cleaning storm water through detention and infiltration, and developing at a sufficiently high density to

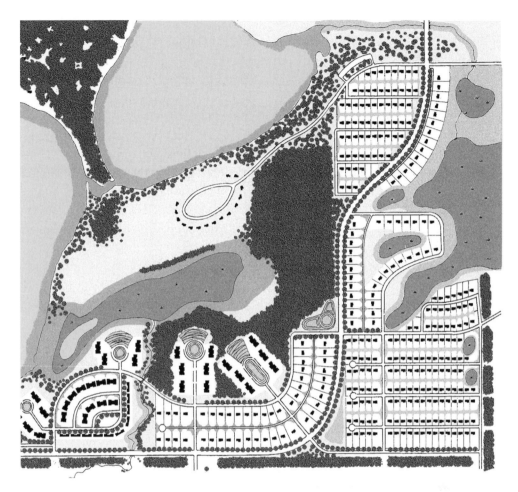

FIGURE 27.1.
Ecological corridor neighborhood design plan: reconnects heterogeneous
ecosystems, cleans storm water before it reaches wetlands, and includes affordable
housing within a mix of types of sewered residential development.

make extension of the municipal sewer system economical, this plan produces higher water quality than the large lots on septic systems and wells
(Fig. 27.2).

Will the extended and connected patches of woodland, storm-water wetlands, "natural" wetlands, and lake shown in Fig. 27.1 support greater
biodiversity than the "present trend" development shown in Fig. 27.2?
Could we have hypothesized a different pattern that would have had a greater
landscape ecological benefit? Will developers and homebuyers recognize the
familiar cues to cultural values that were built into this design? This example
demonstrates the necessity for both biophysical and cultural knowledge to
inform new landscape actions. It also implies the wide-ranging possibilities

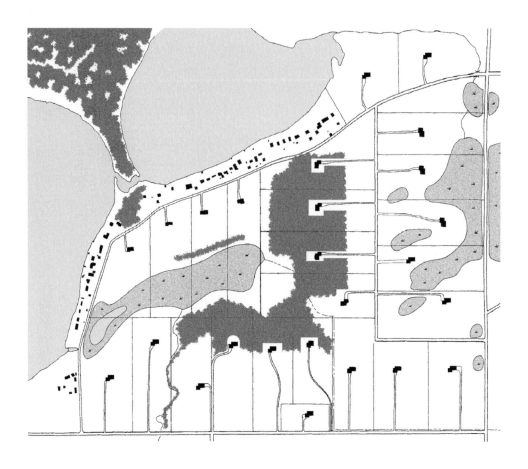

FIGURE 27.2.
A conventional development alternative: further fragments in situ ecosystems, provides housing for 0.06 the number of households in Fig. 27.1, and does not use available local sewer capacity.

for more generalizable experiments that propose and test new prototypes for culturally recognizable and ecologically beneficial landscape structure.

In her 1998 presidential address to the American Association for the Advancement of Science, Jane Lubchenco called for a redirection of American science – away from the single-discipline basic science that was geared toward national defense in the years immediately following the Second World War and toward a new social contract for science that will "help society move toward a more sustainable biosphere," a science that "exercises good judgment, wisdom, and humility." Such a science should look for strategic intersections with culture, as examined by the humanities and social sciences and also as interpreted by design and planning. Strategy does not need to be a compromise of our sense of ecological integrity or our sense of human satisfactions. However, strategy does imply normative change; it moves us toward goals. By definition, landscape ecology can help to define goals for human interactions with ecological systems. It also can identify strategies for achieving those goals by passing ideas for new landscapes between disciplines, so that each can examine and rework those ideas from particular disciplinary perspectives.

We do need to know more about culture just as we need to know more about ecosystems, but we cannot afford to wait. We can begin to act by looking at what we know now in a different way. We should see culture not as a constraint but as a means for landscape innovation. If culture is the means by which humans achieve our needs (sometimes in convoluted and misguided ways), then it also can be the medium for inventing new forms of human settlement that support ecological function. We should study culture not only to predict what will happen if current trends continue but also to conceive what motivates people to change landscapes. What needs are met by skyscrapers and subdivisions, by factory farms and seaside resorts? Rather than accepting these settings as the inevitable detritus of human frailty, we should study them as the incomplete realization of human aspirations, and use our understanding of human needs and landscape ecological function to propose new landscape patterns.

References

Appleton, J. (1975). *The Experience of Landscape.* New York, NY: Wiley.

Baill, A. S., Brun, P., Lawrence, R. J, and Rey, M. C. (eds.) (2000). *Socially Sustainable Cities: Principles and Practices.* UNESCO MOST project. London: Economica.

Collinge, S. K. (1996). Ecological consequences of habitat fragmentation: implications for landscape architecture and planning. *Landscape and Urban Planning*, 36, 59–77.

Corry, R. C. and Nassauer, J. I. (2002). Managing for small patch patterns in human-dominated landscapes: examples in Corn Belt agriculture. In *Integrating Landscape Ecology into Natural Resource Management*, ed. J. Liu and W. Taylor. Cambridge: Cambridge University Press, pp. 92–113.

Gibson, J. J. (1979). *The Ecological Approach to Visual Perception.* Boston, MA: Houghton-Mifflin.

Gobster, P. H. (1997). Perceptions of the oak savanna and urban ecological restorations. In *Proceedings of the Midwest Oak Savanna Conference, February 20, 1993, Northeastern Illinois University, Chicago, IL*, ed. F. Stearns and K. Holland. Chicago, IL: US EPA. www.epa.gov/ecopage/upland/oak/oak93/gobster.html.

Gobster, P.H. (2001). Visions of nature: conflict and compatibility in urban park restoration. *Landscape and Urban Planning*, 56, 35–51.

Gobster, P. H. and Westphal, L. M. (1998). *People and the River: Perception and Use of Chicago Waterways and Recreation.* Milwaukee, WI: National Park Service. Rivers, Trails, and Conservation Assistance Program.

Golledge, R. G. and Stimson, R. J. (1987). *Analytical Behavioural Geography.* Beckenham, Kent: Croom Helm.

Grove, A. B. and Cresswell, R. (1983). *City Landscape: a Contribution to the Council of Europe's European Campaign for Urban Renaissance.* London: Butterworths.

Hough, M. (1995). *Cities and Natural Process.* London: Routledge.

Kaplan, R. and Kaplan, S. (1989). *The Experience of Nature.* Cambridge: Cambridge University Press.

Kaplan, S. (1995). The restorative benefits of nature: toward an integrative framework. *Journal of Environmental Psychology*, 15, 169–182.

Kaplan, S. and Kaplan, R. (1982). *Cognition and Environment: Functioning in an Uncertain World.* New York, NY: Praeger.

Lowenthal, D. and Prince, H. C. (1965). English landscape tastes. *Geographical Review*, 55, 186–222.

Lubchenco, J. (1998). Entering the century of the environment: a new social contract for science. *Science*, 279, 491–497.

Lynch, K. (1960). *The Image of the City.* Cambridge, MA: MIT Press.

Nasar, J. (1993). Proximate physical cues to fear of crime. *Landscape and Urban Planning*, 26, 161–178.

Nasar, J.L. (1998). *The Evaluative Image of the City.* Thousand Oaks, CA: Sage.

Nassauer, J. I. (1988). Landscape care: perceptions of local people in landscape ecology and sustainable development. *Landscape and Land Use Planning*, 8, 27–41. Washington, DC: American Society of Landscape Architects.

Nassauer, J.I. (1995). Culture and changing landscape structure. *Landscape Ecology*, 10, 229–237.

Nassauer, J.I. (1997). Cultural sustainability. In *Placing Nature: Culture and Landscape Ecology*, ed. J. Nassauer. Washington, DC: Island Press, pp. 65–83.

Nassauer, J. I., Bower, A., McCardle, K., and Caddock, A. (1997). *The Cambridge Ecological Corridor Neighborhood: Using Ecological Patterns to Guide Urban Growth.* Minneapolis, MN: University of Minnesota.

Nelessen, A. C. (1994). *Visions for a New American Dream: Process, Principles, and an Ordinance to Plan and Design Small Communities.* Chicago, IL: Planners Press, American Planning Association.

Ness, G. D. and Low, M. M. (2000). *Five Cities: Modelling Asian Urban Population–Environment Dynamics.* Singapore: Oxford University Press.

Peck, S. (1998). *Planning for Biodiversity: Issues and Examples.* Washington, DC: Island Press.

Risser, P. G., Karr, J. R., and Forman, R. T. T. (1984). *Landscape Ecology: Directions and Approaches.* Illinois Natural History Survey, Special Pub. 2. Champaign, IL: IHNS.

Romme, W. H. (1997). Creating pseudo-rural landscapes in the mountain west. In *Placing Nature: Culture and Landscape Ecology*, ed. J. Nassauer. Washington, DC: Island Press, pp. 139–161.

Schroeder, H. W. and Anderson, L. M. (1984). Perception of personal safety in urban recreation sites. *Journal of Leisure Research*, 16, 178–194.

Strong, A. (1965). *Open Space for Urban America.* Washington DC: US Urban Renewal Administration, Department of Housing and Urban Development.

Westphal, L. M. (1999). Growing power: social benefits of urban greening projects. Doctoral dissestation, University of Illinois at Chicago.

The critical divide: landscape policy and its implementation

Forecasts made in planning policy are rarely achieved in the practicalities of local application, and the case for landscape conservation is no exception. The critical divide between landscape policy developed by upper-tier government agencies and the implementation of those conservation measures at a local level is a phenomenon common to many locations. A specific case of this divide was studied in Ontario, Canada over a span of time between the passing and defeat of one planning act and the introduction of another. Through a series of interviews conducted with both the creators and the future implementers of the landscape policy in those acts, central issues that contribute to conservation resistance were examined. This qualitative study compares the responses, identifies the differences, and in the end suggests strategies that may be useful to other jurisdictions to help foster a better land-use planning environment for landscape interpretation, use, and protection in the development process.

The concept of landscape: theory and application

"Landscape" is an idea that has a long tradition in academic literature (Sauer, 1925; Hartshorne, 1939; Hoskins, 1969; Meinig, 1979; Cosgrove, 1984; Schama, 1995). Interest in the concept's utility for planning has grown in the last decade (Mitchell *et al.*, 1993; Maines and Bridger, 1992; Watson and Labelle, 1997; Cardinall and Day, 1998; Rydin, 1998; McGinnis *et al.*, 1999). It has been acknowledged that it can serve as a basis from which planners can integrate natural and cultural elements and issues – historically, two realms polarized from each other (Olwig, 1996). And as a ubiquitous resource it exists as the common ground between various interests in land-use development decisions (Stilgoe, 1982; Jackson, 1984). In this Canadian study, "landscape" was explored in its broadest interpretation from natural

Issues and Perspectives in Landscape Ecology, ed. John A. Wiens and Michael R. Moss. Published by Cambridge University Press.

system (McHarg, 1969; Stilgoe, 1982; Forman and Godron, 1986) to cultural heritage (Daniels and Cosgrove, 1988; Hunt, 1991); from aesthetic experience (Barrell, 1972; Rapoport, 1982; Schauman, 1988; Bourassa, 1991) to economic resource (Gold and Burgess, 1982; Fram and Weiler, 1984; Bolton, 1992); and finally as a place of diverse inhabitants with divergent expectations for the landscape's future (Relph, 1976; Pocock, 1981; Duncan and Ley, 1993).

It was the 1995 introduction of this term "landscape" into land-use planning legislation for the Province of Ontario, Canada, that presented an opportunity for study. In that year, for the first time in Ontario's planning history, landscapes were defined as significant visual and cultural resources by virtue of a proclaimed "Provincial Interest" that was attached to the new Planning Act. For a brief nine months the policy remained as a potentially powerful component in a newly drafted planning act.

A subsequent provincial election with a resultant change of government and a radical shift in ruling ideology had the effect of emasculating the new planning legislation. The new, more conservative government wanted to "streamline" provincial development; planning policy was now forged with business concerns paramount. Landscapes, along with other environmental resources, were given less protection.

In the revised 1996 Planning Act (Ontario Legislative Digest Service, 1996) landscape protections were transformed from powerful "Provincial Interests" to advisory guidelines. In addition, protection for the landscape no longer included conservation of visual resources, making provisions only for cultural landscapes. These vestiges of the government's landscape policy were also moved from the compulsory items in the legislation – the "shall" items – to the best-practice suggestions – the "should" items. The result has been a diminishment of scope and influence for the landscape protections in the land-use legislation.

Study method

This qualitative research was centered on three rural municipalities located along the Grand River corridor in southwestern Ontario. It was based on a total of 40 in-depth interviews with the provincial authors of both versions of the acts (planners, administrators, architects, and landscape architects) and with local planning agents who were to be responsible for the implementation of that policy (politicians, developers, heritage conservationists, municipal planners, and citizen advocates). The same three areas of inquiry were pursued with all participants: the nature of landscape (how they defined it) with the question, "What do you mean by 'landscape'?";

NATURE

Landscape as nature

Natural elements were understood at both levels but the most profound understanding was at a local level.

Landscape as culture

Cultural elements also understood at both levels with the deepest meaning for locals.

Landscape as aesthetic

Expression of Landscape Aesthetic more articulate and emotional at local level with upper tier more concerned with evaluating, defining, and inventorying views for protection.

Landscape as resource

Landscape variously identified for its value from mineral extraction, tourism, waste disposal, residential, industrial, and agricultural uses at both levels. If land is not designated it is considered "blank," ready for development.

Landscape as place

Especially distinctive for locals but difficult to define even though it has great potential to motivate planning efforts.

PERCEPTION

Schism exists between the natural and cultural at both levels. Landscape represents their unification.

Cultural landscapes valued for uniqueness and integrity, effected by local conservation traditions.

Landscape views variously valued, making it difficult to reach consensus about importance and what action to protect.

Different objectives between levels regarding profits and sustainability. Locally the vision tends to be shorter-range economic benefits but they are the ones to deal with impacts of resource development, and the balancing of "progress" and conservation.

Places are elusive yet distinct for locals. Once identified it is clear who exists "inside" and "outside" this place.

REPRESENTATION

Upper-tier conservation policy is necessarily abstract; locally it must be more concrete where boundries drawn and environment is balanced with the economy.

Landscape's potential as a planning tool is recognized at a provincial level but locals are less convinced of its efficacy.

Protection of views brings the battle between public rights and private property to the fore with planning's role to reach a satisfactory compromise.

Use assigned to landscape influenced by perceived capacity to "absorb" use within acceptable parameters of change. Promised technological interventions, lobbying, and market forces affect the level of acceptance for this change.

Landscape places and their protection are a powerful motivation for local people to get involved in local planning initiatives.

FIGURE 28.1
Interpretation of landscape idea.

their perception of landscape (how they valued it), with "What landscapes need protection?"; and the representation of landscape (what actions should be taken, in both the private and public realm), with "What measures do/ would you use to protect these landscapes?"

The transcripts were coded, and through grounded theory (Glaser and Strauss, 1967; Strauss, 1987; Strauss and Corbin, 1990; Ely, 1991; Mitchell *et al.*, 1993; Silverman, 1993; Neuman, 1994; Lincoln and Denzin, 1995; Rubin and Rubin, 1995) patterns emerged that were interpreted within two frameworks that dealt with the landscape idea (Fig. 28.1) and landscape planning (Fig. 28.2). From these codes and themes a final narrative was constructed – a story about the conservation of this complex landscape heritage.

The divide expressed in this interpretation helps to explain, in part, the ultimate demise of the 1995 Planning Act and the subsequent diminishment of landscape regulations in the second act. Beyond this conceptual division, the study points to policy and planning actions that could help foster a better environment for this slippery but potentially powerful planning concept. In the final analysis, it becomes clear that the human dimensions of conservation are paramount and must be understood if any protections in development are to come to fruition.

NATURE

Policies

Subjective experience of landscape requires flexible policy that locals reluctant to embrace as hard to defend and enforce as "softer" planning item.

Scope

Planning must accommodate spatially larger and temporally dynamic nature of landscape for effective stewardship.

Connections

Landscape has great promise for holistic planning that links jurisdictions, communities, and different government levels.

Planning roles

Landscape planning requires increased valuing of local knowledge, and new planning roles for planners, non-governmental organizations, administrators, and developers.

Change

"Good" landscape planning requires change to landscape valuing and who a landscape "expert" really is when a landscape is being planned (scientists and bureaucrats to individual residents).

PERCEPTION

Policy seen to be written by urbanites for rural situations causing "us/them" resistance. Acceptance of policy affected by legislative precedent, public review, and track record of past policies.

Must recognize the potential of landscape as a "home" to unite people across geographic, economic, and societal barriers (political, income, age, cultural, etc.).

Connections by confronting institutional schisms in natural/cultural conservation; fear of conservation; perceived threats to private rights and elitism of past conservation; and governmental mistrust.

Localizing of planning fits with trends to smaller governments, increased volunteerism, and professed commitment to local empowerment.

Political "will" constantly changing dependent upon ideologies of governments in power and the importance placed by community on landscapes. Policy can move scales toward the larger public good.

REPRESENTATION

Policy can be too specific and exclusive, or too loose and meaningless to truly represent and protect landscapes. Policies also must achieve balance between environmental and economic agenda.

Planning Act is not the proper tool for this broad landscape concept; other forms and combinations of policy and action should be sought.

Still role for upper and lower tiers: one for broader scope and connections in landscape and other to address landscape specifics. An intermediary regional jurisdiction may be best.

Local players must play a more significant role in landscape planning through effective and innovative public participation.

Changes needed to policy strength, admissibility of "soft" landscape issues, and an improved recognition of the emotional/spiritual as well as physical aspects of landscape – the cultural and natural.

FIGURE 28.2
Dimensions of effective landscape planning.

The landscape idea divide

From this research it became evident that the biggest challenge of using the term "landscape" in planning is that people cannot agree upon what it is (Schama, 1995). If something is to be protected in the land-use planning process – a process that deals with physical units and zones – something must be identified, bounded, and measured. These are concepts which landscapes confound. Interviewees' views on landscape ranged from positivistic Cartesian notions of mathematical abstractions and Linnaean classifications (McHarg, 1969; Brooke, 1994) to those perhaps more in line with the belief of the Romantics, who saw the landscape as an aesthetic (Laurie, 1975; Crandell, 1993; Schein, 1997), or humanists like Tuan (1979), Pocock (1981), and Crouch (1990), characterizing it as a subjective (Levi-Strauss, 1970; Kaplan, 1987), symbolic (Rowntree and Conkey, 1980; Penning-Rowsell and Lowenthal, 1986), and metaphysical experience (Porteus 1990; Brassley, 1998).

This diversity was also found in the manner in which study participants expressed their ideas of landscape (see Fig. 28.1).

Nature: natural and cultural

The interviewees described a basic division that occurs intellectually and institutionally around the landscape idea – a gap between culture and nature. A division characterized by John Sheail as the "Great Divide" (Sheail, 1988) is also frequently referred to as the dualism of science and humanism (Karetz, 1989), subjectivity and objectivity (Sandercock and Forsyth, 1992), and the country and the city (Pugh, 1990). This polarity also leads to separate fields of research institutions (arts and sciences), different valuing of knowledge ("softer" social issues and "harder" scientific facts), and a governmental organization and programming that fragments into separate silos, one for natural conservation and the other for cultural heritage (e.g., Britain's Sites of Special Scientific Interest and Areas of Outstanding Natural Beauty). The act that was the focus of this study also reflected this division: one policy for landscape views (B13 – "Significant Landscapes, Vistas, and Ridgelines"), and one for the cultural dimensions of the landscape (B14 – "Cultural Heritage Landscapes and Built Heritage"), both written by the Ontario Ministry of Citizenship, Culture, and Recreation. However, the policy on landscape views was written by this ministry with reluctance ("because it's not particularly focused on human heritage"; Pollock-Ellwand, unpublished study transcripts). It was believed that the Ministry of Natural Resources should have authored it. This debate testifies to the persistence of the natural–cultural divide in landscape (Sheail, 1988; Olwig, 1996).

The study participants' comments also revealed another divide that reflects the perennial power struggle between economic and environmental forces – another kind of natural and cultural divide. One local developer in the study put it succinctly, saying he saw landscape protections as impediments to making money: "They're [provincial planners] taking this unilateral approach ... it shall not be developed ... they just fight it ... and industry just shuts down." A provincial participant had a different view on landscape protections: "I feel just because a piece of property won't grow corn ... it doesn't mean it should grow houses" (study transcripts).

In fact, it is difficult to use the fuller meaning of landscape in a scientific model of management and in a land-use planning process that demands bounded ideas. Experts aligned with the rational, scientific, and objective point of view are most often called upon in the decision-making process. In the study, however, there was an impression expressed by participants that non-experts, with their irrational, emotional, and subjective perspective, best understand cultural aspects of the landscape (Pollock-Ellwand, 1997). Yet their richer, subjective, and "softer" knowledge of landscape is devalued against the "harder" scientific and economic measure. One must consider what chance landscape conservation has in a land-use planning system dominated by a strong economic imperative, even though it is clear that landscape conservation carries many more benefits than just economic ones – environmental, genetic, aesthetic, psychological, recreational, and social.

Perception: insiders and outsiders

The second major divide in landscape understanding revolved around the position of the observer as either the "insider" (Cosgrove, 1984) or the "outsider". This was expressed in the study as the "insider" long-term resident and the "outsider" newcomer. Tensions exist. Outsiders, typically ex-urbanite, contend that they have a greater appreciation for the rural place they have come to live in than the complacent locals (Seamon, 1981). Locals in turn say that the newcomers' expectations are inappropriate in rural settings in regard to level of servicing; and that outsiders want to preserve the pastoral ideal at all costs (Bunce, 1994) instead of promoting agricultural and industrial opportunities. These exurbanites are portrayed as, "Lord and Lady Plush Bottom who have free time…don't work" (study transcripts). They force their own conservation agenda over that of the longer-term residents who make their living in the environment and whose sustained welfare may be dependent on landscape change. Stereotypically, the insiders were aligned with pro-development and the call for lower taxes; the outsiders, as the elite, were concerned more with private amenities, pleasant vistas, and arcadian settings (Pollock-Ellwand, 1997).

The comparison of local and provincial study participants shows that local people are knowledgeable and connected with the specifics of landscape. They would quickly pick up a pen and paper to draw a map of *their* landscape, describing in rich detail the landscapes that they intimately know.

However, it must be noted that, even at a distance and removed from the specific local landscape, many provincial participants also eloquently expressed their connections to their own landscape memories. One bureaucrat talked emotionally about the loss of landscapes; he felt it was an "assault on your fantasy world" (study transcripts). In essence, all landscape experience is subjective. Policy makers, out of necessity, have adopted the mantle of rational respectability that comes from the long-entrenched traditions of classification criteria, GIS mapping, and rational planning analysis.

This study underlines this divide – "insiders" and "outsiders" at odds – both believing vehemently in their own reality. Yet, in spite of these fundamental differences, one common theme did emerge – the need to transcend these divisions to build community so landscape could be more effectively conserved.

Representation: theory versus practice

The division between the theoretical and the applied in landscape planning also caused much angst. Practical concerns were often expressed at the local level. One municipal planner, seeing the difficulties of working with the ambiguous landscape concept, deemed it an "extra" in the planning process – an "information item" placed into documents to satisfy government bureaucrats. After that, one gets on with the real job of development.

The meeting of the concrete with the abstract in conservation presents many obstacles. In the study, this was variously described with the difficulties of drawing a line around a classified landscape, reaching community consensus around designations, protecting areas that residents consider to be ordinary and everyday, and having to say "no" to your neighbors who want to develop within a significant landscape. Herein lies the critical divide – seeing landscape as a superfluous planning piece or a new and substantial horizon in development and conservation. The literature and this study show that landscape does represent an opportunity, but before the "promise" of this resource can be embraced, the gap in understanding around the concept must be bridged.

Bridging the divide for effective landscape conservation

Exposing the conceptual divide, this study also revealed some strategies to improve the status of landscape in land-use decision making.

Policy

Conservation policies are official expressions of intent, created to guide protection and development. Often the language employed in these important documents is complicated and can distance the common person from the conservation act. To facilitate local action these policies have to be written in an accessible manner so that those who live in and experience these areas are not alienated by bureaucratic language or terms that are too precise and exclude their own particular landscape interpretation. The best approach would be to represent these landscapes in conceptual terms, describing values that people may invest the landscape with. Terms such as identity, security, pride, and continuity represent more effective and inclusive language (Young, 1990).

Legislation that is descriptive yet succinct is most effective when supported by regulations that are not too voluminous. Too much information can also discourage action. This was the case in Ontario for the first act, which was accompanied by over 400 pages of guidelines. A municipal planner bluntly said at the time, "I think that they all should be burned!" (study transcripts). Policy therefore, can equally damage or nurture the conservation of landscapes.

Scope

Both local and provincial participants questioned the efficacy of the act in dealing with landscape intangibles, stewardship management, and the bluntness of zoning tools for comprehensive landscape conservation. Ironically, the study concluded that "landscape" is too ambiguous to be used in land-use planning. The reality is that this kind of planning policy is only one of many avenues to conservation.

Any planning legislation should be viewed as one part of a group of multi-faceted, community-based landscape conservation strategies. Raymond Williams (1973) went even further, seeing the challenge to landscape conservation as much more than a mere alteration of policy. He saw the real challenge as being the economic system that pits the tangible against the intangible.

Connections

Regardless of the conceptual divide, landscape's potential to connect different jurisdictions, communities, and physical areas was recognized as a holistic approach to planning. A landscape view of the world thwarts the "islands of green" mentality. Landscape theorists speak of the appeal of a larger landscape or regional perspective in ecological health terms (Forman and Godron, 1986) as well as in social equity dimensions (Bookchin, 1992; Plant and Plant, 1992; Sales, 1992). Landscape, in fact, embodies the antithesis of an elitist agenda (Lowenthal, 1985) where injustices inherent to existing land-use planning practice can be addressed. It is a common resource, a habitat, where diverse groups have a vested interest (Relph, 1976).

The impediments to this regionalism are that jurisdictional boundaries are normally aligned to political idiosyncrasies, not natural divisions. One interviewee talked about the advantages of regional administration for landscapes as an intermediate scale between local and upper levels, avoiding planning duplication between smaller municipalities, consolidating development, and avoiding fragmentation of tax structures and tourism efforts. Study participants went further, saying that these new landscape divisions should be based on watersheds such as those that already exist with Ontario's conservation authorities. However, for such a dramatic transformation to occur, both natural and social benefits have to be recognized and changes need to come to both planning roles and societal attitudes.

Planning roles

A shift in scope, connections, and policies means a change for the planner to become more of a facilitator and less of a "doer." Reflecting trends now apparent in planning literature, from Arnstein (1969) to Innes (1998) and Innes and Brooher (1999), interviewees felt that landscape planners should foster all voices in the community and building capacity. With this shift of power, from the "expert" to the citizen, more effective ways must be found to engage the public – from the beginning of the conservation process, when information can be gathered from visioning exercises, cognitive mapping, and oral histories (Sheail 1988; Innes and Brooher, 1999). Clearly, it is the local people who have the deepest knowledge of the landscape and whose input must be given equal weight to scientific studies and technical reports.

Early public involvement in conservation will also result in less contentious land-use decisions (Yaro et al., 1990). The value of conservation should be presented as an enhancement, not a diminishment, highlighting the economic advantages of keeping a resource intact and capitalizing upon it within the monetary return that can be brought to a development proposal (Fram and Weiler, 1984; Cardinall and Day, 1998). In turn, developers would enthusiastically greet early information about significant landscapes. As one interviewee put it, "If you're a developer, what you want is certainty" (study transcripts).

The "civilizing" of planning needs a proper forum (Friedmann, 1987; Forester, 1989), providing intervenor funding to balance development proponents with effective opposition, distributing information about landscape resources equally to all sides of a community debate, and aiding local areas in how to write more detailed landscape policy. Ultimately, all this presupposes an accessibility where language is understandable, schedules are not too tight or prolonged, proceedings are well advertised, and open attitudes are expressed by all involved in the process (Young, 1990).

Transformations

These kinds of changes to landscape-planning approaches necessitate profound transformations of public and professional attitudes and governmental agendas. Foremost, landscape protection is dependent on the goodwill of a community to come together for the common good. Therefore, conservationists must be in touch with the foibles and pettiness of human beings as well as the potential for greatness and the generosity of spirit fundamental to successful conservation.

Study participants suggested that the first task is to contextualize proposals in past land-use decisions. It is essential to tell government and public alike that what they are doing is not unprecedented – landscape conservation, in many guises, has a long history in most locations. As a result, people realize that the task does not seem so unfamiliar and risky. In Ontario, for instance, it would have been useful to remind the detractors of the legislation that landscape protections already exist in more familiar forms such as environmental ease-ments, heritage districts, and natural-area designations.

Comfort levels also rise with examples from other locations. There are landscape conservation success stories to be found in many other jurisdic-tions. Notable initiatives are found in both the United States, with programs such as the Cultural Landscape Initiative and Natural Heritage Areas (Yaro et al., 1990; Keller and Keller, 1994), and the United Kingdom, with the long-established Countryside Commission (now called the Countryside Agency) (Lucas, 1992).

The ultimate resistance to the landscape idea will be presented in a judicial or quasi-judicial forum within a development appeal process where lawyers, traffic engineers, biologists, marketing analysts, and other "experts" argue points. However, when it comes to the defense of landscape it is usually left to impassioned citizens to argue the case. And often the argument is not well organized and too "subjective" for such a court of sober second thought.

Study participants suggested that citizens should enlist "experts" who can speak to the "softer" qualities of a landscape, people such as historians, artists, and psychologists. The appeals court needs to give equal weight to evidence that is typical of landscapes – evidence that can be expressed in dispassionate facts as well as emotional testimony.

This study concluded that landscape is the ideal stage upon which these struggles can occur, landscapes that are known in both a subjective and a collective manner. There is no guarantee that such knowledge will influence land-use decisions. Landscape is where Michel Foucault's triad of power, knowledge, and subjectivity are constantly in flux (Cook, 1993). One can only be cognizant of the underlying power relations and be prepared to engage in a struggle to tell one's own landscape story.

The challenge is to involve all "experts," local implementers of conserva-tion action, and upper-level policy makers. Only then will those efforts be well rooted in the landscape.

As they now exist, landscape policies in the revised Act serve as a toothless reminder of what could have been fully considered in land-use planning decisions. A chance to adopt a new, connected basis of planning was not embraced. It is clear that local municipalities and developers feared the term and citizens did not understand or support the concept. Some planning theorists are left saying it was a good idea and wondering how it might achieve its "promise" some day. The greater lesson to be learnt, from the specifics of this case, is that human dynamics are the ultimate arbiters of a landscape's future.

References

Arnstein, S. (1969). A ladder of citizen participation. *American Institute of Planners Journal*, 35, 216–224.

Barrell, J. (1972). *The Idea of Landscape and the Sense of Place*. Cambridge: Cambridge University Press.

Bolton, R. (1992). "Place prosperity vs people prosperity" revisited: an old issue with a new angle. *Urban Studies*, 29, 185–203.

Bookchin, M. (1992). The meaning of confederalism. In *Putting Power in its Place: Creating Community Control*, ed. C. Plant and J. Plant. Gabriole Island, BC: New Society Publishers, pp. 59–66.

Bourassa, S. (1991). *The Aesthetic of Landscape*. London: Bellhaven Press.

Brassley, P. (1998). On the unrecognized significance of the ephemeral landscape. *Landscape Research*, 23, 119–132.

Brooke, D. (1994). A countryside character programme. *Landscape Research*, 19, 128–132.

Bunce, M. (1994). *The Countryside Ideal: Anglo-American Images of Landscape*. London: Routledge.

Cardinall, D. and Day, J. C. (1998). Embracing value and uncertainty in environmental management and planning: a heuristic model. *Environments*, 25, 110–125.

Cook, D. (1993). *The Subject Finds a Voice: Foucault's Turn Toward Subjectivity*. New York, NY: Peter Land.

Cosgrove, D. (1984). *Social Formation and Symbolic Landscape*. Totowa, NJ: Barnes and Noble.

Crandell, G. (1993). *Nature Pictorialized: "The View" in Landscape History*. Baltimore, MD: Johns Hopkins University Press.

Crouch, D. (1990). Culture in the experience of landscape. *Landscape Research*, 15, 11–14.

Daniels, S. and Cosgrove, D. (1988). *The Iconography of Landscape*. Cambridge: Cambridge University Press.

Duncan, J. and Ley, D. (1993). *Place/Culture/Representation*. London: Routledge.

Ely, M. (1991). *Doing Qualitative Research: Circles Within Circles*. London: Falmer Press.

Forester, J. (1989). *Planning in the Face of Power*. Berkeley, CA: University of California Press.

Forman, R. T. T. and Godron, M. (1986). *Landscape Ecology*. New York, NY: Wiley

Fram, M. and Weiler, J. (1984). *Continuity with Change: Planning for the Conservation of Man-Made Heritage*. Toronto: Dundurn Press.

Friedmann, J. (1987). *Planning in the Public Domain: from Knowledge to Action*. Princeton, NJ: Princeton University Press.

Glaser, B. and Strauss, A. (1967). *The Discovery of Grounded Theory*. Chicago, IL: Aldine.

Gold, J. and Burgess, J. (1982). *Valued Environments*. London: George Allen and Unwin.

Hartshorne, R. (1939). *The Nature of Geography*. Lancaster, PA: Association of American Geographers.

Hoskins, W. G. (1969). *The Making of the English Landscape*. London: Hodder and Stoughton.

Hunt, J. (1991). The garden as cultural object. In *Denatured Visions: Landscape and Culture in the Twentieth Century*, ed. S. Wrede and W. H. Adams. New York, NY: The Museum of Modern Art, pp. 19–32.

Innes, J. E. (1998). Information in communicative planning. *APA Journal*, 64, 52–63.

Innes, J. E., and Booher, D. E. (1999). Consensus building and complex adaptive systems: a framework for evaluating collaborative planning. *APA Journal*, 65, 412–423.

Jackson, J. B. (1984). *Discovering the Vernacular Landscape*. New Haven, CT: Yale University Press.

Kaplan, S. (1987). Aesthetics, affect and cognition: environmental preference from an evolutionary perspective. *Environment and Behaviour*, 19, 3–32.

Karetz, J. D. (1989). Rational arguments and irrational audiences. *JAPA*, 55, 445–456.

Keller, T. and Keller, G. (1994). *How to Evaluate and Nominate Designed Historic Landscapes*. Bulletin 18. Washington, DC: US Department of the Interior, NPS.

Laurie, I. C. (1975). Aesthetic factors in visual evaluation. In *Landscape Assessment: Values, Perceptions, and Resources*, ed. E. H. Zube, R. O. Brush, and J. G. Fabos. New York, NY: Dowden, Hutchinson and Ross, pp. 102–118.

Levi-Strauss, C. (1970). *The Raw and the Cooked*. London: Jonathan Cape.

Lincoln Y. and Denzin, N. (1995). *Handbook of Qualitative Research*. Thousand Oaks, CA: Sage.

Lowenthal, D. (1985). *The Past is a Foreign Country*. Cambridge: Cambridge University Press.

Lucas, P. H. C. (1992). *Protected Landscapes: a Guide for Policy-Makers and Planners*. London: Chapman and Hall.

Maines, D. R. and Bridger, J. C. (1992). Narratives, community and land use decisions. *Social Science Journal*, 29, 363–380.

McHarg, I. (1969). *Design with Nature*. New York, NY: Natural History Press.

McGinnis, M. V., Woolley, J., and Gamman, J. (1999). Bioregional conflict resolution: rebuilding community in watershed planning and organizing. *Environmental Management*, 24, 1–12.

Meinig, D. W. (1979). The beholding eye: ten versions of the same scene. In *Interpretations of the Ordinary Landscapes: Geographical Essays*, ed. D. W. Meinig. New York, NY: Oxford University Press, pp. 33–48.

Mitchell, M. Y., Force, J. E., and Carroll, M. S. (1993). Forest places of the heart: incorporating special spaces into public management. *Journal of Forestry*, 91, 32–37.

Neuman, W. L. (1994). *Social Research Methods: Qualitative and Quantitative Approaches*, 2nd edn. Toronto: Allyn and Bacon.

Olwig, K. R. (1996). Recovering the substantive nature of landscape. *Annals of the Association of American Geographers*, 86, 630–653.

Ontario Legislative Digest Service (1996). 1st Session, 36th Legislature, *Bill Number* 20 *(G)*, Release 20, April 26.

Penning-Rowsell, E. C. and Lowenthal, D. (1986). *Landscape Meanings and Values*. London: Allen and Unwin.

Plant, C. and Plant, J. (1992). *Putting Power in its Place: Create Community Control!* Gabriola Island, BC: New Society.

Pocock, D. (1981). *Humanistic Geography and Literature: Essays on the Experience of Place*. London: Croom Helm.

Pollock-Ellwand, N. (1997). Planning for the landscape idea. Unpublished Ph.D. thesis, University of Waterloo, Canada.

Porteus, D. C. (1990). *Landscapes of the Mind: a World of Sense and Metaphor*. Toronto: University of Toronto Press.

Pugh, S. (1990). *Reading Landscape: Country–City–Capital*. Manchester: Manchester University Press.

Rapoport, A. (1982). *The Meaning of the Built Environment: a Non-Verbal Communication Approach*. London: Sage.

Relph, E. (1976). *Place and Placelessness*. London: Pion.

Rowntree, L. B. and Conkey, M. W. (1980). Symbolism and the cultural landscape. *Annals of the Association of American Geographers*, 70, 459–474.

Rubin, H. J. and Rubin, I. S. (1995). *Qualitative Interviewing: the Art of Hearing Data*. Thousand Oaks, CA: Sage.

Rydin, Y. (1998). Land use planning and environment capacity: reassessing the use of regulatory policy tools to achieve sustainable development. *Journal of Environmental Planning and Management*, 41, 749–765.

Sales, K. (1992). "Free and equal intercourse:" the decentralist design. In *Putting Power in its Place: Create Community Control!* ed. C. Plant and J. Plant. Gabriola Island, BC: New Society, pp. 20–27.

Sandercock, L. and Forsyth, A. (1992). A gender agenda: new directions for planning theory. *JAPA*, 58, 48–59.

Sauer, C. (1925). *The Morphology of Landscape*. University of California Publications in Geography 2. Berkeley, CA: University of California Press

Schama, S. (1995). *Landscape and Memory*. Toronto: Random House.

Schauman, S. (1988). Scenic value of countryside landscapes to local residents: a Whatcom County, Washington case study. *Landscape Journal*, 7, 40–46.

Schein, R. H. (1997). The place of landscape: a conceptual framework for interpreting an American scene. *Annals of the Association of American Geographers*, 87, 660–680.

Seamon, D. (1981). Newcomers, existential outsiders and insiders: their portrayal in two books by Doris Lessing. In *Humanistic Geography and Literature,* ed. C. D. Pocock. London: Croom Helm, pp. 85–100.

Sheail, J. (1988). The Great Divide: a historical perspective. *Landscape Research*, 13, 2–5.

Silverman, D. (1993). *Interpreting Qualitative Data: Methods for Analyzing Talk, Text and Interaction*. London: Sage.

Stilgoe, J. (1982). *Common Landscapes of America, 1580 to 1845*. New Haven, CT: Yale University Press.

Strauss, A. (1987). *Qualitative Analysis for Social Scientists*. New York, NY: Cambridge University Press.

Strauss, A. and Corbin, J. (1990). *Basics of Qualitative Research: Grounded Theory Procedures and Techniques*. Newbury Park, CA: Sage.

Tuan, Y. (1979). Thought and landscape: the eye and the mind's eye. In *The Interpretation of Ordinary Landscapes: Geographical Essays*, ed. D. W. Meinig. New York, NY: Oxford University Press, pp. 89–102.

Watson, A. E. and Labelle, J. M. (1997). An introduction to planning and land use management in the United States, with comparisons to Canada and England. *Environments*, 24, 66–83.

Williams, R. (1973). *The Country and the City*. London: Chatto and Windus.

Yaro, R., Arendt, R., Dodson, H., and Brabec, E. (1990). *Dealing with Change in the Connecticut River Valley: a Design Manual for Conservation and Development*. Amherst, MA: Lincoln Institute of Land Policy.

Young, I. M. (1990). *Justice and the Politics of Difference*. Princeton, NJ: Princeton University Press.

29

Landscape ecology: principles of cognition and the political–economic dimension

It is the view of scientists, and of the public in general, that landscape ecology is a science of landscapes and humans. Landscape is a part of the earth's surface – a region perceived by humans (Hartshorne, 1939; Zonneveld, 1988). However, humans are also inhabitants and users of the landscape. Landscape is their immediate home but it is also a territory of broader political and economic interest. It is the space where humans live, travel, work, and rest. This relationship between humans and the landscape has acquired a special meaning, especially in relation to negative phenomena, even conflicts, which have originated as responses to human activities.

Humans were never on the earth as impartial visitors but from earliest times perceived landscape as their environment. Consequently, environmental problems were those that called for a solution. Humans not only perceived landscape pattern as scenery but they also started to evaluate land-use arrangements by using economic and ecological principles. The impact of humans on the landscape resulting from their activities became the subject of public supervision, decision making, and planning. At the same time, tools useful in acquiring knowledge were activated and scientific research was oriented toward forecasting the consequences of land use and of understanding the potential, or the limits, of a conflict-free functioning of landscape. The theory and methodology of geography, landscape ecology, and also biology (especially geobotany) became the foundation for this reasoning and for the resolution of practical problems.

Well before its formal recognition in the West, landscape research and its applications had been an established part of the state planning procedures in Slovakia and much of central and eastern Europe from the 1950s. For example, the research activities of the institutes of the Slovak Academy of Sciences (SAS) were controlled by the requirements of government agencies and were also directed at a national level in the former Czechoslovakia. Examples of these types of linked activities between landscape research and

 Issues and Perspectives in Landscape Ecology, ed. John A. Wiens and Michael R. Moss. Published by Cambridge University Press.

environmental planning are: the potential vegetation cover maps of Slovakia produced by the Institute of Botany for the State Water Management Plan in the 1950s (Michalko *et al.*, 1986); the spatial analyses for Slovak urbanization projects and for the location of the East Slovak Ironworks by the Institute of Geography; and the collaborative work of the Institutes of Geography, Experimental Biology, and Ecology for the Gabcikovo dam on the River Danube. At regional and local scales, SAS landscape research teams contributed to local government and regional planning institutions on projects related to urban zonation, highway routing, and design for protected natural areas. Here the Institutes of Geography, Landscape Biology, and later the Institute of Experimental Biology and Ecology collaborated to provide input to the planners. After 1989 several private agencies also emerged to provide source materials for planning and decision making. Such projects include environmental impact assessment for motorway construction, industrial parks, shopping centres, etc. Currently, landscape ecological teams of the SAS institutes and departments of the Faculty of Natural Sciences at Comenius University in Bratislava are dealing with issues of ecological stability in territorial systems in the context of regional development (landscape potential) for different administrative units. More recently, work has also been focused on implementing sustainability principles in territorial planning.

This concept of landscape as a research objective in geography was a motivation for scientists in central Europe in the 1960s and 1970s. Here rivalry between the spatial disciplines, such as geographic landscape research and geobotanic research into plant communities, which includes the mapping of potential vegetation, is worth mentioning. These theoretical–cognitive disciplines, which stressed the analysis of singularities and a knowledge of the functioning of spatial wholes, in addition to defending their research results also had to propose practical applications. These disciplines directly influenced the conception of landscape ecology by their methodology and the spatial nature of results obtained. Methodological procedures, analyzing relevant relationships and the mechanisms of the functioning of spatial systems in particular, were developed in landscape ecology. Apart from landscape diagnosis, these procedures also outlined preventative/therapeutic directions. Social order was demanding scientists to bring forward solutions and to identify alternative forms of remedy and regeneration. The scientific approach was expected to present a particular proposal, which together with landscape planning and landscape architecture would envisage the optimum landscape arrangement for a particular problem. Such proposals would outline potential (adequate) functions and their spatial organization and were expected to provide an ideal, conflict-free functioning solution. The theory of landscape as a whole would be verified by applying it to the solution of everyday problems in landscape ecology, which would simultaneously acquire a political–economic dimension.

Landscape ecology: principles of cognition

Limiting environmental problems to landscape systems then becomes a practical matter. This is the logical outcome of the complexity of both concepts – environment and landscape (Weichhart, 1979; Zonneveld, 1988). Weichhart's explanation expresses the broad content and logical structure of the concept of environment. A parallel to ecology can be identified by stressing the *relationships* of environmental reality to aspects of the environment but with a different form of that relationship. The other explanation emphasizes the *differentiation* of the "environmental pivot" (the individuals, small and large social aggregates, mankind) from aspects of the environment (the physical, built, socioeconomic, ideological–cultural environment). This latter differentiation defines the breadth of the term "environment" and limits the scope of the material basis of the environment that can be related to our interpretation of "landscape." An analysis of the relationship of humans (the social aggregates) to landscape points to the focus of the arrangement and to the organizational aspects of landscape. Landscape ecology, by analyzing this human–landscape relationship, focuses on the landscape. It analyzes the geo-elements and their interacting properties that are critical to this relationship. However, humans have always been implicitly considered to be a part of the landscape. Landscape ecology investigates and evaluates landscape "for" humans.

Landscape is the core of landscape ecology. Landscape is represented by a real system consisting of geo-elements (rock, landform, water, soil, vegetation, fauna) and the noospheric dimensions of human beings (Zonneveld, 1988). Its structure is the result of the composition of these elements, their properties, and their interaction. The interaction of natural conditions with human influences as a result generates processes, of which landscape pattern is the result. Hence, landscape function depends on the processes of the natural landscape and processes controlled by humans in relation to that landscape. These are the political and economic principles of land use. The subject of landscape ecology must recognize this fact (Wiens, this volume, Chapter 35). Landscape research should be oriented to understanding the functioning of the natural part (i.e., the relevant elements), where biota are the focus (ecology), and simultaneously the functioning of the cultural landscape (i.e., the driving forces of land use) as it is organized by humans (i.e., the environment). Our understanding of the environment, based on the complexity of landscape, determines our ability to understand the relationships and functions of this human–landscape system.

It is no surprise that aerial photographs and their visual interpretation initiated integrated analysis of the landscape in landscape ecology. In fact, these photographs opened up the possibility of developing an integrated perception of recorded spatial objects on the earth's surface. The discipline of landscape ecology emerged to incorporate three aspects of our general, integrated knowledge: the visual, the chorologic (i.e., spatiotemporal), and the perspective of landscape as an ecosystem (Zonneveld, 1988). Above all, the chorologic and ecosystem aspects make it possible to identify the landscape as a three-dimensional entity with vertical and horizontal heterogeneity changing in time. One of the main characteristics of landscape ecology is that this vertical and horizontal heterogeneity is understood as a holistic object of study. The chorologic (spatiotemporal) aspect of landscape ecology has been applied to the classification of areas with homogeneous, self-regulative mechanisms and consequently homogeneous responses to human input. This aspect is relevant in spatial or territorial planning. The ecosystem aspect emphasizes a landscape's self-regulating mechanisms. It also reminds us of the importance of biota in the interaction of landscape geo-elements, their dynamics and sensitivity. It also indicates the central position of humans as the highest biotic and social entity.

In order to secure the functioning of the landscape system it is necessary to understand the processes, and their regimes, operating in the natural part of the landscape. The natural subsystem has its own self-regulating, functioning mechanisms. All human inputs and interventions will affect this mechanism and will either partially modify or completely alter it. Then, the original self-regulatory matter–energy mechanisms must be regulated or even controlled by man. In an urban and highly technical landscape intervention by man represents the highest and consequently the most costly regulator. In an agricultural, semi-natural, or natural landscape, natural (self-regulating) mechanisms prevail. The essence of a solution to environmental problems lies in knowing these mechanisms and their functioning. The analysis of natural subsystems in real landscapes is, however, an abstraction. It implies the identification of a hypothetical state of the landscape with an emphasis on its substance-energetic content and on the processes of the natural component of the landscape. As a matter of fact, it represents a reconstruction of landscape, which might have existed free of human impact and use, yet under current climatic conditions. Being analogous to the mapping of potential natural vegetation, this obviously reflects the synergetic effects of the functioning (i.e., the processes) of the natural (abiotic) subsystem. Cognition of the mechanisms of such a complex subsystem as the natural landscape is facilitated by the integration of the scientific approaches of geography, geobotany, and landscape ecology. The result of this cognition of the natural part of the landscape is the identification of the relatively homogeneous areas (landscape types) noted above.

The functioning of the landscape system is determined to different degrees by human influence and the interests of society. Its nature depends on the mechanisms of social regulation, or socioeconomic processes, which ensue from the objectives of land use. Primary land-use aims are connected with food and with the satisfaction of providing the basic needs of society. Progressively, land-use planning has been determined by economic and political principles until environmental conflict eventually points to the necessity of understanding the ecological dimensions of landscape. The state of the landscape – that is, landscape pattern – then defines the cognition of both natural and socioeconomic processes in the context of human culture and science (Wiens, this volume, Chapter 35).

Natural conditions, analyzed and identified as a hypothetical state or structure of the landscape, in fact are, to various extents, influenced and used by humans. This natural and human content is materialized in individual components – the landscape objects. These objects, along with (geo)-relief, modify the third dimension of the landscape and humans perceive this through its morphostructural and physiognomic properties. Simultaneously, these visually perceived properties are among the decisive ones used for identification of the real state of the landscape. By means of these properties the content of landscape, thus interpreted, also comes closest to the cognition of its physical state as objective reality. Identification of land cover is considered a suitable integrator of both the visible and content-related landscape qualities. Land cover represents the biophysical state of the real landscape; that is, the natural and also the human-cultivated and created (artificial) material in the landscape (Feranec and Ot'ahel', 2001). The pattern of land cover also indicates the spatial organization of this real (cultural) landscape. However, analysis and identification of these functions, in the context of land use, is necessary for gaining a comprehensive knowledge of the real state of the landscape. Urban and agricultural land cover correspond with land use in a regional dimension. Analysis of land-use functions are, however, indispensable in the case of forest and semi-natural landscapes where economic interests are less visually distinguishable and where land is used for nature conservation, recreation, water management, military purposes, etc. Cognition of these functions in particular areas is also important with regard to the hierarchical assessment of their ecological importance.

These principles of cognition of landscape structure are the preference of those practising the geoecological approach to resolving the priorities and intentions of rational landscape organization. Harmonious landscape organization is, however, not only connected to human beings and their environment, but also to other living organisms. Landscape (land cover) pattern is perceived by humans to have a certain quality. Likewise, animals perceive this quality and their behavior depends on this quality. It is the bioecological orientation of landscape ecology that treats these problems and formulates them according to particular principles. Apart from the principles of structure and diversity, the principles of process and change are also important in the context of biota. The principles of energy flow and nutrient redistribution are also important to the geoecological analysts of the natural landscape. The principle of species flow, however, is a concern of the specialized biologists, even though this principle is connected to land-cover pattern and is significant from the viewpoint of the natural flow of animals in cultural landscapes. The nature of landscape change is mainly connected with disturbances provoked by human activities. The principle of stability is related to the amount and quality of biomass which is able either to resist disturbances or to balance them. Then again, the essence of the principle lies in a cognition of the mechanism of a landscape's inherent natural properties.

Landscape ecology developed its subject matter with various degrees of emphasis on the three aspects noted above. While landscape research along chorologic and ecosystem lines was highly productive and relatively objective, the visual aspect of landscape was generally only implied or was too limited in scope. Knowledge of this visual aspect of landscape is also a matter of perception and its objectives are therefore a matter of aesthetics. It is little wonder, then, that landscape ecological research has been influenced by the behavioral sciences and landscape architecture. Landscape pattern is important with regard to its perception. Such external properties of landscape are, however, closely connected to the quality of the content of landscape, although their cognition and interpretation results from perception. The identification of land-cover pattern is, from this point of view, also efficient in an assessment of the visual qualities of landscape, especially if the assessment respects the general conventions of aesthetics as accepted standards of visual landscape quality. Such standardized approaches are adequate for landscape design and planning. We must also pay attention to the broader significance and complexity of landscape perception. Its cognition is connected to the noospheric aspect of landscape research that includes questions relating to the perception of life and the spirit and identity of landscape. This is the point where the geoecological and the sociological approaches converge.

Landscape ecology: planning and management

Political–economic solutions to environmental problems are naturally relevant to landscape ecology. The solution of particular practical problems in landscape ecological research has helped produce important methodological procedures that emphasize a multidisciplinary approach and consequently stress closer communication between research and the decision-making or planning spheres. The language of communication has simultaneously been influenced by demands from the public arena. The biocentric (ecosystemic) and spatial aspects of landscape research have found an application in the delimitation of areas with homogeneous environmental properties and self-regulatory capacities for land-use planning. The visual aspect is connected to both territorial planning and landscape architecture; landscape architecture and design, in particular, respect aesthetic principles.

An emphasis on knowing a landscape's potential, on the one hand, and the limitations or restrictions imposed by the spatial development of human activities, on the other, is manifest in the principles and procedures used in landscape synthesis (Drdos *et al.*, 1980). The output from this research approach yielded important sources of information for landscape management and planning. The concept of landscape synthesis led successfully to the integration of approaches to landscape research that were oriented toward practical societal issues. The scientific basis of this approach was coordinated by the "Landscape Synthesis – Geoecological Foundation of Complex Landscape Management" working group of the International Geographical Union (IGU). It was only natural that this scientific team has now continued its development within the framework of the International Association of Landscape Ecology (IALE), particularly within the working group "Landscape System Analysis in Environmental Management."

M. Ruzicka, in the former Czechoslovakia, established strong professional ties between landscape ecologists and designers. This cooperation had a significant impact in terms of the methodology of landscape planning, and the procedures developed for the analysis, synthesis, and evaluation in this landscape-ecology-based planning methodology (LANDEP) have come close to becoming a standard approach (Ruzicka and Miklos, 1982). It was applied to actual situations at various hierarchical levels and was applied extensively by public administration bodies in planning and design in Czechoslovakia and beyond. Similar principles of landscape cognition have been developed elsewhere in the formulation of scientifically based landscape-planning procedures. In this respect, the academic status of landscape ecology is important for its practical application and further development (Moss, 1999). In other words, the science seeks the truth and attempts to interpret it to administrators. This interpretation and communication may be achieved by various graphical schemes, graphical spatial models, or maps. These means of communication are the tools also required for spatial understanding. Proposals for environmental planning procedures to generate particular practical solutions originated from this interaction of research with design and administrative institutions (Ot'ahel' *et al.*, 1997). These planning procedures, first of all, identify a potential interest in searching for suitable solutions. This step should precede all actual data acquisition and inputs as a preventative analysis to accompany alternative solutions within any given territory. Interest in such landscape-based solutions requires a considerable knowledge of the technical parameters and spatial properties involved. The parameters of the technical goals, the analyses of the self-regulating capacity of the natural part of landscape, and a diagnosis of the actual state of the landscape (land cover/land use) are the key elements needed for an assessment of the vertical and horizontal conditions existing in a landscape.

A direct interest in landscape, as an economic and political space, is also connected to territorial planning and public administration. Landscape is a resource with potential for regional development. An analysis of the hierarchy of the spatial relationships in landscape organization is a part of the planning and decision-making process (Ot'ahel', 1995). Designers apply analysis of spatial relations in at least two scales: the local and the broader regional scales. The geoscientists usually discern three dimensions for understanding natural and social systems: local (city), regional (nation/state) and global. The differences depend on particular cases, scales, and preferred criteria.

The potential threat for negative environmental impacts is reflected in legal standards and in the control exercised by decision-making bodies. The methods of assessing such impacts on the environment are found in environmental impact assessment (EIA) procedures. The results of such assessments should provide answers to the stated project intentions, their realization and their ongoing operation. Further monitoring of the operation and environmental impact of activities makes possible a post-project assessment and, if necessary, further corrections to the inputs.

In the years of socialism the research conducted by the SAS was centrally managed and controlled, and pursued under what were referred to as the "state plans for fundamental research." SAS was then part of the Czechoslovak Academy of Sciences with statutes recognizing it as an independent organization within the Ministry for Research and Science. The individual institutes were involved in tasks to generate results applicable to social policy. Fulfilling these scientific and practical tasks, and their approval, involved consultation and debate with, and by, the government or regional users of the project. Naturally, the projects of those institutes dealing with landscape research, in the context of landscape ecology (the Institutes of Geography and Landscape Biology), were always oriented to users in the fields of territorial planning (Institute of Regional Planning), agriculture, forest and water management, nature conservation and the environment (each a sectoral institute or department of corresponding ministries). The research of involved geographers was focused on the analysis of patterns in the natural environment, on natural resource use, and the consequences of this use as conflict situations occurred.

A major product, and important data source, derived from the geoecological aspects of this environmental landscape research was a set of more than 30 maps, published at a scale of 1 : 500 000 in the *Atlas of the Slovak Socialist Republic* by, among others, E. Mazur and J. Drdos (see Drdos *et al.*, 1999).

After 1989, some senior administrators of the SAS institutes promoted the idea of focusing their institutes' scientific programs on fundamental research only. This, however, resulted in a considerable reduction in the number of research workers and in the overall research capacity of the SAS. Nevertheless, the institutes of the SAS also had to justify their existence to the government and political circles of the new Slovak Republic, on the basis of social demand. The Act to approve the (new) Slovak Academy of Sciences only came into being in 2002. Now the various institutes obtain credit by their publications and, above all, by their participation in PHARE international projects or in the 5th and 6th European Union Framework Programs. The Institute of Geography, for example, obtained international credit by land-cover mapping and participation in several PHARE–CORINE Land Cover Projects. Examples of these include applications to environmental planning and to the travel industry (for example, the Slovakia CORINE Land Cover Tourist Map). The results of numerous case studies in landscape ecological planning, conducted by the Institute of Landscape Ecology, are summarized in the *Landscape Atlas of the Slovak Republic*. (See Feranec and Ot'ahel', 2001).

Assets and outlooks

The aim for deriving practical outputs from landscape ecological research is to find adequate solutions and alternatives and to prepare resource material for planners and designers. Such outputs should present an ideal option or determine the spatial possibilities and limitations of the proposal for development. It means the presentation of a set of options for the input of human activities to the landscape. The assets of landscape ecology are contained in a distinct ecosystematic or biocentric aspect. Likewise, it is necessary to refer to the assets of the geoecological approach. We can talk about a distinct convergence, as called for by Moss (1999), of both approaches that have been used in the landscape ecological analysis of the natural landscape. A synthesis of both approaches was undoubtedly started by the geobotanical mapping of potential natural vegetation where both approaches were applied in the reconstruction of the areas as homogeneous units. Reconstruction presumed a knowledge, not only of the processes of the abiotic systems, but also of the ecological relations of the mapped vegetation unit.

The bioecological approach is even more desirable for identifying solutions to landscape stability problems. Here analysis of biota is oriented to issues of origin, size, shape, continuity, and neighborhood in terms of determining ecological importance. Initial spatial analysis of biota, within the conceptual framework of preservation of the landscape's ecological stability, is oriented to the identification of a hierarchical system of ecological significance in the landscape. A system of biocenters and biocorridors presents a framework for ecological quality which, with the activities of eco-stabilizing functions, is able to transfer gene-pool information. Concepts of the ecological stability of landscape also include analysis of negative or stress elements. Hierarchical systems of positive and negative landscape elements predetermine the natural linkages (conduits) of biota. These natural linkages can be analyzed in the context of ecological significance and suitability of positive elements, including porosity, intensity, and number of barriers, and the limits imposed by negative elements. Consequently, Moss's (1999) invitation for a synergy of the bio- and geo- approaches in landscape ecology is fully justified, especially in the context of the identification of land cover (habitat types) pattern and its significance.

The traditional ways of understanding the natural (geoecological) part of landscape represent the basis for the correct identification of key eco-systemic relationships and self-regulating mechanisms. Respect for self-regulating principles is central to the concept of sustainability, which recognizes spatial development of socioeconomic activities in harmony with a landscape's character and potential. Comparison between a hypothetical state and an actual real one is an adequate approach to understanding natural conditions and land-use assessment. Sound spatial development of any activity requires, first of all, a knowledge of the real state of the contemporary landscape. Remote-sensing data and their processing in Geographical Information Systems (GIS) may be valuable tools. Satellite images and aerial photographs also make possible the spatial identification of positive and negative landscape objects and may suggest more efficient spatial relations for the synthesis of the landscape as a whole. Higher spatial coherence of landscape objects recorded in these images provides a better solution for developing compatibility between human intentions and the functioning of these landscape elements.

The appropriate presentation of such results to design and administrative institutions has increased the importance of landscape ecology in terms of its social value. Changes to these values, and the criteria used, have been reflected in legislation on nature conservation, environmental impact assessments, and territorial systems of ecological stability at local and regional levels, as well as in the foundation of environmental boards and offices of planning and regional development. These results have been achieved while solving particular environmental problems. Such positive results have helped to increase education and to promote the significance of landscape ecology while simultaneously increasing the ecological awareness of the public. The political–economic dimension of landscape ecology has stimulated methodological progress and has fostered the application of landscape ecology and its increased importance among the geosciences, and in the spatial dimensions of the social and economic sciences. This is evident from the introduction of landscape ecology departments in technical and other universities and in the teaching of ecological and environmental subjects in primary and secondary schools. I am convinced that extended education in the geosciences and the ecological disciplines will contribute to the promotion of landscape ecology itself, and to society in general, by preparing new experts in the sphere of landscape planning and management.

References

Drdos, J., Mazur, E., and Urbanek, J. (1980). Landscape syntheses and their role in solving environmental problems. *Geograficky Casopis*, 32, 119–129.

Drdos, J., Bezak, A., and Podolak, P. (1999). A landscape-ecological approach to sustainable regional development: the case of Slovakia. In *Landscape Synthesis: Concepts and Applications*, ed. M. R. Moss and R. J. Milne. Guelph, Ontario: University of Guelph, pp. 157–184.

Feranec, J. and Ot'ahel', J. (2001). *Land Cover of Slovakia*. Bratislava: Veda.

Hartshorne, R. (1939). *The Nature of Geography*. Lancaster, PA: Association of American Geographers.

Michalko, J., Berta, J., and Magic, D. (1986). *Geobotanical Map of Czechoslovakia*. Bratislava: Veda.

Moss, M. R. (1999). Fostering academic and institutional activities in landscape ecology. In *Issues in Landscape Ecology*, ed.

J. A. Wiens and M. R. Moss. Guelph: International Association for Landscape Ecology, University of Guelph, pp. 138–144.

Ot'ahel', J. (1995). Spatial relationships and their hierarchy in environmental planning. *Ekologia (Bratislava)*, 14 (Suppl. 1), 29–36.

Ot'ahel', J., Lehotsky, M., and Ira, V. (1997). Environmental planning: proposal of procedures (case studies). *Ekologia (Bratislava)*, 16, 403–420.

Ruzicka, M. and Miklos, L. (1982). Landscape-ecological planning (LANDEP) in process of territorial planning. *Ekologia (CSFR)*, 1, 297–312.

Weichhart, P. (1979). Remarks on the term "environment". *GeoJournal*, 3, 523–531.

Zonneveld, I. S. (1988). Landscape ecology and its application. In *Landscape Ecology and Management*, ed. M. R. Moss. Montreal: Polyscience, pp. 3–15.

Integration of landscape ecology and landscape architecture: an evolutionary and reciprocal process

Landscape architecture is a professional field that is significantly focused on landscape pattern – the spatial configuration of landscapes at many scales. Landscape architecture is informed by scientific knowledge and aspires to provide aesthetic expressions in landscapes across a range of spatial scales. Landscape ecology has been defined as the study of the effect of landscape pattern on process, in heterogeneous landscapes, across a range of spatial and temporal scales (Turner, 1989). The logical reasons for integrating these two fields are clear and compelling, with a great potential to support sustainable landscapes through ecologically based planning and design.

The integration of landscape ecology and landscape architecture holds great promise as a long-awaited marriage of basic science and its application; of rational and intuitive thinking; of the interaction of landscape pattern and ecological process over varied scales of space and time, with explicit inclusion of the "habitats," activities, and values of humans. To the optimistic, this integration promises to provide a robust and appropriate basis for planning and design of sustainable environments. The focus on application is integral to most definitions of landscape ecology but has been slow to gain complete acceptance, or to demonstrate widespread success in "real world" landscape architectural applications. Unfortunately, the promise of integration remains more of a goal than a reality at this time.

I believe it is instructive to see the integration of landscape ecology and landscape design as an evolutionary, three-stage process (Fig. 30.1). I define key concepts and characterize the three stages including a discussion of the potential benefits and challenges of realizing a full, informed, and reciprocal integration (stage three). In this essay, "landscape architecture" denotes all those activities relating to the planning and design of landscapes, across a range of scales and landscape contexts. I submit that the three stages I describe have evolved uniquely in different parts of the

Issues and Perspectives in Landscape Ecology, ed. John A. Wiens and Michael R. Moss. Published by Cambridge University Press.

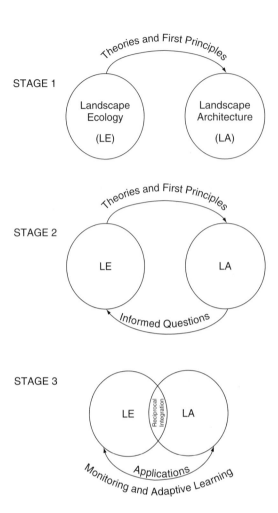

FIGURE 30.1
The three evolutionary stages of integration of landscape ecology and landscape architecture.

world. In Europe, for example, the integration of landscape ecology in landscape design is generally more advanced than in North America (Schreiber, 1990; Forman, 1990).

Stage 1: theory and principles

The first stage of the integration of landscape ecology and landscape design is the articulation of basic theory and first principles – robust statements of knowledge that transcend a particular cultural, temporal, or environmental circumstance. First principles synthesize the knowledge base, frame questions for future research, and build an intellectual basis for application. Defining contributions in this area have been made by Isaak S. Zonneveld, Karl F. Schreiber, Zev Naveh, Michel Godron, and Richard T.T. Forman, among others. Monica Turner's seminal paper "Landscape ecology: the effect of pattern on process" (1989) synthesized the discipline's knowledge into a clear and compelling statement which defined, from a scientific perspective, the potential of applications of landscape ecology. Richard Forman (1995) proposed 10 "first principles" that provide insight into landscape pattern or process. These ideas, principles, and theories, among others in the literature, have focused primarily on biological and physical resources and processes; for example, nutrient flow, landscape pattern change in response to disturbance, species response to landscape pattern change, and species movement and survival in heterogeneous landscapes (Hersperger, 1994). As a complement to the physical–biological focus, Nassauer (1995) proposed four "broad cultural principles" for landscape ecology to address culture–landscape interactions in the context of landscape ecology. The addition of these cultural principles to the previous physical and biological "first principles" represents a working theoretical base for an applied landscape ecology.

What distinguishes the landscape ecological principles from other established principles in ecology, cultural geography, and other physical and social sciences is the assertion that they are useful for application or, more specifically, to inform the planning, design and management of landscapes. These landscape ecological principles aim to integrate physical, biological, and cultural knowledge. They identify the potential for future experiments, and suggest a basis for informed application. I argue that these principles represent a sound foundation upon which an intellectual basis for informed application in landscape architecture can be built.

Stage 2: questions and dialogue

In the second stage of the evolution of the integration, planners and designers begin to ask intelligent questions of scientists that arise from their understanding of landscape ecology theory and principles. The questions concern issues of scale, landscape process(es), disturbance, and human—landscape interactions. The questions include:

- What is the proper spatial scale for understanding ecological patterns and processes?
- How does a particular place constrain or support an ecological process?
- What timescales are appropriate for planning? For which processes?
- Which species or species groups should be planned for? Can a particular species represent the habitat needs of larger species groups?

- How should disturbance be understood in landscapes? What are the intensity, duration, and spatial extent of disturbances?

The dialogue has evolved to more specific questions, for example:

- How large a forest patch is required to support a given species, or ecological process?
- What configuration of corridors is needed to sustain species interactions and buffer nutrient flows across a heterogeneous and fragmented landscape?
- How can the benefits and values of "ecological corridors" be tested to determine their value and appropriateness in conservation planning?
- How can landscapes be planned to accommodate specific disturbance regimes?
- What types of monitoring are appropriate to learn if landscape ecological applications achieve their intended results?

In this second stage, landscape architects also began to examine the implications for the new landscape-ecology paradigm on aesthetic expression at the scale of human experience and perception in the landscape. The quest for full integration of ecology and design transcends that of biological, physical, and cultural knowledge and principles. It requires a "consilience" of rational and intuitive thinking (Wilson, 1998). Landscape ecology, as a scientific discipline, is appropriately based on rational and empirical thought and research. Landscape architecture and environmental engineering are engaged in solving problems, mitigating impacts, and accommodating human activities. Landscape architecture, as distinguished from environmental engineering, strives to produce original combinations of science and art that express cultural meaning and inspire intellectual reflection and aesthetic expression. As the late John Lyle argued, this cannot be achieved solely through rational thought:

> In reality, however, nature is silent, ambivalent, and contradictory. We know now that she will not tell us what to do. In any given situation, any numbers of different plans are possible. The recognition of diverse possibilities is the all-important element missing from the four-step (scientific) paradigm and from so many other efforts to define design process. Recognizing possibilities takes creative thought, and creativity tends to be stifled by a rigid framework of logic. When we stifle creativity, we shut out a great many possibilities, and in a world that so desperately needs better solutions, that is something that we cannot afford to do.
>
> (Lyle, 1985: 127)

I submit that the second stage of landscape ecology–architecture integration is a self-limiting model. Because it is a one-way flow of knowledge and information, from science to application, it denies the possibility of a reciprocal integration in which new knowledge and modes of thinking can be learned through design and then examined or "applied" in the science of landscape ecology.

Stage 3: reciprocal integration

In the third stage of integration, landscape ecology and landscape design are engaged in a reciprocal integration in which theory, principles, knowledge, and applications flow in both directions: science informs design, and design informs science. Rational and intuitive thinking are integrated. The third stage of integration is more of a challenge than a reality at this point in time, with some notable exceptions (Hulse *et al.*, 2000). I believe it is the stage at which the application of landscape ecology can reach its potential. I propose five issues and challenges that must be understood and engaged as a prerequisite to realizing a full and reciprocal integration.

The paradox of time

Change and uncertainty are fundamental in natural and cultural systems. In ecology, economics, and in other natural and social sciences, change is understood as a fundamental process rather than an aberration. Landscapes are not different. Change is also fundamental and uncertainty is a "given." Natural processes occurring in landscapes need time and certainty in some places, yet cultural and economic forces demand flexibility to change in others. This is the paradox of time in landscape planning (Sijmons, 1990). Landscape ecology can help to define or design a durable/sustainable landscape framework that supports the long-term ecological processes (e.g., the "slow turning wheels," groundwater and nutrient flows, species survival and evolution). By implication, the "interstices" within the landscape framework are available to accommodate change, specifically the intensive uses and landscape types (agriculture, urbanization, transportation) that contribute little or that degrade ecological functions. The contemporary landscape architect is challenged with designing the framework and its interstices to simultaneously sustain long-term ecological processes and accommodate contemporary needs, while also being mindful of cultural needs, values, and aesthetics (Van Buuren and Kerkstra, 1993). The challenge presented by the paradox of time is familiar to designers: to artfully accommodate and balance complementary and competing land uses. The paradox presents challenges that are new to most ecologists: to think strategically, to make intelligent compromises, and to understand the place of dynamic land uses within a more stable framework.

The positive potential of landscape change

To resist landscape change unilaterally is like "putting on the brakes" against unstoppable ecological and global economic forces in defense of a historically and continually diminishing "nature." Resisting change is a defensive position that maintains a polarization between the "doers" and the "protectors" and denies opportunities for more creative and proactive solutions, in both landscape planning and design (Vroom, 1997). While many changes are undisputedly negative, an acceptance of the inevitability of change and recognition of its positive potential is essential to achieving a full integration of ecology and design.

The power of spatial concepts

A spatial concept expresses through words and images an understanding of a planning/design issue and the actions considered necessary to address it. Spatial concepts are related to the proactive or anticipatory nature of landscape design, in that they express solutions to bridge the gap between the present and some desired future situation. Spatial concepts are often carefully selected metaphors; for example, "Green Heart" or "Stepping Stones," which communicate the essence of the concept clearly to build consensus for an overarching planning policy and to form a clear basis for more specific design decisions.

Although scientific input from landscape ecology is essential to conceive spatial concepts, its potential is limited. Many scientists are reluctant to make the "leaps of faith" that are essential to conceive spatial concepts. There is an essential element of creativity in the design of spatial concepts. They represent an interface of empirical and intuitive knowledge. Through spatial concepts, rational knowledge is complemented with creative insights. A well-conceived spatial concept represents a powerful tool to guide, inspire, and support landscape design. Figure 30.2 presents an example of several spatial concepts often used in landscape architecture.

Physical expression of landscape processes

The idea of making natural processes visible through design is a common theme in the literature of ecological aesthetics (Olin, 1988). Indeed, the pattern–process dynamic, fundamental to landscape ecology, offers a

CONTAINMENT
To control the enlargement or expansion of a core resource area, or an area of land use change.
Example: urban greenbelt

FIGURE 30.2
Spatial concepts for landscape architecture and planning.

INTERDIGITATION
A spatially integrated pattern based on an intrinsic resource distribution pattern.
Example: ridges and valleys

SEGREGATION
A strategic concept to benefit from concentration, or to minimize the impacts of selected land use(s).
Example: framework concept, zoning

CONTROLLED EXPANSION
To direct land use change or expansion in a prefered direction, as along a corridor.
Example: urban highway corridors

PROTECTED CORE
A defensive strategy to maintain a core resource area in a threatening or non-supportive environment.
Example: "the Green Heart," habitat patch

LINEAR NETWORK
A simple system of linkage in which discrete elements can form an integrated system, may be heirarchical.
Example: road network, hedgerows, canals

DENDRITIC HIERARCHICAL NETWORK
A system of linkage, caused by or emulating the most efficient means to accomodate flows or movements.
Example: drainage network

NODE AND CORRIDOR NETWORK
A system of core areas combining the benefits of large core areas with advantages of connectivity.
Example: ecological network

compelling challenge to designers to give visible form to landscape function(s). Some notable success has been realized in this area when designers have engaged, for example, the ecology of storm-water hydrology, plant succession, and fire as an ecological disturbance. In this way, people can "see" where the rainwater goes, how a meadow can become a forest, and how a landscape responds to fire. When successful, such designs engage the public, raise awareness and understanding, and contribute to a new aesthetic sensibility. When these expressions remain in the domain of "high art," they have been criticized as being remote from the culture or elitist. I see this as a valid challenge, and one that offers tremendous opportunities for collaboration between scientists and designers.

The dilemma of uncertainty

As professionals operating in the real world, landscape designers are often confronted with a mandate for action. Projects operate in response to short-term economic or politically driven goals and objectives. Inevitably, the knowledge on which to base these actions is incomplete and uncertain. The designer can't afford to plan through trial and error, and inaction is, in itself, a management decision with its own negative consequences. Scientists are justifiably uncomfortable making specific recommendations in the face of uncertainty. Adaptive management offers a strategy to address this dilemma. It explicitly acknowledges uncertainty and develops a range of possible actions, conceived as experiments. Hypotheses are formulated and design actions are proposed following accepted principles of experimental design. With an appropriate monitoring protocol, the experiments yield results, which contribute to new knowledge. The objectives, assumptions, decisions, and outcomes are documented so that new knowledge and understanding are gained through the process of application (Peck, 1998).

Conclusion

I have attempted to articulate three stages of integration of landscape ecology and landscape design, each characterized by specific activities and issues. The final stage, which may be elusive, promises a full reciprocal integration with a two-way flow of information and knowledge. It would be descriptive and prescriptive. Through empirical research, designs would be more informed of their ecological consequences, and through monitoring, implemented plans and designs would yield new empirical knowledge for ecology. The challenges to achieve such an integration have proven to be significant in terms of the modest successes to date in applied landscape ecology. The reward and motivation for a successful integration should be progress toward sustainability – hopefully a sufficiently noble goal to motivate ecologists and designers to seek deeper integration.

References

Forman, R. T. T. (1990). The beginnings of landscape ecology in America. In *Changing Landscapes: an Ecological Perspective*, ed. I. S. Zonneveld and R. T. T. Forman. New York, NY: Springer, pp. 35–41.

Forman, R. T. T. (1995). *Land Mosaics: the Ecology of Landscapes and Regions*. Cambridge: Cambridge University Press.

Hersperger, A. M. (1994). Landscape ecology and its potential application to planning. *Journal of Planning Literature*, 9, 14–29.

Hulse, D., Eilers, J., Freemark, K., Hummon, C., and White, D. (2000). Planning alternative future landscapes in Oregon: evaluating effects on water quality and biodiversity. *Landscape Journal*, 19, 1–19.

Lyle, J. T. (1985). *Design for Human Ecosystems*. New York, NY: Van Nostrand Reinhold.

Nassauer, J. I. (1995). Culture and changing landscape structure. *Landscape Ecology*, 10, 229–237.

Olin, L. (1988). Form, meaning, and expression in landscape architecture. *Landscape Journal*, 7, 149–168.

Peck, S. (1998). *Planning for Biodiversity: Issues and Examples*. Washington, DC: Island Press.

Schreiber, K.-F. (1990). The history of landscape ecology in Europe. In *Changing Landscapes: an Ecological Perspective*, ed. I. S. Zonneveld and R. T. T. Forman. New York, NY: Springer, pp. 21–33.

Sijmons, D. (1990). Regional planning as a strategy. *Landscape and Urban Planning*, 18, 265–273.

Turner, M. G. (1989). Landscape ecology: the effect of pattern on process. *Annual Review of Ecology and Systematics*, 20, 171–97.

Van Buuren, M. and Kerkstra, K. (1993). The framework concept and the hydrological landscape structure: a new perspective in the design of multifunctional landscapes. In *Landscape Ecology of a Stressed Environment*, ed. C. C. Vos and P. Opdam. London: Chapman and Hall, pp. 219–243.

Vroom, M. J. (1997). Images of ideal landscape and the consequences for design and planning. In *Ecological Design and Planning*, ed. G. F. Thompson and F. R. Steiner. New York, NY: Wiley, pp. 293–320.

Wilson. E. O. (1998) *Consilience: the Unity of Knowledge*. New York, NY: Knopf.

ROB H. G. JONGMAN

31

Landscape ecology in land-use planning

When you see the geese fly south or you suddenly get a glimpse of a badger, you do not easily realize that they have a target to go for. The geese fly south to migrate from their breeding grounds in the north of Europe, Asia, or America to their winter biotope. The badger goes along his usual route for foraging. Common toads migrate in large groups from their hibernation shelter to the water, where they have been born, to deposit their eggs. Salmon try to find their way up the streams to their spawning grounds. Storks return to their nests from Africa just like people return home from their holidays. It sounds very human, for in this behavior there is not much difference between wild species and mankind. As long as the migration routes are available and without too much danger for the species, we do not notice it, because they come and go. The birds fly over, the badger passes in the night just like the toads, and the only thing most people notice are the toad eggs in the water and the stork when it has returned to its nest.

Under the influence of changes in human food demands, caused by demographic trends, the cultivated area of North America and Europe has shown considerable fluctuations. Agricultural areas move from one region to another, forests are removed in one part of the world and forests of exotic species are planted elsewhere. At present, the agricultural productivity in Canada, the USA, and the EU, measured in kg dry matter per unit of acreage, continues to rise thanks to ongoing advancements in agronomic knowledge. Through changes in agriculture and forestry practices, landscapes have suffered rapid and often irreversible changes. These changes can be classified into two groups (Fry and Gustavsson, 1996):

- Those resulting from the marginalization of farmland and forests and consequent abandonment of earlier practices.

Issues and Perspectives in Landscape Ecology, ed. John A. Wiens and Michael R. Moss. Published by Cambridge University Press.
© Cambridge University Press 2005.

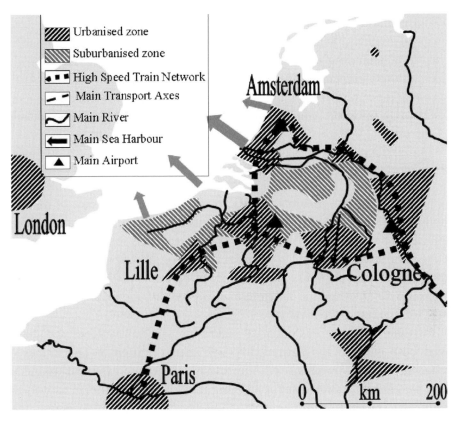

FIGURE 31.1
The Northwest-European Delta Metropolis as depicted in the Second Structure Plan for Benelux (Secretariat General Benelex Economic Union, 1996) and the Fifth Dutch National Policy Paper on Spatial Planning (Ministry of Housing and Environment, 2001).

- Those arising from the intensive use of highly productive land. Such processes have resulted in increased urban areas, less land being farmed, but farming and forestry are done more intensively, more specialized, and at larger scales.

In western Europe, urban land use is more and more dominating spatial structure and spatial developments (Fig. 31.1). In the urban fringe of western Europe, intensive agriculture used to be an important land use. Now its role is strongly diminishing and changing into other functions such as horse keeping, garden centers, and recreation facilities (Lucas and van Oort, 1993). Many people live in a totally urban environment but prefer the combination of urban and natural environment. As it might have been in earlier times, urban dwellers want to enjoy the countryside. They show that they have a need for rural landscapes, because these landscapes provide:

- aesthetics related to the identity of an area
- an attractive living environment
- understanding and experiencing nature
- outdoor recreation close to the living environment
- richness in species
- water transport, climate regulation, purification of water, air, and soil.

Land-use planning problems

The history of inhabited landscapes is different from that of natural landscapes (Meeus *et al.*, 1990). In the European agricultural landscapes, long traditions have caused recognizable patterns that are regionally different. They have become cultural landscapes consisting of characteristic land-use and urbanization patterns. The intensity of disturbances is greater than in natural landscapes and the decisions made by humans are the main determining factors of land-use patterns.

Increasing agricultural intensity makes land monofunctional and takes away both cultural and natural diversity. Intensification by one farmer – reducing production costs – will improve his position on the market. Also, here we have to realize that the farming market is international as well as within the European Union or in America. The farmers in the Paramo of the Andes have to compete with the large-scale potato farmers in Canada, and the small Greek farmers have to compete with the industrial Dutch and Danish farmers on the cheese market. If the market is not regulated, the farmers in the less-favored regions will be marginalized. Both intensive and extensive land use are expressed in the landscape: the structure of the land, the size of the parcels, and the area of natural and semi-natural vegetation that is present. Regulation of these land-use changes therefore becomes an international question, but land-use planning is still a national or regional activity that can hardly be expanded to continental dimensions.

In the Netherlands, the claim for urbanization until 2020 has been estimated to be 500–900 km^2, 2–3% of the total area of the country. The influenced area will be much larger. It will be comparable elsewhere in the world. In the competition with urban functions, the rural functions mostly cannot survive. That causes a number of problems for nature, agriculture, and outdoor recreation. In all countries, even the most industrialized ones, nature is needed for the above-mentioned functions. The Nature Policy Plan of the Netherlands (Ministry of Agriculture, Nature Management and Fisheries, 1990) contains a long-term strategy plan, the National Ecological Network (NEN). It must lead to a coherent network of (inter)nationally important ecosystems consisting of core areas, nature restoration areas, buffer zones, and ecological corridors. This, however, requires international cooperation with neighboring countries.

Fragmentation of the landscape has many causes. Increasing traffic and intensifying agriculture have caused many barriers in the European cultural landscape (Jongman, 2000). Transport infrastructure in Europe (roads, waterways, and railways) intersects habitats of species and thereby decreases the possibilities of species to disperse between different habitats that are divided by traffic lines. In the Netherlands, urbanization, agriculture, and industry have put increasing pressure on the total area that has been reserved for landscape and nature. The remaining natural area is fragmented due to a dense network of motorways, railways, and waterways that covers the country. This process of fragmentation has been going on for several centuries (Ministry of Transport, Public Works and Water Management, 1999) and this has resulted in loss of habitats, faunal casualties, barrier effects, disturbance (noise and light), and local pollution (IENE, 2003). These negative impacts influence many animal species in the Netherlands (Ministry of Transport, Public Works and Water Management, 1999).

Because it is impossible to prevent a confrontation between nature and urban developments, the Dutch Ministry of Agriculture, Nature Management and Fisheries (1995) proposed in its report "Urban Landscapes" an integrated approach for urban–rural relationships. Increasing road density, building of railroads, and the intensity of use lead to an increase of barriers in the landscape. Species can be hampered in their living space through land use, because the space needed for living depends on dispersal. For small species, roads are often inaccessible barriers, which means that the animals must find living space within the areas. Some animals like amphibians in spring take the risk of crossing roads toward breeding ponds. Larger animals will be hampered in their movements by urban areas, roads, and unattractive land.

It is not only urban planning that influences ecological processes in the landscape. Running waters are far more than just longitudinal river corridors, and modern ecology recognizes them as complex ecosystems (Jungwirth 1998). According to Townsend and Riley (1999), the science of river ecology has reached a stage where explanations for patterns rely on links at a variety of spatial and temporal scales, both within the river and between the river and its landscape. The links operate in three spatial dimensions:

(1) longitudinal links along the length of the river system, such as the river continuum (Vannote et al., 1980) or downstream barriers to migration
(2) lateral links with the adjacent terrestrial system, such as the flood-pulse concept (Junk et al., 1989)
(3) vertical links within and through the riverbed.

Many linkages occur between the river and its environment, so the river continuum must be considered within broad spatial and temporal scales (Roux *et al.*, 1989). Water flows are changed in quantity and quality and many animal species are sensitive to fragmentation. Through water relationships in a river catchment, agricultural and urbanization developments can have an impact over long distances, through both quantitative and qualitative changes (Alterra, 2004).

Fragmentation of natural areas is a spatial problem that has been defined by Forman (1995) as the breaking up of a habitat or land type into smaller parcels. In an ecological sense it is the dissection of the habitat of a species into a series of spatially separated fragments. Fragmentation leads to a diminishing habitat area and an increase in barriers or an increase in spatial discontinuity. Fragmentation is caused by barriers such as roads, urban areas, inaccessible land in both time and space, or by a decrease of landscape elements (connectedness: small forests, hedgerows, riparian zones). A consequence can be that the effect of external negative impacts on habitats increases and the number of suitable and reachable habitat sites decreases. The effects are species-specific and depend on the needed functional area, species mobility, and isolating effects of the landscape (roads, urban areas, and canals). Both decrease of functional area of a habitat site and isolation increase the chance of local extinction of populations and diminish the chance of spontaneous return of species. The spatial effects (Mabelis, 1990) are:

- decrease in suitable area of the original ecotope
- increase in landscape heterogeneity and land use
- landscape fragments with subpopulations
- source–sink relationships in natural populations (larger natural areas become increasingly important).

The early role of landscape ecology in land-use planning

Landscape ecology has had a mutual relationship with spatial and land-use planning. Landscape ecology made ecologists look beyond the species level and beyond ideal ecosystems. It made the scientific world realize that the landscape is the reality where we have to deal with humans and all wild species, and that ecological science for practical application is not only done in laboratories and reserves but especially in living landscapes. Already in the first Landscape Ecology Congress in Veldhoven different theoretical frameworks were presented, such as the LANDEP approach for integrated planning (Ruzicka and Miklos, 1982) and the functional approach for nature-reserve planning (van der Maarel, 1982). Both approaches had in common that they considered the whole landscape and tried to apply principles from biogeography, vegetation classification, and material fluxes into complex planning models.

Planners did not understand all these complex ecological models. They already had to deal with complex economic models, traffic models, and trends in land use, production, and urbanization. They were not pleased to be confronted with yet another player in the field who told them to have an overall concept based on ecological processes as well. This is accepted only when landscape ecology is not only a problem-stating but also a problem-solving science (Naveh, 1991). Specialists in several aspects of landscape ecological science have carried out fundamental research in hydro-ecological modeling, population dynamics, and landscape modeling. The generalists among landscape ecologists translated these principles into land-use planning concepts and applied them in the reconstruction of wetlands, development of ecological networks, catchment approaches in water planning, and new approaches in monitoring landscapes. In this way, nature is more and more accepted as an issue for land-use planning. Nature can provide principles on which plans can be built and also can deliver criteria for constructing patterns and managing processes.

Landscape ecological principles

Landscape ecology supplies important concepts that can be applied in land-use planning. These can be ordered in a hierarchy from more or less general and holistic to more specific landscape- or population-oriented ones.

Sustainability is the capacity of the earth and its landscapes to maintain and support life and to persist as a system. The concept of sustainability is not only fundamental to the earth as a whole, but also to smaller systems within it. This parallels the approach of landscape ecology, in that it is essential to maintain ecosystems, which are dynamic but also self-reproducing, without spoiling nutrients and species. Sustainability implies that it is necessary to maintain a resource, whether it is wildlife, amenity, or agriculture. The good-husbandry concept of farming as considered in the nineteenth century is in many respects reflected in some of the recent landscape-ecological work on modern agro-ecosystems. Landscape ecology develops the concepts that make it possible to find a balance between land use and ecology.

Landscapes operate at different levels involving complexes of different elements. Urban *et al.* (1987) provide an important perspective on landscape ecology, as they discuss the *hierarchical relationships* between elements within the landscape and their interdependence as well as the role of humans in their management and manipulation (Mander *et al.*, 2003; Wassen and Verhoeven, 2003). On the one hand, one can study a whole catchment such as the Mississippi or the Rhine. The Rhine catchment consists of mixtures of whole landscapes, from Alpine to mountainous landscapes with large-scale forestry and mixed farming through to alluvial and lowland landscapes characterized by intensive dairy farming. On the other hand, within that landscape one can examine structures such as woodlands and the surrounding land and their relationships. Planning also takes places at these different levels. The use of the Rhine is coordinated by the Rhine Commission in which all countries are represented; its land use is planned within countries and regions and its water use and management are taken care of in the water management systems in the different countries.

A wider basic principle is that landscapes involve *gradual changes and ecotones* (Naiman and Décamps, 1990). It is recognized that many ecological elements do not show sharp boundaries between each other, but rather grade together in time and space. The stability and dynamics of such systems are based on physical parameters rather than biological ones. This concept has been used in planning and nature conservation but is not yet well supported by research.

With the increased pressure on semi-natural habitats there has been much concern about *biodiversity*. It is a basic concept in the management of landscapes and in planning. Policy objectives for national parks and nature reserves are often formulated with the objective of maintaining an existing high biodiversity. Biological diversity is the outcome of historic processes and therefore refers to both time- and space-related processes (Pineda, 1990). Biodiversity is dependent on the natural richness but it is also dependent on the impact of humans and the way they have changed nature into cultural landscapes (Jongman *et al.*, 1998)

A very important landscape-ecology concept for land-use planning concerns population dynamics in manmade landscapes: the *metapopulation* (Opdam, 1991). This represents the concept of interrelationships between subpopulations in more or less isolated patches within a landscape and helps one understand the impact of progressive isolation of individual areas of vegetation and their associated animal populations in modern agricultural landscapes. Temporary extinction and recolonization are characteristic processes in metapopulations. In this respect the following aspects are important:

- The dynamics of the subpopulations (extinction and immigration rate). If a patch is small and highly isolated, the extinction rate might exceed recolonization and a subpopulation becomes extinct.
- The connectivity between patches. Important landscape variables in this respect are the absence of barriers and the presence of corridors.

- The spatial and temporal variation in habitat quality. This is introduced by the absence or presence of disturbances in agricultural landscapes represented by land-use practices.

Applications and questions

The most important contribution of landscape ecology to landscape planning has been to focus attention on natural spatial and temporal dynamics. In promoting a broader-scale view than traditional site-based conservation, we are more likely to be successful in maintaining a high biodiversity, even in urbanized areas. In addition, landscape ecology has an integrating role, linking human and ecological aspects of countryside management. Current moves toward a greater integration of human and social needs in conservation planning have resulted in the inclusion of landscape conservation in national and international programs (Council of Europe, 1995). The underlying landscape ecological principles can be expressed in relation to nature conservation and human needs as follows (Fry, 1996):

- The spatial configuration of landscape elements affects the survival and distribution of species of plants and animals
- The spatial configuration of landscape elements affects human landscape preferences.

These premises seem not only to be intuitively correct, but are also backed by an increasing body of scientific literature (Forman, 1995). Landscape ecology offers exciting new prospects for planning whole landscapes, but there are problems. For example, despite the enormous amount of ecological research during the past decades we still lack detailed knowledge about the impacts of different land-use intensities and landscape configurations both in space (pattern) and time (change). Much has been claimed about the importance of movement corridors in a landscape. Unfortunately, we do not yet understand well how to design these most effectively, whether they act as corridors or as barriers, or if they are more important for the introduction of predators or disease-spreading species.

The spacing of woodlots in the countryside is also likely to be important as planned new plantations throughout Europe change the pattern of forest cover. Work in The Netherlands has shown that isolation of woodlots can be very important and can lead to regional extinctions (Opdam, 1991). To a mobile group such as birds, woodland in the landscape probably has to be less than 10–20% cover before isolation becomes an important ecological factor (Andrén, 1994). In several European agricultural landscapes, we have reached this point for woodland, and an even lower percentage of the previous meadow and pasture cover remains. River systems and their species show an even worse picture, because no natural rivers still exist; most rivers are dammed and fish migration is an illusion in nearly all major rivers in Europe and most rivers in North America. Landscape ecology is the integrating field of science that can help to repair the damaged landscape connectivity.

Problems arise when trying to generalize landscape ecological principles from one species to another or from one type of landscape to another. It may well be that each specific landscape–species interaction is unique. The big question is, "What general rules exist?" In most planning situations, landscape ecological questions will be integrated with other land-use questions. This underlines the need for a deeper understanding of landscape processes and interactions, rather than trying to find answers that will give the optimal landscape solution from the point of just one or a few species.

Until empirical evidence is available to refine our understanding of landscape dynamics we need rough generalizations. These may still be useful if they can rank planning options in the form of "option A is better than option B for species/ function X." The following questions are typical of those asked of landscape ecologists by planners of agricultural landscapes (Fry and Gustavsson, 1996):

(1) Is habitat fragmentation a major threat to wildlife and amenity and, if so, can we compensate by adding new habitat patches or corridors?
(2) Are large habitat blocks better than several small ones, and are there critical minimum sizes?
(3) Is linking habitats together better than not doing so?
(4) Which landscape elements are barriers to species dispersal?
(5) Are edge effects good or bad and under what circumstances?
(6) At what scale should we plan farm and forestry landscapes?
(7) How do we include farming/forestry systems and their rotation dynamics in planning?
(8) How to coordinate efforts between land owners to enable planning at the landscape scale?
(9) How can landscape ecological concepts best be presented to planners?
(10) How do we measure success in landscape planning?

This all leads to planning and implementation of ecological networks in many parts of the world. The questions that need to be answered for urban landscapes or river systems are not yet formulated, but will be much more complex.

Land-use planning and design

We especially need principles to give good advice now when so many opportunities for designing and managing new landscapes exist. Ecologists need to communicate to planners about design principles from a landscape ecological point of view (Dramstad *et al.*, 1996). When translating these questions into real-world problems we mostly have to deal with landscapes where other functions for society exist as well. In landscapes where multifunctional land use is required, for instance where outdoor recreation and nature use the same space, a well-designed structure including physical barriers for people can help to construct quiet ecological corridors alongside trails. The trail should be close to nature to allow walkers to enjoy nature, but the shelter of the natural species should not be affected. In the Dutch lowlands this is done by designing trails and ecological corridors with eye contact but preventing physical contact (Fig. 31.2).

Design does not only mean the development of a multifunctional corridor. It also can mean the crossing of a barrier. Barriers can be of all kinds, but they are often species-specific. Increasing traffic and intensifying agriculture have caused many barriers in the European cultural landscape. Canalization of waterways and the building of motorways have disturbed both the habitat of species and their possibility to disperse. Planning of ecological corridors is a method for compensation of a long-term fragmentation process in agricultural landscapes.

Roads are made as technical infrastructure to help human society in its transport needs. Natural infrastructures such as streams and rivers have been adapted to drainage and water transport. Both structure and intensity of use make it impossible for animals to cross these. The structure of roads consists of a wide strip of asphalt or concrete, often with ditches and fences. The structure of waterways consists of straight deep water, weirs and locks, steep shores and lack of shallow-water areas and islands. That makes the manmade infrastructure difficult to cross and for many species it is impossible to reach the other side. Most fishes never get through the maze of locks and weirs in the Dutch delta area.

Planning an ecological network means also mitigation and compensation of the manmade infrastructure. Fish ladders have to be built to make it possible for fish to cross weirs and locks. Road crossings can be tunnels or flyovers. Flyovers or ecoducts are meant for larger species (Fig. 31.3). In all cases the landscape in its surrounding has to be adapted to its function; hedgerows and small forests for guidance and shelter have to be planted. For those animals using water as a corridor (e.g., otter, *Lutra lutra*), bankside waterway crossings have to be developed. Natural banks must be maintained,

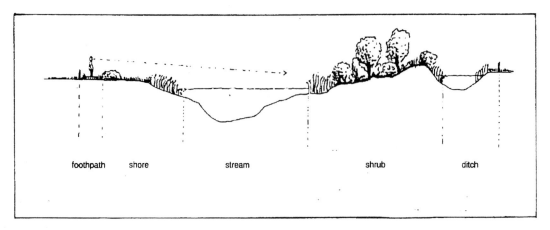

FIGURE 31.2
Combination of a trail and an ecological corridor in an agricultural landscape (Elzinga and van Tol, 1994).

FIGURE 31.3
Ecoduct over the Amsterdam–Germany motorway in the Veluwe (A1), the Netherlands. (Ministry of Transport, Public Works and Water Management, 1998).

and where roads cross waterways, tunnels have to provide both a dry and a wet passage possibility for fauna.

References

Alterra (2004). The Pantanal–Taquari project. www.pantanal-taquari.alterra.nl.

Andrén, H. (1994). Effect of habitat fragmentation on birds and animals with different proportions of suitable habitat: a review. *Oikos*, 71, 355–366.

Council of Europe, UNEP, and European Centre for Nature Conservation (1995). *A Vision on Europe's Natural Heritage: the Pan European Biological and Landscape Diversity Strategy*. Strasbourg: Centre Naturopa.

Dramstad, W. E., Olson, J. D., and Forman, R. T. T. (1996). *Landscape Ecological Principles in Landscape Architecture and Land-Use Planning*. Washington, DC: Island Press.

Elzinga, G. and van Tol, A. (1994). Groene netwerken voor natuur en recreatie. Otters en natuurgerichte wandelaars, kanoërs en toerfietsers in het Groene Hart. Unpublished M. Sco. thesis, Wageningen, Agricultural University.

Forman, R. T. T. (1995). *Land Mosaics: the Ecology of Landscapes and Regions*. Cambridge: Cambridge University Press.

Fry, G. L. A. (1996). A landscape perspective of biodiversity. In *The Spatial Dynamics of*

Biodiversity, ed. I. Simpson and P. Dennis. Aberdeen: IALE, pp. 3–13. .

Fry, G. L. A., and Gustavsson, R. (1996). Testing landscape design principles: the landscape laboratory. In *Ecological and Landscape Consequences of Land Use Change in Europe,* ed. R. Jongman. Proceedings of the first ECNC seminar on land use changes and its ecological consequences. ECNC Man and Nature series 2, pp. 143–154.

IENE (2003). Infra Eco Network Europe website. www.iene.info.

Jongman, R. H. G. (2000). The difficult relationship between biodiversity and landscape diversity. In *Multifunctional Landscapes: Interdisciplinary Approaches to Landscape Research and Management,* ed. J. Brandt, B. Tress, and G. Tress, Roskilde, Denmark: Centre for Landscape Research pp. 72–83.

Jongman, R. H. G., Bunce, R. G. H., and Elena-Rossello, R. (1998). *A European perspective on the definition of landscape character and biodiversity. in Proceedings of the European IALE Seminar, Myerscough College,* pp. 21–36.

Jungwirth, M. (1998).River continuum and fish migration: going beyond the longitudinal river corridor in understanding ecological integrity. In *Fish Migration and Fish Bypasses,* ed. M. Jungwirth, M. S. Schmutz, and S. Weiss. Malden, MA: Fishing News Books, Blackwell Science, pp. 19–32.

Junk W. J., Bayley P. B., and Sparks, R. E. (1989). The flood pulse concept in river-floodplain systems *Canadian Special Publication of Fisheries and Aquatic Sciences,* 106, 110–127.

Lucas P. and van Oort, G. (1993). *Dynamiek in een stadsrandzone-Werken en wonen in de stadsrandzone van de agglomeratie Utrecht.* Utrecht: Faculteit Ruimtelijke Wetenschappen, Rijksuniversiteit Utrecht.

Mabelis, A. (1990). Natuurwaarden in cultuurlandschappen. *Landschap,* 7, 253–267.

Mander, Ü.,Külvik, M., and Jongman, R. H. G. (2003). Scaling in territorial ecological networks. *Landschap,* 20, 113–127.

Meeus, J. H. A., Wijermans, M. P., and Vroom, M. J. (1990). Agricultural landscapes in Europe and their transformation. *Landscape and Urban Planning,* 18, 289–352.

Ministry of Agriculture, Nature Management and Fisheries (1990). *Nature Policy Plan of the Netherlands.* The Hague: Ministerie Van Landbown, Natuurbeheer en Visserij.

Ministry of Agriculture, Nature Management and Fisheries. (1995). *Discussienota Visie stadslandschappen.* Den Haag: LNV Directie Natuurbeheer.

Ministry of Housing and Environment (2001). *Vijfde Nota over de Ruimtelijke Ordening* 2000/ 2020. Den Haag: Ministerie van Volkshuisvesting, Ruimtelijke Ordening en Milieubeheer.

Ministry of Transport, Public Works and Water Management (1998). *Ecoduct over A*1 *bij Kootwijk.* Information Bulletin.

Ministry of Transport, Public Works and Water Management (1999). *Ontsnippering.* Delft.

Naveh, Z. (1991). Some remarks on landscape ecology as a transdisciplinary ecological and geographical science. *Landscape Ecology,* 5, 65–74.

Naiman R. J., and Décamps, H. (1990). *The Ecology and Management of Aquatic–Terrestrial Ecotones.* UNESCO Man and the Biosphere series 4. London: Parthenon.

Opdam, P. (1991). Metapopulation theory and habitat fragmentation: a review of holarctic breeding bird studies. *Landscape Ecology,* 5, 93–106.

Pineda, F. D. (1990). Conclusions of the international symposium on biological diversity, Madrid 1989. *Journal of Vegetation Science,* 1, 711–712.

Roux A. L., Bravard, J. -P., Amoros, C., and Patou, G. (1989). Ecological changes in the upper Rhône River since 1750. In *Historical Change of Large Alluvial Rivers: Western Europe,* ed. G. E. Petts, H. A. L. Möller, and A. L. Roux. Chichester: Wiley, pp. 323–350.

Ruzicka, M. and Miklos, L. (1982). Methodology of ecological landscape evaluation for optimal development of territory. In *Perspectives in Landscape Ecology,* ed. S. Tjallingii and A. A. de Veer. Wageningen: Pudoc, pp. 99–108.

Secretariat General Benelux Economic Union (1996). *Tweede Benelux Structuurschets.* Brussels.

Townsend, C. T. and Riley, R. H. (1999). Assessment of river health: accounting for perturbation pathways in physical and ecological space. *Freshwater Biology*, 41, 393–405.

Urban, D. L., O'Neill, R. V., and Shugart, H. H. (1987). Landscape ecology. *BioScience*, 37, 119–127.

Van der Maarel, E. (1982). Biogeographical and landscape-ecological planning of nature reserves. In *Perspectives in Landscape Ecology*, ed. S. Tjallingii and A. A. de Veer. Wageningen: Pudoc, pp. 227–236.

Vannote, R. L., Minshall, G. W., Cummins, K. W., Sedell, J. R., and Cushing, C. E. (1980). The river continuum concept. *Canadian Journal of Fisheries and Aquatic Sciences*, 37, 130–137.

Wassen, M. and Verhoeven, J. (2003). Upscaling, interpolation and extrapolation of biogeochemical and ecological processes. *Landschap*, 20, 63–78.

PART VII

Retrospect and prospect

32

The land unit as a black box: a Pandora's box?

Modern landscape ecology developed in the first half of the twentieth century – before the computer age. As a marriage between geography and biology, its essence is the idea of land or landscape as a system, which means a *correlative complex of relations at the earth surface*. Models (hypotheses) of such a relational complex were originally either in the form of written metaphors, sometimes very simple algorithms, or often in the form of diagrams or graphical models (J. I. S. Zonneveld, 1985). Examples of written metaphors have been used in describing land as a device in cybernetic (dynamic) equilibrium (van Leeuwen, 1982) or land as a pseudo-organism, as in Lovelock's Gaia hypothesis (Lovelock, 1979). An example of a simple algorithm is:

$$L = F(r, \ w, \ s, \ c, \ p, \ a, \ m)\Delta t$$

in which L = land or landscape and F is a function of r = rock; w = water; s = shape of the terrain, landform, relief, topography, c = climate (atmosphere), p = plants, a = animals, m = man, and t = time. Examples of graphic representation are those showing relations as connecting lines between boxes representing the (supposed) building blocks and forming factors. Such metaphors, algorithms, and graphical representations were, and are still, commonly used in the sciences that fall under the umbrella of landscape ecology.

From my perspective, the dawning of landscape-ecological systems thinking began by trying to integrate the graphical webs (systems) of the soil scientist with those of the vegetation scientist and the geomorphologist. Each of these specializations appears to have many common elements or factor boxes (I. S. Zonneveld, 1987, 1995). But, as at the present time, even in those early days a kind of schism divided scientists within both the soil and vegetation disciplines as well as between reductionists and holists. On the one hand, the reductionists believed that the whole could only be understood from its details. On the other hand, the "pragmatic holists" concentrated on the whole or on the complexity of the entities, at least at preliminary stages. This latter group often classifies its models in this way and uses input/output data without knowing the details of processes involved and their interrelationships. Many of them consider output data as being sufficient for land evaluation, land planning, and land management for development and conservation.

Often in a scientific activity one is unable to define exactly, or to delineate initially, one's object of study. So too in landscape ecology the subsystems encountered may only be recognized initially by certain superficial characteristics and by input and output rather than by the actual processes and properties of the systems themselves. Such (sub)systems may be, at least temporarily, described as *black boxes* and may even be classified by a limited number of easily perceptible properties and recognizable characteristics. This is done in order to incorporate them as parameters in a model, or even in a final application, when there has been neither the opportunity nor the possibility of analyzing them in more detail. It is apparent that this type of generalist, and especially surveyors, may have more familiarity in applying the black box approach than do the reductionists. Such reductionists are, by their character, mainly interested in concentrating on detailed studies of basic processes, without regard to the time or the cost required. The concept of a black box, to this group, has a bad taste. To these scientists this black box could, without too much exaggeration, be reminiscent of Pandora's box – the source of evil in classical Greek mythology. For two and a half millennia this old Greek tale told of the beautiful, but evil, Pandora who carried around a box with potentially disastrous contents. By opening the box all the evil it contained escaped – disease, misery, greed, corruption, injustice, lies, and uncertainty. Hesiod's story goes further – Pandora closed the lid of the box before Elpis (Hope) – also in the box – could escape.

Reductionism versus holism

In my country, the Netherlands, just before and during the Second World War, competition developed between soil chemists and survey-minded soil scientists for government funding and contracts for applying their science in the evaluation of land (in fact soil) for agricultural use. The differences mentioned above were the basis of this competition. The soil chemist sought to determine, as directly as possible, "single values" such as the availability of certain minerals and, in case one wanted to know the distribution of these data in space, to put these as single values on topographic map sheets. The soil surveyors of those days, for example Oosting and Edelman, advocated the mapping of "soil bodies." These three-dimensional basic units of the landscape were determined by using morphometric, but easily perceived, properties of relief and the soil profile as classificatory characteristics. As indicators of boundaries between these map units and their land pattern, they applied scenery, landform, and vegetation/land use. The later introduction of the use of stereo aerial photography fitted ideally with this approach and perfected its development. The basic philosophy of this so-called "regional soil science" is the idea of a "correlative complex" including its spatial pattern. This represents a system of all soil factors that synergistically determine the capability, quality, and suitability for a particular use of the land, as far as the soil is concerned. To proceed from the large to the small is a general principle in any survey. This fits perfectly with the spatial, that is the landscape, orientation of "regional soil science" as the basis for survey. This approach to soil survey gradually became widely adopted internationally, especially through the FAO in developing countries.

Vegetation scientists, in those days, also argued that vegetation maps could, even more comprehensively, indicate the ecological quality of land. Their reasoning was that vegetation reflects the actual climate, water, animal, and human action in an integrated manner, in addition to the output of all soil and soil-forming factors.

For various, often political, reasons, the soil surveyors succeeded in obtaining the necessary financial support and contracts. This culminated in a systematic soil survey of the Netherlands. Later, vegetation surveys were also initiated and generally applied to grassland, forest, and (semi)natural areas, as well as arable fields based on weed communities, partly in combination with, or even integrated with, soil maps. In this context, landscape ecology as an applied discipline for region-wide ecological studies developed. The object became the region-wide study of relations in the geosphere, that being the total system at the earth surface or global landscape or "Gaia," but concentrated at region-wide scales, the appropriate order of magnitude for evaluation for land use. Within the geosphere several sub-spheres or "land-attributes" are distinguished. These are the lithosphere (rock and its influence on other spheres), the atmosphere (air/climate and its influences on the other spheres, including the main influence of the cosmosphere), the hydrosphere (water and its influences), the biosphere (plant and animal life and its influences), and the noosphere (humans and their cultural influence, artefacts, etc.). These land attributes are reflected and recognizable by landscape features such as scenery, landscape pattern, vegetation composition, relief, and, after some digging or augering, by the properties of soil and rock, as well as animal life and human artefacts.

Similar developments to these in the Netherlands took place in other parts of the world such as Canada, East and West Germany, Czechoslovakia, and other European countries including Russia in particular. In Australia, and to some extent in Britain, the well-known "land system" survey methodology developed and became widely applied in Australia and in various forms of "ecological," "biocenological," and "integrated" surveys elsewhere. The Toulouse conference on integrated surveys held in 1964 (UNESCO, 1968) was the first meeting point for the various approaches in applied land survey after the Second World War. Within the correlative complex of relations, studies demonstrated the geospheric (Gaia), the chorologic (horizontal), as well as the topologic (mainly vertical) dimensions.

Developments in the landscape paradigm

Various scientists working within the landscape paradigm had different approaches and also sought different applications. Consequently, they also named their foci differently, using terms like "integrated survey," "land survey," "land evaluation," "regional soil survey," "ecosystem survey," and "landschaftskunde." But here we encounter an interesting phenomenon, the influence of language on the development of science. As with many other words, the use of the German term *Landschaft* (in Dutch *landschap*) has gradually changed. Before the seventeenth century it still meant a more general region (in German also *Erdgegend*). It meant a specific part of the land defined according to certain selected criteria. It could be synonymous with either the Greek *chore* – referring mainly to spatial characteristics – or to *topos* – identified by content. This content could refer either to its natural, cultural, national, or administrative aspects. It may also have a somewhat narrower meaning when used as the very general term *Land* (land) which can be used for "area," irrespective of its use, or size, or character. In the seventeenth century the use of the term *Landschaft* (*landschap*) started to shift under influence of painting and art in the sense of a "picture" or "scenery" (in Dutch *stadsgezicht*, *landgezicht*). With such a background, in an aesthetic sense, a landscape could neither be good nor bad, suitable or unsuitable, but one just appreciated on a scale from beautiful to ugly. Both of these perspectives, the limited aesthetically based one and the neutral, more functionally based one, coexisted for quite some time – about two centuries. When the Germanic–Dutch term became anglicized into "landscape," after it was imported from the continent with painting and art, only its more limited, mainly aesthetically loaded meaning was associated with it, while on the continent both meanings persisted for at least two more centuries. For example, Carl Troll introduced the term *Landschaftsökologie*, originally as a subdiscipline of *Landschaftskunde* (landscape science), meaning the "region-wide study of the functional aspect of the correlative relational complex," alternatively described as the system of land at the earth surface. Hence, he used the term *Landschaft* in the original sense, as a region including its more superficial aspects like shape and even aesthetics, but did not concentrate on these items. Similarly, Edelman named his landscape-oriented soil science *regionale bodemkunde* (regional soil science) to correspond with the original, neutral, functionally loaded meaning of *landschap*.

These two tenors of the term "landscape" are also reflected in the world of landscape architecture. There we have two schools, one mainly focused on aesthetics (the design of the environment) and the other that is purely functional and which may even include exploitation of land resources.

When "landscape ecology," in its English translation, became generally accepted as an academic field the two meanings caused some considerable confusion. This effect is even enhanced by the fact that for these two different meanings two totally different words may exist in the non-Germanic languages. Indeed, it would have been better if Troll, in translating his *Landschaftsökologie,* had used the more neutral term "land" or "region" as in "regional ecology" or "land ecology." The latter term I have used at the request of my international students who are mainly interested in its application in land evaluation. In our international courses at ITC (the International Institute for Geo-Information Science and Earth Observation) and Wageningen University, it appeared that students from developing countries assumed the term "landscape ecology" referred to aesthetics and was an unnecessary luxury to consider for use in developing countries. By contrast, students from Japan and China, where there is a strong tradition of landscape art, were disappointed when lectures appeared to concentrate only on the physical/biological aspects of the "correlative complex of relations." And an even more negative reaction came from university administrators, who understood the term only in its more luxurious sense and consequently gave it little financial support or priority.

But what is in a name? In spite of this confusion a wide variety developed in region-wide approaches to studies of the "correlative land complex." These ranged from studies of mainly biological interest into the effects of spatial heterogeneity, such as the importance of metapopulations to comprehensive landscape development, and from purely scientific studies to applications in land planning – for development, management, and production, as well as conservation. Some of these studies and projects included a recognition of the aesthetic aspects of landscape as a necessity for an optimum human environment.

Landscape ecology as it developed in the United States in the last quarter of the twentieth century, when compared to the classic European, Canadian, and Australian approaches, very quickly developed in the chorological dimension;

that is, spatial pattern and its influence on life. Such studies of heterogeneity and the relation between patches of land in this dimension were even declared by some to be *the* core of landscape ecology! It is here that the confusion about the term landscape may have stimulated rather than hampered a fruitful co-evolution of landscape ecology on both sides of the Atlantic Ocean. It did lead to the emergence of IALE and consequently to a worldwide, flourishing transdisciplinary science.

This essay concentrates on the application of survey for the evaluation of land for management. Of greatest importance for the development of landscape ecology in this context has been the overwhelming rise in the use of the computer and its influence on methodology. Models can be treated much more comprehensively; classification using multivariate analyses can become a realistic option. Geo-information systems and other forms of geo-mathematics have become powerful methods for research and application. It has also increased the possibility of studying pattern as an end in itself. But it has lead to a hyping of pattern indices, some of which still have to demonstrate their value in increasing our knowledge of landscape and its interrelationships. There has been a tendency for this "patternology" to be overblown in value. However, the contributions to the methodology of the ecological meaning of pattern, within the correlative complex of relations in all dimensions (topological, chorological, and geospheric), has been a most important development for landscape ecology as a transdisciplinary science.

Computer techniques also appear to strengthen the reductionists in their conviction that one should study the wholes in their finest detail. Is it not logical to suppose that even the almost unlimited capacity of the modern computer can open the possibilities of analyzing and integrating these finest details and then combining them using GIS? This would be the opposite approach to that of the holists who can use GIS to separate the inferred land attributes from the total land entity! Would this therefore mean that intuitive thinking, partly based on "farmer's wisdom" and "the vague approach of the black box," would from now on be totally outdated and superfluous – yes, even dangerous?

Let us see now how far modern developments indeed justify the opinion of the reductionists in stating their somewhat exaggerated malcontent about "holism," that the black box is comparable with Pandora's box as the source of all evil.

Content and function of the black box and its hidden factors

I started this essay by stating that the correlative complex of relations at the earth surface is so intricate that it is practically impossible to handle without a reduction of its complexity. How can this be done? Where to begin? How far may one go? A first step in any science is the sampling and/or the description of the object under investigation. At the landscape scale we use rock formations, characterized by an association of minerals rather than individual mineral components, soil bodies rather than individual soil components, and plant communities characterized by structure and/or a certain association of species rather than individual plants. Even if we are interested in the role of one specific plant species in the correlative complex we may start with the more complex unit (community) to which that species appears to belong as a component. The same holds for certain rock and soil components. (For the much more dynamic animal-species component different approaches are also needed.) Such vegetation, soil, or rock subsystems of the total correlative land(scape) complex are really nothing more than a kind of *black box*. Apparently Nature, according to systems theory, allows us to bring order into our thinking by describing (classifying) it in hierarchical wholes. At the stage before more detailed study, these "wholes" necessarily represent hierarchically arranged black boxes. In fact, any complex parameter we put into our computer is, in itself, at least representing a complex of factors. The fact that many parameter values are a result of rather indirect measurements enhances this statement. We have even begun giving them names in order to categorize them and handle them systematically. This is common in vegetation and soil science and in geology and geomorphology prior to further detailed study of these units. Rather than using the bright ideas of modern professional philosophers, in this respect we follow millennia-old wisdom, derived from the common practices of pre-technological land users like hunters, farmers, and herdsmen who invented this principle at the dawn of humanity's struggle for life in the landscape. A major testimony to this is the wealth of information represented by the ecologically inspired land toponyms; that is, land names, in all languages, used for detailed land units up to regions of larger scale (Oba, 2001). A *chore*, which is just a patch or space at a certain location, is such a "toponym" but one raised to the level of a *topos*; that is, a particular individual landscape ecological unit recognized by its content and function. These black boxes, which may be "black" (more correctly "opaque") as far as internal processes are concerned, appear, however, to represent an entity – a "Gestalt" body – that can be recognized and hence named and classified. They may even be colorful and beautifully structured Gestalts (scenery!) with aesthetic and sometimes even emotional (territorial!) values.

For certain applications, sufficient knowledge may be obtained from such black-box descriptions or mapping. This may include some output data but without too much knowledge of the driving forces and processes of the total entity. Evaluation can then be done on the basis of input and output.

Improvements can be made on the basis of such empirical knowledge. Agriculture, for example, has been based for nearly 10 000 years on the empirical wisdom of farmers about these black boxes. This is long before the last 150 years when knowledge of the processes gained by modern science started to contribute to management. In a large part of the world, subsistence farming is still managed in this empirical way. Nowadays, more transparency through research into the internal processes of the black box is required, especially in cases where non-traditional management methods have to be introduced. The objective for such changes used to be to increase output (production), to enhance system stability by conservation, and to improve management in general – all required as a consequence of dramatically increasing human population size. But this requires, first of all, a knowledge of the factors of the correlative complex of relations at the earth surface. These factors may be subdivided in three categories: operational, conditional, and positional factors (van Leeuwen, 1966).

Operational factors

Operational factors are the actual physical and chemical processes that directly determine material abiotic and biotic reality. Variation and nuances in physicochemical processes inside organisms (plants and animals), which result in the products of assimilation and respiration, dominate the relationship between organisms and other attributes of the landscape. These processes are guided by fluxes of water inside the organisms, as well as in the surrounding landscape attributes, and carry minerals, nutrients, and waste products. According to the ideals of reductionism, these processes should be directly measured. The precise and direct measurement of such chemical and physical reactions and fluxes requires, however, that as far as possible, sophisticated methodologies, demanding expense in both time and money, be employed. In landscape studies, certainly in the applied sphere, in developing countries – which is a main source of my experience – these may rarely be available. But even under laboratory conditions such measurements depend more on input and output measures than on direct observation. Instead, one has often to be satisfied with inferring the process from the results of a registration of more robust phenomena like the behavior of organismic black boxes (plants and animals) within the biosphere, and from patterns in other landscape attributes like geomorphology, soil, or rock. These phenomena and patterns may have, more or less, the character of the synergetic output of the integrated operational factors resulting from assimilation, respiration, erosion, sedimentation, etc. These outputs may appear as static patterns or, in the case of robust fluxes, dynamic features. Using such data to represent operational factors can be an application of the black-box principle as long as the verification – or the falsifying – of the actual correlation between pattern and operational process remains to be done.

Observed phenomena may, instead of being the result of certain processes, also typify those conditions which themselves cause, maintain, and stimulate specific operational factors. This brings us, therefore, to "conditional" factors.

Conditional factors

Conditional factors are those phenomena which create, determine, and condition the operational factors. An example is "soil texture," which, in itself, is not an operational ecological factor. It does, however, determine various processes by conditioning, for example, the absorption of minerals and water and hence the availability of these factors, and these in turn condition various basic bio-processes of plant growth. This holds also for those abiotic processes concerned with stability, plasticity, and porosity of the soil in relation to its permeability to air and water. This will then affect sensitivity to erosion and subsequently to other land-forming processes, and so on. Vegetation cover conditions the availability of light and moisture for the soil surface and the organisms living there. A special form of such conditional factors, especially in the landscape context, are the "positional" factors.

Positional factors

Positional factors refer to the position in the landscape in relation to energy and information fluxes, in both the vertical and horizontal directions. They depend totally on the three-dimensional pattern of the landscape. Low-lying places obviously receive fluxes of water, minerals, sediments, etc. from higher areas. Neighboring land patches (units) of equivalent elevation are only influenced from the neighbor if the flux comes from that direction. This may be by atmospheric action (wind) carrying materials and diaspores, or tracks of animals or manmade fluxes. If the flux direction oscillates 180 degrees, an intensive, mutually connective relationship exists. Here we touch upon a core item of landscape ecology – that is, the importance of landscape heterogeneity as an influence on the structure and composition of pattern as spatial phenomena in the correlative system at the earth surface. One of the most important phenomena or principles, discussed in "relation theory" (van Leeuwen, 1966), is "to separate or to connect" or "closure versus openness" on any scale from membranes in living organic processes to chorological relationships between landscape pattern elements or land units.

A special aspect of positional factors is that they are conditioned by and can be read by the physiognomy of the earth surface – its pattern and topography (relief). These are the items used also in classification, survey, and mapping and thus facilitate the relationship between survey and the classification of the land's black box, on the one hand, and research into the factors determining the quality of land, on the other.

Depicting the black box using stereoscopic aerial photo interpretation

Carl Troll came to his *Landschaftsökologie* through observation of stereo pairs of early aerial photographs. He later declared how he achieved, in one glance, an impression of the *correlative complex* from the three-dimensional pattern image (exaggerated as it was in the *z* axis) that was immediately revealed. Indeed, in any comprehensive study of the landscape, or even of one of its main attributes like soil or vegetation, stereointerpretation of aerial photographs and landscape ecological thinking cannot exist without each other.

Photo interpretation, using stereo pairs of aerial photographs, is an art requiring initially both a deep understanding of the subject being interpreted and a knowledge of photogrammetry. That means, in this case, understanding the land and the landscape as the correlative complex of relations at the earth surface. Photo interpreters per se do not exist, but in this context landscape ecologists, soil surveyors, vegetation surveyors, geomorphologists, and other types of specialist do. They use aerial photographs as just one of the variety of tools any surveyor must master. This is somewhat different from other types of remote sensing, such as the (satellite) scanning of radiation, where the methodology and instrumentation is considerably more complicated than it is for using a simple stereoscope. The production of such remotely sensed images not only allows, but requires, processes of enhancement which demand both extra effort and time. This may tend to compete with critical "land ecological" thinking. Observing features of the landscape using stereo aerial-photointerpretation is, however, much more intense and also more integrative than any other remote-sensing method because of the very realistic image presented by the stereo image. The process of aerial-photo interpretation is a combination of both observation and recognition by integrating our conscious with our unconscious hidden knowledge which results from experience and intuition. The wealth of detail, even on small-scale (global) photographs, is integrated and structured in the interpreter's brain by a convergence of evidence. It stimulates insight, correlation, recognition, and discovery of features related to *conditional* or *positional* factors, as well as to the delineation of the observed piece of land as a mapping unit. Black and white, wide-angle photographs are far superior, in this respect, to any kind of satellite imagery currently available. These other means of remote sensing may, however, have special advantages: for example, in certain wavelengths radiation can be used for special purposes in providing both multi-spectral and multi-temporal imaging. Therefore, a combination of both types of remote sensing is advocated for the identification as well for further analysis of the area – or the "region-wide black box."

Classifying the black box

Since the landscape, as an entity, can be an object for study and evaluation, a systematic ordering of that object is also required. Fundamental reductionists may look down with contempt on this procedure – an activity known to both soil and vegetation scientists. In biology it was even the first major scientific activity in the days when organisms were still seen to be excellent examples of black boxes with their own clear, individual identity. So why should landscape not be classified as an entity? Classification is an ordering of the object of study in a practical, retrievable system and for that reason has a hierarchical form. In classification of spatial objects that cover the earth as a continuum, like soils and vegetation, two kinds of classification exist: abstract *typing* by agglomeration and *chorological*, partly by subdivision.

Typing or typifying is the common form of classification of organisms and other discontinuous individual items. It can, however, also be used by the continua covering the earth like soil, vegetation, and also landscape. Within these continua individual entities can be distinguished. Any reasonable morphometrically described properties can be used as (diagnostic) characteristics in an abstract system. The guiding principles to select these characteristics, and especially the hierarchical structure, may vary. Often ontogenetic criteria are used for this purpose. It may, however, also be that the properties relevant for application are used. An example of this, in the case of soil or vegetation classification, may be "fertility."

The properties used in vegetation classification as characteristics are usually derived from species composition and canopy structure, in soils from texture, horizonation, and chemical composition, and for landforms from shape, relief, etc.

The individual units that are abstracted at the lowest-order unit level, in any kind of classification system, show a relatively high degree of similarity according to the characteristics chosen. At one level higher the lowest-level units in the typology are grouped into units of higher rank that differ more in characteristics than those of the lower rank but less than units at the next higher rank, and so on.

Chorological classifications differ from "typing by agglomeration" principally in their hierarchical structure. The most common form of a chorological classification is a map. The lowest categories on detailed maps are identical to those used in typing by agglomeration. A worldwide chorological system is necessarily global and this perspective is reflected in any accompanying map and legend. Maps at an intermediate scale show units that exist as a composition of nested units of lower rank. The elements in these nested complexes do not need to be related in properties but only in their location. Unit composition and hierarchical arrangement tend to be regionally unique. This makes generalizations difficult. It means that a worldwide chorological classification, independent of maps, has hardly any added value beyond that of being a regional map. These classifications add mainly superfluous complications by generating more nomenclature. The doubtful results, from a comparable design of a chorologically based general classification in vegetation science, the so-called Sigma systematics, confirm this. A general, worldwide classification for land units (other than in the form of ad hoc maps with a special purpose) would have even more of a disadvantage than those mentioned above for soil and vegetation surveys. But, then, would a purely abstract typology (hence by agglomeration) for land units as such be useful?

As we have seen, land units can be considered as entities; hence, it would indeed seem logical to advocate the design of a general, abstract landscape typology in the same way as typologies exist for soils and vegetation. The design of an abstract landscape typology would certainly stimulate interesting scientific activity, as it has in other disciplines. A major point of discussion would be the selection of the criteria for determining the guiding principles. Any type of practical application, however, may demand different solutions causing huge complications. Moreover, ordering of basic data in the applied sphere, at the landscape level, is not strictly necessary because landscape units can be characterized by a combination of their building blocks – the land attributes (landform, soil, vegetation, etc.) – for which excellent abstract, regional, and general classification systems already exist. This latter procedure appears to be suitable also for the practice of land evaluation, this applied discipline being the main subject of this essay. So far, I have never felt the need for a worldwide, general landscape typology, whether it be in the humid tropics, the savannas, the arid zone, or the Arctic, and certainly not in the densely cultivated temperate landscapes (I. S. Zonneveld, 1995; van der Zee and Zonneveld, 2002).

Classifications using land attributes give the regional landscape ecologist freedom to choose the characteristics needed. My good friend the late Henk Doing classified the coastal dune region of the Netherlands very elegantly with a land-unit system based on an existing vegetation classification typology in combination with an existing geomorphic one. To this he occasionally added the occurrence of some individual plant species and other mosaic forms, including land and settlement features.

The use and misuse of the black box

Reductionists may abhor an acceptance of this indirect approach. They may, however, risk concentrating on only one, or a few, relatively easily measured factors in the system, possibly even combined in a simple mathematical model, and use that as though it would work in isolation. This can, however, have unscientific consequences. It is well known from development history that, when applied, such consequences can be quite disastrous. It should be remembered that incomplete knowledge about supposed operational factors has caused significant environmental damage. Opening the *black box* before a reasonably comprehensive knowledge is acquired about the balance between the positive and negative effects of intervening in a natural *complex of relations* may make it into a *Pandora's box*. The failures of large schemes in developing countries worldwide are well known and continue to produce, for example, accelerated erosion, the disastrous effects of misuse of pesticides and artificial fertilizers, and the incorrect manipulation of water in a non-integrated way. Beyond that, the single-minded management focus on vegetation as necessary land cover in arid zones has influenced climate and induces worldwide environmental change, the consequences of which still cannot be fully predicted. The same holds true for the new techniques being used in the most basic elements in the biosphere, the manipulation of DNA in genes.

So, if we use as a metaphor for these problems the Greek fable about a beautiful, enticing woman, evil is not necessarily caused by the content of the box since these can be manipulated with patience and wisdom before opening it. The power of evil is the woman herself, appropriately named *Pandora*, who represents the worst characteristics of humankind – stupidity, lack of wisdom, shortsightedness, irresponsibility, even criminal negligence by removing the lid before the negative consequences have been studied. A more prudent "pragmatic holistic" approach to black-box situations will produce more advantages and prevent negative impacts. Processes or factors exist which we never suspected, like the old grandmother of the Neolithic village who supervised agriculture in her territory for decades but had no idea of P, K, and N! Studying the input and output of pure black boxes is the necessary first step in discovering such factors. Powerful modern computers and cunning software designers may open new ways to unravel systems that were, until the present time, too complex or too difficult to analyze. For the fundamental reductionist reader, who still distrusts the black box, this may be the Elpis (Hope) that Pandora conserved in her box.

Conclusions

In discussion it has been argued that a prudent application of the black-box principle is not a source of evil. The black-box principle can contribute to, and can even enhance, the characteristics of land ecology as an applied science by:

- stimulating awareness of its complexity and risks
- providing an efficient methodology to study and survey the landscape
- providing a proper base for land evaluation
- stimulating vision in the field of management and conservation of land and landscape

Through use of the black-box principle it is possible to direct an approach to the land as an entity in itself whose properties and characteristics can be measured and, by extension, can be mapped and registered as changes in four dimensions. Land evaluation based just on input and output provides in an efficient way, in low-budget circumstances, reliable estimations. The black-box approach provides a first step in tackling the analysis of complex systems as research objects. The land unit, delineated as a black-box by using only superficial characteristics representing conditional and positional factors, can be used as a vehicle for knowledge concerning input and output. Even more important is its use for storing and integrating local wisdom and any other empiric knowledge or personal experience. It is a source for increasing vision about management and for enhancing the need and direction for further research. The land unit can be used for stratifying analysis and integrating the positional factors of the land attributes (soil, vegetation, water, landform, relief, climate, etc.). It can be used also as the basis for mapping these factoral attributes. Electronic geo-information systems (GIS) can be used to separate these data, in cartographic form, from the holistic reality of the land. The pattern of land units that are individually opaque boxes at the survey stage may reveal, when used in combination with empiric knowledge gained from both local knowledge and common (scientific) sense, important positional and conditional factors. And possibly, with good judgment, they may even lead to inferences about the operational factors.

Most of all, the accumulated and integrated knowledge concerning the *land-unit black box* will stimulate an awareness of the complexity of land and landscape. This awareness will include a consciousness about the danger of destroying the balance of its intricate web and of triggering unforeseen, even non-restorable, destruction – which has been the consequence of so many development projects in the past. If the *black box* is damaged in this way it may turn into a *Pandora's box* from which the evil of loss of diversity, erosion, and other devastating and impoverishing processes of resource depletion may spread. The hope is that, for the time being, there remains the possibility of wise management by the careful monitoring of the input and output of these opaque black boxes, without the full knowledge and the precise working of all their operational factors. In the meantime, a deeper, more detailed landscape-ecological research agenda must evolve with a good balance between modest reductionism and holism. This may then lead to an improved knowledge base for the better management of the intricate *correlative complex of relations at the surface* of our increasingly over-populated planet.

References

Lovelock, J. E. (1979). *Gaia: a New Look at Life on Earth*. Oxford: Oxford University Press.

Oba, G. (2001). Indigenous ecological knowledge of landscape change in East Africa. *IALE Bulletin*, 19, 1–3.

UNESCO (1968). Aerial surveys and integrated studies. In *Proceedings of the Toulouse Conference on Principles and Methods of Integrating Aerial Survey Studies of Natural Resources for Potential Development* 1964. Paris: UNESCO.

Van der Zee, D. and Zonneveld, I. S. (2002). *Landscape Ecology Applied in Land Evaluation, Development, and Conservation: Some Selected World-wide Examples*. Enschede: ITC/IALE.

van Leeuwen, C. G. (1966). A relation theoretical approach to pattern and process in vegetation. *Wentia*, 15, 25–46.

van Leeuwen, C.G. (1982). From ecosystem to ecodevice. In *Perspectives in Landscape Ecology*, ed. S. P. Tjallingii and A. A. de Veer Wageningen: Pudoc, pp. 29–34.

Zonneveld, I. S. (1987). Landscape ecology and its applications. In *Landscape Ecology and Management*, ed. M. R. Moss. Montreal: Polyscience, pp. 3–16.

Zonneveld, I.S. (1995). *Land Ecology: an Introduction to Landscape Ecology as a Basis for Land Evaluation, Management, and Conservation*. Amsterdam: SPB Academic.

Zonneveld, J. I. S. (1985). *Graphical Models Used in Landscape Ecology*. Utrecht: VCMr, University of Utrecht.

33

Toward a transdisciplinary landscape science

In the current period of transformation from an industrial to a post-industrial, information-rich age with its severe ecological, socioeconomic, and cultural crises, it has become very obvious that a critical point has been reached in the earth's capacity to support both nature and the growing consumption and expectations of its rapidly growing human population. For the first time in the history of the earth, one species – *Homo sapiens* – has acquired the power to eradicate most life in our natural and semi-natural landscapes, threatening not only their vital life-support functions but also human life itself. To divert the present evolutionary trajectory, which is leading toward breakdown, collapse, and extinction, to a breakthrough toward the sustainable future of nature and the highest attainable quality of human life, there is an urgent need for a far-reaching revolution of environmental and cultural sustainability (Laszlo, 2001). This is imperative in order to reverse global biological and cultural degradation and for dampening the dangerous effects of global warming and the elimination of the scourge of poverty. According to Brown (2001) this *sustainability revolution* will be driven by the widespread adoption of technological innovations in regenerative and recycling methods and in the efficient utilization of solar and other clean and renewable sources.

There are already many encouraging indicators that this is not an unrealistic Utopia. For example, the use of wind turbines and photovoltaic cells is growing now at over 25% annually, and will very soon be competitive with fossil fuels. Organic farming has become the fastest-growing sector in the world agricultural economy. However, these achievements must be coupled with more sustainable lifestyles and consumption patterns, more caring for nature and even investing in nature. This requires landscape ecologists to be morally committed to the solution of the current ecological crisis and its implications for the future of our landscapes, to broaden their disciplinary

 Issues and Perspectives in Landscape Ecology, ed. John A. Wiens and Michael R. Moss. Published by Cambridge University Press. © Cambridge University Press 2005.

and fragmented thinking, and to act with both an integrative and a transdisciplinary outlook. Like other environmentally concerned scientists, they will have to leave behind an obsolete belief in the concept of an objective, mechanistic, and reductionistic science. Instead, they must involve themselves in mission-driven, forward-looking, transdisciplinary research, education, and action, which bridges the gaps between the natural sciences, the social sciences, the arts and humanities.

Some major premises for a transdisciplinary landscape science

In this brief essay I am arguing that, for such a transdisciplinary shift, landscape ecologists will have to achieve more than just a focus on the biophysical and ecological landscape parameters, as suggested by Moss (1999). The science of landscape ecology has to become much more than simply another "normal" academic scientific discipline (*sensu* Kuhn, 1970). According to the IALE mission statement (IALE, 1998), landscape ecology deals mainly with "the study of spatial variation in landscapes at a variety of scales." Landscape ecology needs a much broader holistic, future-oriented conceptual basis with a clearer definition of its theoretical and practical aims. These must include those human ecological aspects which deal with the people – living, using, and shaping the landscape – for good or bad, enjoying them or suffering from them. Instead of reducing them to nothing more than "socioeconomic factors" in their landscape models and interpreting their behavior merely as "*Homo economus*," landscape ecologists will have to take into consideration not only the material aspects of human ecology but also humans' intellectual and spiritual needs, their wants and aspirations. Humans, in a much broader holistic sense, are the ones who have to avoid further landscape impairment and who have to restore the integrity, productivity, and beauty of landscapes and ensure their future sustainability. Therefore, landscape ecologists will have to change their view of landscapes, from a multidisciplinary and interdisciplinary perspective of being composed of physical, chemical, and biotic and abiotic landscape elements and processes, into a more holistic systems view of landscape and its multifunctional natural and cultural dimensions and functions, as an undividable whole. They will have to adopt innovative transdisciplinary principles and methods in research, education, and action, transcending and crossing the disciplinary borders, now restricted to the conventional natural sciences and based in ecology and geography. According to Jantsch (1970), the goal should be to reach out beyond interdisciplinarity to an even higher stage of integration and cooperation with the relevant fields of the social sciences, the humanities, and the arts, and aim toward a common systems goal – in this case that of sustainability.

This does not mean that they will have to neglect their own unique disciplinary expertise, gained as it is from different fields of knowledge and molded to deal with the land as a whole, as discussed by Moss (1999). Rather, they will have to share it with those synthetic "eco-disciplines" which are already successfully integrating their discipline-based social foci with ecological principles and knowledge. These include such fields as ecological economics, eco-psychology, social ecology, urban ecology, and industrial ecology. Rather than being discipline-oriented, this type of transdisciplinary research should be problem-oriented and carried out in close collaboration both with the professionals who deal with land-use planning, management, and decision making, and with the public at large. Thus, by taking an active part in the practical implementation of their research and working together with other scientists and professionals toward this common systems goal – that is, the sustainability revolution – landscape ecology will hopefully become an influential transdisciplinary landscape science.

Space does not allow me to cite examples, from the many encouraging signs, of a recognition of the need for changes in this direction by landscape ecologists. Thus, for instance, at the 1995 World Congress of IALE, Richard Hobbs (1997) pleaded for a more active involvement of landscape ecology in the solution of pressing environmental problems. Probably the most forceful expression of the need to transform landscape ecology into a transdisciplinary landscape science is to be found in the resolutions made at the 1997 conference of the Dutch Association of Landscape Ecology (Klijn and Vos, 2000).

Any trends toward transdisciplinarity are not possible without the acceptance of a holistic concept of landscapes as synthetic nature–culture systems. The foundations of this perspective were laid in Central and Eastern Europe by the end of the Second World War. They are now widely accepted and practiced worldwide. Some of the major theoretical and conceptual cornerstones for such a transformation were outlined by Naveh and Lieberman (1994). These have been updated more recently by Naveh (2000, 2001, 2003), Li (2000), Tress and Tress (2001), Carmel and Naveh (2002), and Bastian and Steinhardt (2003). Here I will focus briefly on a few of the most important issues.

This revolutionary, transdisciplinary landscape paradigm can only be comprehended fully within the broader context of a "scientific revolution" as expressed by Kuhn (1970). Rooted in General Systems and Hierarchy Theory, it is based on a major shift from reductionistic and mechanistic paradigms to a holistic and organismic scientific world-view and to a new scientific understanding of the "web of life" (Capra, 1997). As lucidly shown by Laszlo (1994, 2001), this has led to an all-embracing concept of a synthetic, cosmic, geological, biological, and cultural evolution as a non-linear but coherent evolutionary process. It has far-reaching practical implications for providing solutions to our present crises at the crucial transitional "macroshift" toward the information age.

The total human ecosystem and the total human landscape

A holistic landscape-ecological conception fits very well into this integrative systems view of the world. It culminates in the recognition that humans are a part of nature, but not apart from nature, or above nature. Together with their total environment they form an indivisible co-evolutionary geo-bio-anthropological entity of the "Total Human Ecosystem" (THE), as the highest ecological global micro-level of the macro-level of the self-organizing universe.

Landscapes are the spatial and functional matrix for all organisms, populations, and ecosystems. As such they are also the concrete space-time defined ordered wholes of our Total Human Ecosystem, ranging from the smallest mappable landscape cell or ecotope, to the global human-dominated "Total Human Landscape" (THL).

According to this hierarchical systems view, each landscape unit, regardless of its size, should be treated on its own right as a "holon" of the global THL "holarchy," that is more than the measurable sum of its living and non-living components. Interlaced as spatial and functional networks, landscapes have become entirely new entities of ordered and irreducible whole "Gestalt" systems, which are more than puzzles of mosaics in repeated patterns of ecosystems. As "medium numbered systems" (Weinberg, 1975), neither mechanical nor statistical approaches nor their description and analysis as Archimedian geometric configurations can do full justice to their organized structural and functional complexity. Innovative approaches and methods are required for their study.

Multidimensional and multifunctional landscapes as tangible bridges between nature and mind

Whereas the natural landscape elements have evolved and are operating as parts of the geosphere and biosphere, their cultural artefacts are creations of the "noosphere," namely the sphere of human mind. As described brilliantly by Jantsch (1980), the late, great systems thinker and planner, this is an additional natural envelope of life in its totality that *Homo sapiens* have acquired throughout the evolution of the human mind. It is our "mental space" and the domain of our perceptions, knowledge, feeling, and consciousness, which enables our self-awareness and cultural symbolization and our linguistic and artistic expression. It enabled the development of additional noospheric realms of the info-, socio-, and psycho-sphere that have emerged during our cultural evolution. As a result, our Total Human Landscapes are driven both by geo-spheric and noospheric processes, which are transmitted simultaneously by biophysical and by cultural information, chiefly with the help of our natural and formal scientific language.

For transdisciplinary study, and for the appraisal and management of the natural and cultural dimensions of multifunctional landscapes, we have to tear down the perceptual barriers which view landscapes as either entirely physical or entirely mental–perceptual occurrences. This can be achieved by treating them with a "biperspective systems view" by which single, self-consistent mind events of human cognitive systems and natural, physical space-time events of concrete biophysical systems are observable and manage-able simultaneously as integrated natural-cognitive and psychophysical sys-tems (Laszlo, 1972). This enables us to treat these multidimensional and multifunctional landscapes as the tangible bridges between nature and mind.

Biosphere and technosphere landscapes and their integration in the post-industrial symbiosis between human society and nature

Throughout the period of dynamic, non-linear cultural evolutionary process, characterized by sudden leaps and crucial bifurcations, pristine land-scapes have undergone far-reaching modifications and conversions by human land use and activities. Our present disorganized "Total Human Industrial Landscape" (THIL) is the result of the Industrial Revolution. This caused a major bifurcation between the natural and semi-natural solar-energy--powered biosphere landscapes, operating as self-organizing and autopoietic regenerative systems on one hand, and on the other hand human-made and maintained urban–industrial technosphere landscapes, driven by polluting and high-entropy dissipating fossil energy. As unsustainable throughput systems, they are threatening the future health of both humans and nature. The same is also true for the "hybrid" solar- and fossil-energy-powered, intensive agro-industrial landscapes.

Biosphere landscapes, and their spontaneously developing and reproducing plants and animals, fulfill vital multiple life-supporting functions for human physical and mental health without the need for any external energy or mate-rial inputs. To overcome these antagonistic relations, and to ensure full spatial and functional integration between bioagro- and techno-landscapes in our Total Human Landscape (THL), new symbiotic relations between human society and nature have to be created. One of the most significant contributions of landscape ecologists to this symbiosis, and thereby also to the sustainability revolution, should be their active involvement in the dynamic management, conservation, and restoration of the most valuable and richest biosphere "keystone systems" on which further biological evolution depends.

As shown in a recent multinational European Union modeling project of the Sustainable European Information Society, such a symbiosis could be achieved by the creation of mutual supportive cultural and economic auto- and cross-catalytic networks closely linking natural, ecological, socio-cultural and economic processes for the benefit of both nature and humanity (Grossmann, 2000).

Some important issues for transdisciplinary landscape research

Among the most important practical consequences arising from this transdisciplinary approach to our Total Human Landscape is the need for a much broader, integrative appraisal of their multidimensional landscape functions. The biperspective view enables their evaluation, not only in the anthropocentric dimension of "hard" instrumental and marketable values, but also in the "soft" ecocentric and ethical dimensions, which are not dependent on utilitarian values but are grasped with our cognitive and perceptual dimensions and consciousness. Ongoing exponential landscape degradation cannot be prevented by treating landscapes solely as a commodity to be exploited or as a resource on which we project our economic interest and measure by monetary parameters and products of the "free market play." We have to recognize the intrinsic values by which they become not a means to an end, but an end in themselves. Even the term "natural capital," introduced by ecological economists, cannot account fully for the most vital life-support functions provided by fertile soil, clean air, and water. Nor can this account at all for the intangible aesthetic, cultural, spiritual, and re-creative values of healthy and attractive biospheric landscapes.

The importance of these landscapes for our quality of life and mental well-being in our emerging information society is now greater than ever. Therefore, much greater attention should be paid to "psychotherapeutic landscape functions." These are derived from the restorative experience of nature acting against the multitude stresses of modern life. This is particularly the case with "direct attention fatigue" (*sensu* Kaplan, 1995) after prolonged intensive mental and creative work, such as that performed by computer operators working in the high-tech field.

The biperspective view, and its application for the utilization of multifunctional landscape complexity, is also a precondition for the above-mentioned, integrated ecological, socioeconomic, and culturally sustainable forms of development and their cross-catalytic networks. The preparation of practical strategies, supported by dynamic, transdisciplinary systems simulation models and other interactive methods and tools, can only be realized as a joint transdisciplinary effort by both landscape ecologists and scientists from relevant natural, social, and human disciplines as well as with artists, planners, architects and eco-psychologists, land-use managers and decision makers.

As mentioned above, highest priority has to be given to research and action that ensure further evolution of organic life in our most valuable natural and semi-natural, solar-powered, autopoietic biosphere landscapes and keystone systems. For this purpose we have to maintain and restore their dynamic homeorhetic flow equilibrium, fostering their inbuilt resistance and adaptation capacities to the unexpected, and utilizing their regulation and connectivity functions and their buffering, sheltering, and filtering capacities.

Another, most urgent transdisciplinary challenge is the development of practical tools for the integrated assessment of closely connected biodiversity, cultural diversity, and ecological macro-and micro-site heterogeneity by joint indices of "Total Landscape Eco-diversity" (TLE-d) that can be easily applied by land managers and stakeholders.

All these research activities should be part of the overall effort toward the functional and structural integration of all our natural and cultural landscapes into a more coherent, better organized, and more sustainable post-industrial Total Human Landscape. For this purpose, future-oriented, mission-driven, transdisciplinary landscape ecologists will be much better equipped to help in the conversion of unsustainable, high-input, high-throughput agro-industrial landscapes into sustainable, regenerative, non-polluting but no less productive agro-ecological landscapes, and in the creation of healthier, more livable, and more attractive urban–industrial technosphere landscapes. They will have to shift their focus from the rigid, geometric landscape structures and from theoretical exercises aimed at inventing more and more sophisticated landscape indices, to the understanding of dynamic landscape processes and functions. They will have to be ready to present their work, not only as strictly scientific publications, but also as well-illustrated, non-formal, and easily accessible "pragmatic" information.

Landscape ecologists, planners, and managers will, very soon, also have to find very creative and sound solutions to the consequences of dramatic landscape change, such as the large-scale abandonment of agricultural fields and upland pastures, and the changes caused by establishing solar- and wind-powered installations. They should be ready to deal with uncertainties and surprises using both virtual landscape scenarios and risk models, and with biological and ecological landscape-engineering methods which attempt to avert the catastrophic results of, for example, forest destruction, river damming, and wetland filling. The disasters likely to be caused by increasingly extreme climatic events related to global climate destabilization, such as drought, flooding, hurricanes, sea- and river-level rises, landslides, and erosion are additional objectives of this approach.

This emerging transdisciplinary landscape paradigm cannot be imprisoned by a deterministic and mechanistic predictive scientific theory, for which classical Newtonian physics has served as a model, but which has already been abandoned by many innovative theoretical and quantum physicists. Therefore, instead of trying in vain to mature into a "predictive" science, landscape ecology will have to renew itself as a dynamic, anticipatory, and prescriptive science.

We cannot predict the future of our landscapes and their rapid and sometimes even chaotic changes by simply extrapolating from the past and present into the uncertain future. But we can take part in creating their future by translating our vision into action, realizing that what we will do today will shape the world of tomorrow. With the help of positive scenarios, we can prescribe what, in our opinion, should be done to realize those that are most desirable. We should make every effort to promote the shift from the fossil-energy-driven despoiled, polluted, homogenized, and suburbanized landscapes of the industrial society into more sustainable, healthier, attractive, productive, viable, and livable landscapes.

In concluding, I hope that we will be able to educate a new breed of committed, transdisciplinary landscape ecologists, planners, managers, and restorationists who will respond to all these challenges, as experts in their own field and as integrators, who will be able to combine landscape-ecological knowledge with broad ecological wisdom, and with consciousness and environmental ethics.

References

Bastian, O. and Steinhardt, U. (2003). *Development and Perspectives in Landscape Ecology*. Dordrecht: Kluwer.

Brown, L. R. (2001). *Eco-Economy: Building an Economy for the Earth*. New York, NY: Norton.

Capra, F. (1997). *The Web of Life: a New Scientific Understanding of Living Systems*. New York, NY: Anchor Doubleday.

Carmel, Y. and Naveh, Z. (2002). The paradigm of landscape and the paradigm of ecosystems: implications for landscape planning and management in the Mediterranean region. *Journal of Mediterranean Ecology*, 3, 35–46.

Grossmann, W. D. (2000). Realizing sustainable development in the information society. *Landscape and Urban Planning*, 50, 179–194.

Hobbs, R. (1997). Future landscapes and the future of landscape ecology. *Landscape and Urban Planning*, 37, 1–9.

IALE (1998). IALE mission statement. *IALE Bulletin*, 16, 1.

Jantsch, E. (1970). Inter- and transdisciplinary university: a systems approach to education and innovation. *Policy Sciences*, 1, 203.

Jantsch, E. (1980). *The Self-Organizing Universe: Scientific and Human Implications of the Emerging Paradigm of Evolution*. Oxford: Pergamon Press.

Kaplan, S. (1995). The restorative benefits of nature: toward an integrative framework. *Environmental Psychology*, 15, 169–182.

Klijn, J. and Vos, W. (2000). *From Landscape Ecology to Landscape Science*. Dordrecht: Kluwer.

Kuhn, T. S. (1970). *The Structure of Scientific Revolutions*. Chicago, IL: University of Chicago Press.

Laszlo, E. (1972). *Introduction to Systems Philosophy: Toward a New Paradigm of Contemporary Thought*. New York, NY: Harper Torchbooks.

Laszlo, E. (1994). *The Choice: Evolution or Extinction? A Thinking Person's Guide to Global Issues*. New York, NY: Putnam.

Laszlo, E. (2001). *Macroshift: Navigating the Transformation to a Sustainable World*. San Francisco, CA: Berret-Koehler.

Li, B.-L. (2000).Why is the holistic approach becoming so important in landscape ecology? *Landscape and Urban Planning*, 50, 27–47.

Moss, M. R. (1999). Fostering academic and institutional activities in landscape ecology. In *Issues in Landscape Ecology*, ed. J. A.Wiens and M. R. Moss. Guelph: International Association for Landscape Ecology, University of Guelph, pp. 138–144.

Naveh, Z. (2000). What is holistic landscape ecology? A conceptual introduction. *Landscape and Urban Planning*, 50, 7–26.

Naveh, Z. (2001). Ten major premises for a holistic conception of multifunctional landscapes. *Landscape and Urban Planning*, 57, 269–284.

Naveh, Z. (2003). The importance of multifunctional self-organising biosphere landscapes for the future of our Total Human Ecosystem: a new paradigm for transdisciplinary landscape ecology. In *Multifunctional Landscapes. vol. 1: Theory, Values and History,* ed. J. Brandt and H. Vejre. Southampton: WIT Press, pp. 33–62.

Naveh, Z. and Lieberman, A. T. (1994). *Landscape Ecology: Theory and Application*, 2nd edn. New York, NY: Springer.

Tress, B. and Tress, G. (2001). Capitalizing on multiplicity: a transdisciplinary systems approach to landscape research. *Landscape and Urban Planning*, 57, 143–157.

Weinberg, G. M. (1975). *An Introduction to General Systems Thinking*. New York, NY: Wiley.

34

Toward fostering recognition of landscape ecology

The volume of essays (Wiens and Moss, 1999) produced for distribution at the Fifth World Congress of IALE, the International Association for Landscape Ecology, generated a good deal of interest and comment. What has now emerged from that original collection of essays is this expanded and updated version. The essay I contributed to the original volume (Moss, 1999) contained my personal observations on the status of the field of landscape ecology and the role played by IALE, academic institutions and practitioners in advancing the field. Now, five years later, it is perhaps appropriate to re-examine these comments and to make some reassessment of how the profile of landscape ecology may have changed amongst its adherents, within the scientific community at large, within academic institutions, and amongst those practitioners who apply its ideas to solving environmental problems.

In the 1999 essay my main argument focused on the need for a clear understanding of what "landscape" means to landscape ecologists (see also Moss, 2000). One of the major problems I saw then was the need to bring together into this focus the "two solitudes" within landscape ecology: the geoecological and the bioecological traditions. Since that time this same issue has been raised by several commentators. Bastian (2001) has added a great deal to this debate, starting from a historical perspective, and Opdam *et al.* (2002) expanded the discussion to the context of landscape-ecological input to spatial planning. What I find, however, is that much of my discussion from 1999 can legitimately be repeated. My thesis remains: that landscape ecology has now come of age, but that its healthy, youthful development will be cut off before it matures if it does not recognize and develop its own distinctive core and focus. Furthermore, the many progressive developments now taking

Issues and Perspectives in Landscape Ecology, ed. John A. Wiens and Michael R. Moss. Published by Cambridge University Press. © Cambridge University Press 2005.

place in landscape ecology will become marginalized if some fundamental concepts about landscapes do not emerge to form a clear focus to which the diverse perspectives raised in this volume can contribute. It will be argued also that unless landscape ecologists agree upon such a conceptual core for their field, the fundamental questions about landscapes cannot be asked, and hence no particular body of general theory about landscape ecology will emerge. In other words, the *science* will not develop and the benefits we see from its *application* will not materialize. Unless the scientific community and practitioners of landscape ecology can identify with a clearly defined body of knowledge when looking for solutions to particular landscape problems, then the applications or the *practice* of the science will be limited. And unless this materializes, academic institutions will have little reason to be persuaded to support the development of programs and courses to educate future students in the field. These will be essential in producing a new generation of landscape ecologists. I would argue that these individuals should be educated as landscape ecologists rather than as individuals who see the field as peripheral to some other academic sphere – the situation still typical for many of the first, still dominant, generation of landscape ecologists. My conclusion is that, unless landscape ecology emerges as a disciplinary field in its own right rather than as an inter-, cross-, trans-, multi-disciplinary field, it will never become accepted as an academic endeavor of worth in most institutions of higher education, nor will there be a clear avenue for its applications. To achieve this status, a strong theoretical and methodological base must be developed. Without an academic anchor the interaction of participants from across the field, which now tends to occur only at IALE congresses and regional meetings, will not be able to generate the cross-fertilization needed to advance the subject.

From a somewhat pessimistic standpoint, one can say with some certainty that we can readily recognize the twin origins of the field, which have persisted as two solitudes to the present time. Should these two solitudes remain unreconciled – that is, without their adherents recognizing that their respective sub-fields are part of a broader concept – what will emerge under the umbrella of landscape ecology is likely to be an increasing divergence away from *landscape* as its core.

An optimist would, however, recognize that these two solitudes do appreciate the others' perspectives. After all, did not IALE come into existence in Piešt'any, Czechoslovakia, in 1982 (Ruzicka, 1999) when the bioecological tradition, primarily from the United States, sought to wed the geoecological traditions, primarily from middle and eastern Europe, in a marriage brokered by the Dutch at the First International Symposium held in Veldhoven, the Netherlands, in 1981?

What are the current issues for the field of landscape ecology?

Landscape ecology has developed remarkably over the last two decades but it remains at a critical threshold. It is increasingly recognized as a field of scientific investigation, and some of the results have been put into practice by practitioners such as landscape architects and resource planners; it has established international journals and basic texts; it has a growing cohort of adherents; and it is establishing a foothold in academic institutions throughout the world. What, then, is the problem?

Perhaps to start this discussion it would be wise to state what, in my view, landscape ecology is not. It is not the only field dealing with landscape issues and it certainly is not *the* all-embracing environmental science. It is, however, a field with the potential to make a unique contribution to solving a particular subset of natural-resource-based issues. But to achieve this goal requires answers to three points. First, what fundamental, generic questions does landscape ecology ask about the landscape that differ from those of other fields? Second, what types of information can landscape ecology generate by addressing these fundamental questions? Third, do all adherents to the label "landscape ecologist" subscribe to the same basic focus?

My own response to the last question would be "no." And therein lies something of an answer to the first two questions. Beyond a certain superficial level, most people would recognize the continuing existence of the two founding "solitudes": the bioecological perspective and the geoecological perspective (Fig. 1 in Moss, 1999). Without much doubt the major advances in the discipline in the last 25+ years have been within the bioecological sector, particularly through initiatives from within the United States. The longer-established tradition of the geoecological perspective dates back to the early decades of the twentieth century in Europe, based on either geographic or soil-science traditions. This subfield subsequently advanced in state research institutes and academia, largely in the former Soviet bloc. The bioecological approach is derived from, and based almost entirely within, the biological sciences, particularly ecology, and stems from a recognized need to understand the significance of the spatial dimension in vegetation and animal populations and in community-scale dynamics. The geoecological approach in its early developmental phase sought to define land systems and regional spatial entities on the basis of the systematic interpretation of land-related components such as landforms, soils, vegetation, and human land-use impact. In addition, energy, moisture, and biogeochemical forces, which integrate these landscape elements to produce distinct landscape units, added a dynamic aspect to this work.

Are these two solitudes irreconcilable, or have they merely remained relatively distinct, largely due to their different linguistic and geographic bases? It is perhaps worthy of note that where these two perspectives have been effectively integrated, for example by Dutch and Danish landscape ecologists, the degree of impact of landscape-ecological applications on resource, land, and environmental planning appears to have been most successful. It is along these lines that some minor yet discernible changes in attitude have taken place over the last five years, particularly by applied landscape scientists. By raising the issue, and by generating discussion about the dangers of continuing along existing pathways, the benefits of more collaborative approaches become evident (Opdam *et al.*, 2002). Should the two sub-themes continue to exist independently, they will inevitably become increasingly divergent, obfuscating the real potential of the field.

What is needed is an identification of the unifying goals and critical fundamental questions that will form the *one* focus for both (a) the bioecological theme of ecology *in* the landscape, and (b) the geoecological theme *of* land(scape) system science. The current underlying weakness of (a) for this scenario is that its main justification is the importance of the spatial perspective to plant and animal community dynamics. This inevitably means that the main reason for its existence is to improve our knowledge of plant and animal communities. The landscape merely is the broader context, or the template, in which this takes place. To justify the existence of landscape ecology merely as a spatial science is severely restrictive. Do not most environmental disciplines require a spatial dimension in their approaches? Geographers, for example, have found (to their cost) the limitations of justifying their subject on the basis of the study of spatial distributions only. This became known as the spatial encumbrance. A spatial dimension is critical to any discipline dealing with variations in the character of its objects over areas of the earth's surface. But it cannot be its sole justification.

The underlying weakness of (b), the geoecological theme, has been in making assumptions that the superimposition of individual land-component data generates functional landscape units. In fact, to understand function requires a knowledge of process, and the study of processes, in complete land-unit systems, requires a functional integration, not merely the combining or superimposition of a range of pedological, hydrological, geomorphic, lithospheric, and atmospheric process information. Furthermore, it is a widely held (but often invalid) assumption that the abiotic elements inevitably determine the nature and character of the biotic landscape elements. There is often also the assumption that a given set of abiotic characteristics would result in a predictable set of spatially repetitive biotic characteristics. This viewpoint ignores the ability of biotic elements to modify their own environments. It is an approach that is particularly misleading in regions where human activities have affected landscapes, particularly their biotic components, for long periods of time. The emergence of pertinent process information about land systems, whether modified or relatively untouched by human activity, remains severely limiting, being based on many other disciplines whose objects of study are merely parts of a land system and not the systems themselves. Another serious limitation of this sub-field, to date, has been a lack of knowledge of changes in the spatial interrelationships of land-system data, particularly as these spatial interrelationships will respond differently to a range of management and land-use impacts over time.

What is really needed is a clearly defined, unique approach to landscape system analysis capable of generating a set of analytical tools for landscapes. Ecosystem analysis is not land-system analysis at a finer scale; it is biotically focused. Land(scape) system analysis, on the other hand is both biotically and abiotically focused as well as integrative. Although there is a spatial dimension to each of these approaches, the significance of human impacts and land-use change is still not well understood, either from a temporal or from a spatial perspective or in an integrated or disaggregated investigation of landscape.

Perhaps some degree of mutual understanding has been brought about by the use made by all landscape ecologists of remotely sensed information and geographic information systems. But these techniques are not the preserve of any one discipline. They merely provide and display information as one source for problem solving and for generating further research questions.

Consequently, landscape ecology must reconcile the divide between the two sub-fields before they become too divergent and driven by forces from outside the landscape focus. To achieve this goal requires that some very fundamental questions about landscapes be asked so that the two solitudes can both turn to one common focus – the understanding of landscape.

The organizational framework for landscape ecology: the role of IALE

Given the above discussion, has IALE, *the* international organization for landscape ecology, failed in its mission? Most of us would say emphatically "no!" After two decades, IALE has begun to act as the essential bridge between its own theoreticians and other scientists and between the academics and the practitioners. It must continue to act in a collaborative, leadership role rather than one which merely perpetuates and reflects the existing views of its various constituencies.

One of the major debates within IALE over the past few years revolved around the development of a statement of purpose which would satisfy all its constituents. In 1998 the Executive Committee developed the following statement (IALE, 1998):

> Landscape ecology is the study of spatial variation in landscapes at a variety of scales. It includes the biophysical and societal causes and consequences of landscape heterogeneity. Above all, it is broadly interdisciplinary.
>
> The conceptual and theoretical core of landscape ecology links natural sciences with related human disciplines. Landscape ecology can be portrayed by several of its core themes:

- the spatial pattern or structure of landscapes, ranging from wilderness to cities,
- the relationship between pattern and process in landscapes,
- the relationship of human activity to landscape pattern, process and change,
- the effect of scale and disturbance on the landscape.

This statement should, however, be merely a starting point for further clarification, both for groups within the field and for persons in other fields seeking direction and purpose. The statement discusses "landscape ecology" rather than "landscape." What IALE now needs to address is the development of a short list of critical questions *about* landscapes that the majority of landscape ecologists would find acceptable as guiding principles, and to which they can contribute answers by their own individual initiatives and research. In so doing, landscape ecologists themselves will have a clearer idea of the goals and the context for their work. But of equal importance, the non-landscape ecologist will have a much clearer idea of what landscape ecologists do and can do. In other words, the field needs a focus and a profile. Many would say that we are still in a developmental stage, building from what people bring to the field but without really clarifying precisely what that field is. When you need to know something about plants you ask a botanist. But who do people ask now about landscape issues? What questions should both the scientists and practitioners of landscape ecology be asking? What answers can landscape ecologists give that relate a landscape perspective to broader environmental issues? By having a core, a focus, or a subject into which people see their work fitting, landscape ecologists will avoid the dilemma of geographers, particularly those in much of the English-speaking world. What do geographers do? What is the focus of geography? The usual, somewhat glib and unsatisfactory, answer is that "geography is what geographers do."

Without any clear focus or role it is little wonder that across North America, in particular, universities and institutes of higher education continue to close many departments and programs in that discipline.

By defining and clarifying a landscape focus and by identifying critical questions about this focus we need not narrow the field nor hinder others from related fields from making their contribution. Indeed, much of the strength of IALE, and of landscape ecology in general, has been in bringing together people of diverse interests. But have we really articulated the value and the purpose of this diversity in clarifying the goals and the purpose of landscape ecology?

Landscape ecology and its status in academia

The status of landscape ecology as an environmental sub-field for both instruction and for academic research varies tremendously from country to country. Again, an underlying distinction can be found between those areas with a long, geographically based tradition coupled with a record of application, and those areas where it is striving to gain even a minor foothold within an existing, biologically based academic discipline. The first area is perhaps best illustrated by the former Soviet-bloc countries where strong, traditional geography programs, often directly linked to state planning institutions and to research academies, were the norm. The opposite extreme is to be found, principally in the United States, where landscape ecology, often as only a single course within a degree program, is offered through a biology-based discipline, which may or may not be "ecology". This is yet a further reflection of, and a potential to deepen, the "two solitudes" discussed earlier. In the former Soviet bloc, where the land-system or geoecological approach predominates, early developments in the field of a recognizable landscape ecology have probably not advanced much beyond this foundation during the past 20 years, the period of greatest international growth of landscape ecology. By contrast, the recent major advances and the higher profile of landscape ecology have come, without any doubt, from the theoretical and methodological advances made by those ecologists who relate their work to scalar and spatial ecosystem analysis. However, again one would be negligent if one did not cite particular countries where the interpretation of these two themes has been brought together – or where they were never identified as being distinct. This would include, for example, the work of many of the Dutch landscape ecologists. A fine example of the institutional and academic synergy that can generate both the theoretical and applicable aspects of the field is to be found in Alterra, the Research Institute for the Green World, based at Wageningen University in the Netherlands. The work of the younger Czech landscape ecologists and their engagement in landscape rehabilitation and restoration resulting from the decollectivization of agricultural lands following the fall of communism after 1989 serves as a further example of the benefits of synergy within the field in providing practical solutions.

But the problem that remains is that with the existing distinct approaches the situation tends to be self-perpetuating. This has inevitably arisen because of the discipline backgrounds and geographic location of the people who established landscape ecology curricula, the "first generation." How, then, do we train and educate a new generation of "complete" landscape ecologists? One solution must lie in the need to recognize the unity that can emerge despite the cross-disciplinary origins of landscape ecology and of its protagonists. For example, at the University of Guelph, Canada, three ("first generation") landscape ecologists (one a geographer, one a wildlife ecologist, one a landscape architect) collaborated to produce an introductory "principles of landscape ecology" course. Hopefully, by their efforts to integrate and show connections across the material addressed by these three individuals, the students get *one* basic picture of the field. Based upon this foundation, then other courses covering techniques such as GIS, together with courses from related disciplines such as soil science, community ecology, physical geography, law, and policy, should have greater relevance to landscapes. This should be the case particularly where the content from these other disciplines can be related directly to the core objectives of landscape ecology. In this way, the knowledge base appropriate to the core is developed. This need not include everything about geomorphology, soil science, etc. but should focus on the landscape dimensions of these related disciplines and the need to extract *from* their respective cores the relevant landscape-related information and to place it *into* the landscape core. It is not merely a question of borrowing from existing but related disciplines, but one of utilizing this information by the methodologies and techniques of landscape ecology itself.

In other words, we have to define our academic needs more succinctly as well as justifying the value of our field. If we take the conceptual initiative suggested, then a major obstacle will be overcome. Until that focus is defined and can be justified as a valuable academic endeavor, the training of a future generation of "complete" landscape ecologists will remain problematic and very difficult to achieve under the many prevailing constraints inherent (and inherited) in our academic institutions.

To me there are interesting contrasts between the way geography and ecology have progressed in academic institutions. In many universities, especially in North America, geography departments and institutes have closed. This is at a time when the need for a geographic knowledge in the population has never been greater. Two reasons exist for this. First, geography as a discipline has not developed its own theoretical and methodological base. Second, most geographers practice on the peripheries of their discipline bordering on other fields rather than addressing any unique or individualistic approach at that periphery which relates to core questions that geography might ask about its particular environmental concerns. By contrast, the field of ecology has built a very strong theoretical and methodological base and continues to develop as a field despite being more commonly structured in academic institutions as an inter-disciplinary program rather than as a distinct academic department. There are clearly lessons here for the development of landscape ecology within academia.

Summary: landscape ecology and its societal applications

In several essays in this volume the use of principles of landscape ecology in addressing landscape-scale planning and development problems has been well illustrated. However, these remain relatively few in comparison to the many instances, when dealing with landscape problems, that many other "specialists," without any knowledge of these principles, have failed to address a particular problem adequately. Most commonly the result has led to further landscape deterioration or even catastrophe when the solution required called for an integrated landscape approach.

Society in general and governments in particular continually ask questions, raise issues, and identify problems. More frequently than not, these problems and questions focus on issues that either cross, or are totally unrelated to, the artificial boundaries which typify many academic administrative structures. How well do our traditional academic education and training systems support the provision of solutions to such emerging problems? Do the traditional academic disciplines enable us to address these problems? Can the disciplines, either singly or in combination, enable us to respond to the types of issues raised by society? Experience tells us that virtually all environmental issues are those that transcend single discipline bounds. They often go well beyond the scope of interdisciplinarity. They often require quite novel approaches to be developed. The distance between the raising of issues and the training of individuals also opens up an increasingly wide gap in the ability of science to respond. The only solution to this dilemma is for new problem-solving foci to emerge. Landscape ecology has been developing as one of these, developing, in part, from *within* several existing disciples or interdisciplinary fields (see Fig. 1 in Moss, 2000).

In some ways, many people see an evolving interdisciplinary approach to the solution of separate problems as a strength of landscape ecology. But it is of limited value to society because it requires a constant coming together of separate disciplines and specialists to address these individual problems. Once addressed, that particular problem focus is lost, and the team(work) falls apart. No advances have been made in building a system of general principles relating to the field. It is from the weakest link in the continuum between societal needs and traditional academic structures that problem-solving initiatives have to be taken. With the constant shift between issues and disciplines there are tremendous opportunities for new fields to develop, and with the increasing societal demand for solutions to problems of landscapes the time now is most opportune for landscape ecology to crystallize its thinking. Given the advancements in the field over the past few decades, and by coming together as a discipline, or at least with a discipline-like focus, the connections between societal demand, training and education, and institutionally led research initiatives, environmental problems requiring a landscape-scale focus can be much more effectively addressed.

The three major points, then, are:

- the need to define the core of the field – the landscape
- the need to explain the conceptual uniqueness of landscape ecology
- the need to consider this uniqueness from a set of fundamental, conceptual questions and problem-based issues about landscapes

To achieve these objectives, IALE can play a major role in identifying and enhancing this core and in elaborating a research agenda. Academic institutions without a tradition of landscape ecology will only begin to support initiatives from an interdisciplinary base once the goals of that endeavor are clearly articulated. The field will only advance as a body of knowledge if it works outward from a common conceptual base rather than from the individualized, peripheral, single-problem-based approach that it has tended to employ up to now.

References

Bastian, O. (2001). Landscape ecology: towards a united discipline? *Landscape Ecology*, 16, 757–766.

IALE (1998). IALE mission statement. *IALE Bulletin*, 16, 1.

Moss, M. R. (1999). Fostering academic and institutional activities in landscape ecology. In *Issues in Landscape Ecology*, ed. J. A.Wiens and M. R. Moss. Guelph: International Association for Landscape Ecology, University of Guelph, pp. 138–144.

Moss, M. R. (2000). Interdisciplinarity, landscape ecology and the "Transformation of Agricultural Landscapes." *Landscape Ecology*, 15, 303–311.

Opdam, P., Foppen, R., and Vos, C. (2002). Bridging the gap between ecology and spatial planning in landscape ecology. *Landscape Ecology*, 16, 767–779.

Ruzicka, M. (1999). My role and contribution of Slovak landscape ecology to the development of IALE. *IALE Bulletin*, 17, 1.

Wiens, J. A. and Moss, M. R., eds. (1999). *Issues in Landscape Ecology*. Guelph: International Association for Landscape Ecology, University of Guelph.

Toward a unified landscape ecology

The variety of topics and approaches represented by the essays in this volume testifies to the diversity of landscape ecology as a discipline. Remote sensing, fragmentation, ecological networks and greenways, percolation models, spatial statistics, cultural perceptions, metapopulation dynamics, land-use planning, experimental model systems, watershed hydrology, individual-based modeling – landscape ecology is all of these, and more.

This diversity is at once the great strength and the potential weakness of landscape ecology. Landscape ecology can gain strength from the sharing of problems, perspectives, and procedures that are derived from different research traditions and cultures. "Interdisciplinary" has become a fashionable label, and while many interdisciplinary approaches are simply traditional disciplines dressed in new clothes, landscape ecology truly *is* interdisciplinary. It is this convergence of different avenues of thought and practice that gives landscape ecology its tremendous vitality and that offers the promise of new insights into the ecology of land (and water; see Wiens, 2002) systems. But this diversity also carries with it the threat of fragmentation and polarization. As landscape ecology continues its explosive growth, there is a risk that subdisciplines will seek their own identity and will look inward rather than outward, splintering rather than consolidating landscape ecology.

If landscape ecology is to contribute meaningfully in such arenas as the resolution of land-use issues, the emergence of comprehensive conservation initiatives, or the development of spatially sensitive ecological theory, it must become conceptually and operationally unified. All of the issues addressed in this volume are necessary elements of this unification, but in my mind three stand out. These are, first, the need to determine what we really mean when we talk about "landscape"; second, the need to assess how landscape ecology should be done; and third, the need to consider how human culture affects everything we do in landscape ecology.

Issues and Perspectives in Landscape Ecology, ed. John A. Wiens and Michael R. Moss. Published by Cambridge University Press. © Cambridge University Press 2005.

What do we mean by "landscape"?

If "ecology" is the study of the interrelationships between organisms (including humans) and their environments, then how does the addition of the adjective "landscape" narrow this definition? Standard dictionaries usually define "landscape" in terms of natural scenery or landforms. At the opposite extreme, Forman and Godron (1986) defined "landscape" as "a heterogeneous land area composed of a cluster of interacting ecosystems that are repeated in similar form throughout." Others follow the nineteenth-century geographer von Humboldt in defining landscape as *Der totale Character einer Erdgegend* (the total character of an earth region) or, in more contemporary terminology, the ecology of land ecosystems or what Zonneveld (1995) calls "land ecology."

Although the emphasis in these definitions is on something about the land and its physical arrangement, recent discussions have implied something more. Some proponents of hierarchy theory, for example, have argued that "landscape" refers to a level of biological organization that is more inclusive than an ecosystem but less inclusive than a biome. Others have associated "landscape" with a broad, kilometers-wide spatial scale. As Allen (1998) and King (this volume, Chapter 4) have persuasively argued, "landscape" is neither a level of organization, nor is it necessarily restricted to broad spatial scales. What a landscape *is*, in my view, is a spatially defined mosaic of elements that differ in their quantitative or qualitative properties. Landscapes are characterized by their spatial configuration. It is this locational pattern, and the way it affects and is affected by spatially dependent processes, that is the subject of study of landscape ecology.

"Landscape ecology," then, is ecology that is spatially explicit or locational; it is the study of the structure and dynamics of spatial mosaics and their ecological causes and consequences. This spatially referenced linkage between pattern and process may apply to any level of an organizational hierarchy, or at any of a great many scales of resolution. It is a shared interest in the importance of spatial relationships and interactions, as they are played out over a land (or water) area, that unites landscape ecologists who otherwise ask quite different questions about quite different systems from quite different perspectives.

How should landscape ecology be done?

One way to unify landscape ecology is to recognize the essential sameness of the phenomena we study. Can the same reasoning be applied to the ways in which studies in landscape ecology are conducted? At one level, the answer is clearly "yes." Despite the variety of questions that landscape ecologists ask about a variety of systems at a variety of scales, they can use a common set of tools to obtain the information to answer these questions. These tools – remote sensing, GIS, spatially explicit individual-based models, experimental model systems, spatial statistics, and the like – provide increasingly powerful ways to generate locational data, whatever one's objectives. At another level, however, the answer is "no." Sharing a common set of tools does not make all landscape ecologists alike, any more than a common set of paints and brushes makes all artists alike. Science, like art, involves more than tools and their mastery. *How* the tools are used depends on the questions that are asked and the context in which the results will be interpreted and used.

In most areas of science, the questions and contexts are often segregated into "basic" and "applied" areas. Ecology exemplifies this dichotomy, as evidenced by the way journals divvy up publications (e.g., *Journal of Ecology* versus *Journal of Applied Ecology*, *Ecology* versus *Ecological Applications*, or even *Conservation Biology* versus *Conservation in Practice*). Because of its polyphyletic origins, this tendency is even more apparent in landscape ecology. A large part of landscape ecology, particularly in Europe, is closely associated with human ecology and applied land-use issues. Another, historically separate, theme is rooted in basic ecology and population biology. I contend that the distinction between basic and applied work is as false and counterproductive in landscape ecology as it is in other areas of science. The unification of landscape ecology requires a melding of basic research with practical applications, of science with action.

The relationship between the science and the action of landscape ecology is reciprocal. On the one hand, the science of landscape ecology gains strength by addressing issues that are relevant to society. The answers to the questions posed in basic scientific investigations in landscape ecology are likely to be of broader significance if those questions are framed in the context of applied issues. Moreover, because most of the world's landscapes bear the imprint of human actions, it would be naive to conduct basic scientific investigations of those landscapes without considering the anthropogenic forces that have shaped them. On the other hand, the action of landscape ecology is likely to make valuable and lasting contributions to such areas as land-use planning, environmental management, or natural-resource conservation only if it has a strong scientific foundation. In the absence of such a foundation, it is all too easy to fall prey to advocacy, and to promote positions that have little support other than intuition. The interactions of patterns and processes in landscapes are complex, however, and our intuitions about what might happen as a result of changing land use, mosaic fragmentation, or different land-management practices may often lead us astray. The objectivity and rigor of well-designed science are checks against mistaken intuition and advocacy.

But does the science of landscape ecology have what is required of it to inform enlightened action? Elsewhere (Wiens, 1999), I have characterized landscape ecology as scientifically immature. This judgment is based on the notion that "mature" scientific disciplines are characterized by a unifying conceptual structure or body of theory, which (I argue) landscape ecology lacks. If landscape ecology as a science is to provide a firm foundation for applications, it needs more than an array of disparate findings about, for example, the effects of fragmentation in this or that system, on this or that kind of organism. It requires more than general statements of the form "scale matters" or "all ecosystems in a landscape are interrelated." It requires a core of concepts, principles, methodologies, and predictive theories that generate specifics from generalities. This is the nucleus from which the varied approaches to landscape ecology all radiate, and to which they all contribute.

Landscape ecology now has lots of ideas and "proto-principles" and is generating new data at an accelerating pace. How should all of this coalesce to form this core? That I can't say, but I can suggest some of the key elements of this core. These elements derive from the way "landscape" and "landscape ecology" are defined, and they can be framed as three fundamental questions about landscapes:

- What creates pattern in landscapes? What are the sources of spatial variation in the quantitative or qualitative properties of systems?
- How does landscape pattern affect processes? How do gradients or discontinuities in landscape mosaics affect flows of energy, materials, individuals, or information through space?
- How does scale affect all of this?

These questions are often asked as part of specific studies, and they generate specific answers. What we need is a conceptual framework or set of theories that will consolidate the specifics into general statements. These general statements cannot take the form of "laws"; landscapes are too complex and varied for that. If we set our minds to it, however, I am convinced that we can derive *contingent* generalizations – "if … then … " answers to the above questions.

The effects of human culture

Science is conducted in a cultural context – what we regard as important, or as issues requiring resolution, is conditioned as much by culture as it is by the science itself. With landscape ecology, these roots lie very deep. Humans, and human cultures, evolved in landscapes. Landscapes are at the heart of our perceptions of nature and the aesthetic values we place on scenery. These perceptions and values, in turn, are the basis of legislation regulating land use or of policies governing the establishment of natural parks or scenic areas. Landscapes figure prominently in the art, music, and literature of all aboriginal and civilized cultures. "Landscapes," the focus of study of landscape ecology, are inexorably intertwined with human culture.

This inseparability of landscapes and culture affects the conduct and content of landscape ecology in two ways. First, it affects the ways in which we perceive landscapes. George Seddon put it well in his essay, "The nature of Nature" (1997):

> Whether or not there is a world out there independent of our perceptions of it, we cannot escape the variability of those perceptions. The ways in which we perceive, imagine, conceptualise, image, verbalise, relate to, behave towards the natural world are the product of cultural conditioning and individual variation.

It is no accident, then, that landscape ecologists tend to think of landscapes on scales that correspond with the kilometers-wide scale of scenery, or that landscape ecologists from different cultural backgrounds differ in their views of what landscape ecology is about. Perception is everything, and the challenge of overcoming our culturally conditioned perceptions of landscapes to deal with landscapes at other scales, or to define landscapes using different qualities than those we see or value, is formidable.

The second way that the culture–landscape linkage affects landscape ecology has to do with ethics. There is in most human cultures a deep-seated ethic about landscapes, reflecting the sense of a stewardship over the land. Every world religion contains teachings about how we draw strength from the land and how we have responsibility (or dominion) over it. Ecologists have recently taken up this call under the mantra of "sustainability," but the pragmatism of this term belies the deeper ethical foundations. Here is Aldo Leopold, writing in *A Sand County Almanac* (1949):

> That land is a community is the basic concept of ecology, but that land is to be loved and respected is an extension of ethics. That land yields a cultural harvest is a fact long known, but latterly often forgotten.

Ethics is one of the pillars of human culture, and land ethics affect both the ways in which we perceive landscapes and how we use landscapes. In an ethical sense, then, landscapes are more than mappable spatial mosaics, more than the environmental setting for conservation or units to be managed for sustainability. Landscapes have properties that go beyond science. Because we are products of our cultures, our science at some level reflects these ethical underpinnings, and our concepts and findings are applied within cultural contexts in which land ethics establish priorities and constraints. Doing landscape ecology without recognizing the cultural context is incomplete.

The unification of landscape ecology as a discipline, then, requires that we recognize what is important about "landscapes." It demands that we avoid partitioning the discipline into basic and applied camps and instead bind both the science and the action to a well-developed conceptual core. And it obligates us to recognize that culturally based approaches to landscape ecology (e.g., Nassauer, 1997) lie at the center rather than the periphery of the discipline. The unification of these themes will not be easy, but what a great challenge for the new millennium!

Postscript

I wrote the above essay in the spring of 1999. Reading it over now, in the autumn of 2004, I find that the basic points still ring true. There have been important technological advances during the past several years, some conceptual progress, and many publications and symposia, but the need to unify landscape ecology remains. And it is now more urgent than ever.

Or perhaps I am just more aware of this urgency now. Two years ago I left the hallowed halls of academia – in which discussions about various approaches to landscape ecology or the relative merits of basic or applied research were, well, academic – to join The Nature Conservancy in its efforts to preserve the earth's biodiversity by protecting the lands and waters that harbor that biodiversity. The emphasis in The Conservancy is on *places*, and there is an increasing recognition that these places are parts of landscapes – landscapes that embody the structure, function, and change that landscape ecologists are so fond of talking about (e.g., Hobbs, 1997; Turner *et al.*, 2001). The fight to stem the erosion of biodiversity is well upon us, and landscapes are the battlegrounds.

The relevance of landscape ecology to conservation is therefore clear. And several "principles" of landscape ecology are already moving place-based conservation in new directions, to wit:

- **Landscape elements differ in quality.** Clearly, not all places in a landscape are the same. This is the basis for various site-selection or reserve-design algorithms (see Groves, 2003) or, at a broader scale, the debate over the conservation value of "hotspots" versus "coldspots" (Myers, 2003; Kareiva and Marvier, 2003).
- **Boundaries influence dynamics.** Elements in a landscape are not isolated from their surroundings, and element boundaries are the "filters" that influence what goes where in a landscape (Cadenasso *et al.*,

2003). The conservation value of a particular landscape element may depend on the nature of the boundaries – their permeability or impermeability to movements of focal species, predators, disturbances, and the like. Boundary characteristics must be included in conservation planning.

- **Patch context is important.** The conservation value of a place is also influenced by its surroundings. The high diversity of ant communities in the Argentine Chaco, for example, is a reflection of a varied landscape mosaic that includes both semi-natural areas and areas of intense human use (Bestelmeyer and Wiens, 1996). Although nature preserves are often managed as if they were islands in a sea of human land uses, they are not. Conservation based on protected areas alone will not do the job of preserving biodiversity; the matrix must be managed as well.
- **Connectivity is a key feature of landscape structure.** Conservationists talk (and sometimes argue) incessantly about corridors and their merits in reducing the impacts of habitat fragmentation (Bennett, 1999). Landscape ecologists are increasingly recognizing, however, that the true connectivity of a landscape goes beyond simple corridors to entail how elements of differing quality are arrayed in space, how their boundaries affect movements, and how the dispersal or propagation of organisms or processes of interest is influenced by landscape configuration (Tischendorf and Fahrig, 2000; Wiens, 2001). Whether one's focus is on critically endangered species or ecosystem processes, understanding how the fabric of a landscape mosaic is woven together to facilitate or impede movement is critical to effective conservation.
- **Everything is scale-dependent.** Landscape structure and composition change with changes in scale. Moreover, the organisms, communities, or ecological processes that are the targets of conservation differ in the scales on which they occupy places or respond to environmental conditions, and the factors that threaten their persistence likewise vary in the scales on which they are relatively benign or potentially decimating. As a consequence, the conservation actions appropriate at one scale or for some targets may be inappropriate at another scale or for other targets. Conservation efforts must simultaneously encompass multiple scales; simply saying "bigger is better" won't do.

Traditionally, the focus of conservation has been on species or, less often, on communities, habitats, or ecosystem processes. The Nature Conservancy, along with many other non-governmental organisations and government agencies, has sought to preserve this biodiversity by protecting the places they occupy. The overriding message of landscape ecology, however, is that *conservation of context is just as important as conservation of content*.

One final point. It should be clear that conservation *must* incorporate the principles and practices of landscape ecology to be effective. What is perhaps less evident is the role that conservation can play in reconciling the disparate approaches to landscape ecology followed in different parts of the world. To many European landscape ecologists, for example, humans and human activities are inseparable from landscapes, and landscape ecology must therefore be "transdisciplinary" (see Zonneveld, this volume, Chapter 32; Naveh, this volume, Chapter 33). To a good many North American landscape ecologists, on the other hand, such holism is unscientific, and they pursue a (arguably) more rigorous approach to measuring landscape spatial patterns and assessing their effects on ecological systems, at multiple scales. Although both perspectives are ultimately right, bringing them together has proven to be difficult. But one of the emerging insights of conservation is that *effective conservation must include rather than exclude human activities*. This is the essence of The Nature Conservancy's "working landscapes" approach and of Rosenzweig's (2003) "win–win ecology." If this view is followed, it means that both humanistic/holistic landscape ecology *and* more strictly ecological/reductionist landscape ecology will make important contributions. Conservation may be the catalyst that finally unifies landscape ecology.

References

Allen, T. F. H. (1998). The landscape "level" is dead: persuading the family to take it off the respirator. In *Ecological Scale: Theory and Applications*, ed. D. L. Peterson and V. T. Parker. New York, NY: Columbia University Press, pp. 35–54.

Bennett, A. F. (1999). *Linkages in the Landscape: the Role of Corridors and Connectivity in Wildlife Conservation*. Gland, Switzerland: International Union for Conservation of Nature and Natural Resources (IUCN).

Bestelmeyer, B. T. and Wiens, J. A. (1996). The effects of land use on the structure of ground-foraging ant communities in the Argentine Chaco. *Ecological Applications*, 6, 1225–1240.

Cadenasso, M. L., Pickett, S. T. A., Weathers, K. C., and Jones, C. G. (2003). A framework for a theory of ecological boundaries. *BioScience*, 53, 750–758.

Forman, R. T. T. and Godron, M. (1986). *Landscape Ecology*. New York, NY: Wiley.

Groves, C. (2003). *Drafting a Conservation Blueprint: a Practitioner's Guide to Planning for Biodiversity*. Washington, DC: Island Press.

Hobbs, R. (1997). Future landscapes and the future of landscape ecology. *Landscape and Urban Planning*, 37, 1–9.

Kareiva, P. and Marvier, M. (2003). Conserving biodiversity coldspots. *American Scientist*, 91, 344–351.

Leopold, A. (1949). *A Sand County Almanac*. New York, NY: Oxford University Press.

Myers, N. (2003). Biodiversity hotspots revisited. *BioScience*, 53, 916–917.

Nassauer, J. I. (1997). *Placing Nature: Culture and Landscape Ecology*. Washington, DC: Island Press.

Rosenzweig, M. L. (2003). *Win–Win Ecology: How the Earth's Species can Survive in the Midst of*

Human Enterprise. Oxford: Oxford University Press.

Seddon, G. (1997). *Landprints: Reflections on Place and Landscape*. Cambridge: Cambridge University Press.

Tischendorf, L. and Fahrig, L. (2000). On the usage and measurement of landscape connectivity. *Oikos*, 90, 7–19.

Turner, M. G., Gardner, R. H., and O'Neill, R. V. (2001). *Landscape Ecology in Theory and Practice*. New York, NY: Springer.

Wiens, J. A. (1999). The science and practice of landscape ecology. In *Landscape Ecological Analysis: Issues and Applications*, eds. J. M. Klopatek and R. H. Gardner, pp. 371–383. New York: Springer.

Wiens, J. A. (2001). The landscape context of dispersal. In *Dispersal*, ed. J. Clobert, E. Danchin, A. A. Dhondt, and J. D. Nichols. Oxford: Oxford University Press, pp. 96–109.

Wiens, J. A. (2002). Riverine landscapes: taking landscape ecology into the water. *Freshwater Biology*, 47, 501–515.

Zonneveld, I. (1995). *Land Ecology*. Amsterdam: SPB.

Index

Note: page numbers in *italics* refer to figures and tables